David Sharp

Contributions to an Insect Fauna of the Amazon Valley

Coleoptera-Staphylinidæ

David Sharp

Contributions to an Insect Fauna of the Amazon Valley
Coleoptera-Staphylinidæ

ISBN/EAN: 9783337267926

Printed in Europe, USA, Canada, Australia, Japan

Cover: Foto ©berggeist007 / pixelio.de

More available books at **www.hansebooks.com**

IV. *Contributions to an Insect Fauna of the Amazon Valley.*/Coleoptera—Staphylinidæ. By D. SHARP, M.B.

[Read 2nd February, 1876.]

THERE is probably no part of the world outside of the temperate zones of whose insect fauna we know so much as we do of that of the valley of the Amazon. During a long residence in this interesting region Mr. Bates formed rich collections of its *Insecta*, and since his return to Europe has published numerous important memoirs descriptive of these stores.

Mr. Bates naturally selected for study those groups which are best known, and about which therefore most interest is felt by entomologists; and some few years ago he made over to me the whole of the specimens in his possession of Amazonian *Staphylinidæ*, with the hope that I should be able to examine and describe them. While I was engaged in this task, Dr. J. W. H. Trail, of Aberdeen, made a journey to the Amazon valley in the interests of natural history, and on his return handed over to me, in the most disinterested manner, the *Staphylinidæ* (and some other *Coleoptera*) collected by him, and, as the result, I found myself in possession of an important supplement to Mr. Bates' collection. I have also received through Mr. Janson a few species of the family collected at Pará by Mr. H. H. Smith three or four years ago, and one or two interesting species from the upper portion of the valley, collected by Mr. Hauxwell, have reached me.

I am thus enabled to enumerate a considerable number of species of the family as inhabiting the valley of the Amazon; a large proportion of these species are small, obscure and unattractive to the general collector, but perhaps on this account their importance just now to the genuine student of nature is all the greater; for there is prevalent a generally entertained, but I believe quite erroneous, opinion as to the existence of minute and obscure species of insects in the tropics. It appears to be generally supposed that small and unattractive species of insects which we all know to be so numerous in temperate regions, are comparatively less frequent in the tropics and are there replaced by the brilliant and magnificent forms which at present represent the insects of the tropics in our collections. I am myself, however, of opinion that obscure and minute species of insects are quite as abundant in the tropics as they are in temperate regions, and that the real difference that exists between the tropical and cooler regions of the world in this respect is, that in the tropics these minute insects are accompanied by a large number of brilliant and massive forms, which disappear gradually as the cold regions are approached. The following quotation will show that the removal of such misconceptions is of importance. In Lyell's "Students' Elements of Geology," that very careful author, in alluding to the remains of numerous species of insects found in the limestone of the Lias, in Gloucestershire, says (p. 342):— "The size of the species is usually small, and such as taken alone would imply a temperate climate, but many of the associated organic remains of other classes must lead to a different conclusion."

If my estimate of the abundance of obscure forms in the tropics be correct, the discrepancy alluded to by Sir Charles Lyell, in the passage just quoted, between the evidence from insect and that from other classes, would be considerably reduced, if not entirely removed. I need not, however, insist on this point, for now that we have obtained a considerable knowledge of the more striking insect forms of the tropics, we are constantly having revealed to us glimpses of the enormous number of minute species which probably exist there; I may, however, indicate Mr. Wollaston's important work on the *Cossonidæ* recently published by the Society, as illustrative of the correctness of my estimate.

Turning now to the number of species from the Amazons, it will be seen that the number here enumerated is 487, of which 463 are described as new. The proportion of new species to those previously described is therefore about as 19·1. This very large proportion of new species suggests forcibly how nearly complete is our want of knowledge of the *Staphylinidæ* of tropical America; nevertheless a good number of Mexican species have been described by Fauvel and Solsky, and latterly several species from Peru have been made known by the latter of these savants; the most important contribution to a knowledge of the South American *Staphylinidæ* hitherto obtained, however, is the descriptions contained in Erichson's "Genera et Species Staphylinorum," of species collected by Moritz in Columbia, and by Beske and others in Brazil. From these and from some other sources, we have had altogether just about 600 species of South American *Staphylinidæ* previously described, and, as will be seen from this paper, I have been able to identify very few of these described species with my Amazonian material. There is from this fact reason to suppose that, as a rule, the individual species of *Staphylinidæ* have not a wide range in South America, and this opinion is confirmed by all the other facts I am acquainted with. It is an interesting point, however, that the group *Piestini* seems to contain a much larger proportion of widely-distributed species than does any other group of the family, the proportion of new to previously-described species being in it only as $3\frac{1}{2}\cdot 1$.

This number of 487 species of *Staphylinidæ* inhabiting the Amazon valley, though it may at first appear by no means inconsiderable, is yet, I feel convinced, only a small fraction of the species of the group to be found in this prolific region. Mr. Bates informs me he only collected the species of this family when more important and valuable insects of other families were not to be met with; while out of the seventy-seven species of *Staphylinidæ* brought back by Dr. Trail, no less than fifty-five proved to be new, and not previously found by Mr. Bates. Taking all I know about these insects into consideration, I am unable to estimate the number of species of *Staphylinidæ* at present existing in Amazonia at less than four or five thousand species.

This paper, therefore, lengthy as it is, is but a preliminary contribution to a knowledge of the Amazonian *Staphylinidæ*, and in executing my task I have had constantly to bear in mind that I am only accomplishing a very imperfect work. This has been a great discouragement to me, for recognizing, as I do most thoroughly, how difficult is the task of determining obscure and small species by means of descriptions, however well drawn up, I have been often in doubt as to whether my labour would not be wasted, or, at any rate, rewarded only by very inadequate benefits to the cause of science. The certain discrimination of species by means of descriptions has proved as yet unattainable, and this must, in my opinion, continue to be the case until all or nearly all the actually existing species are known to us, and until descriptive terms are used with much more precision and definition than is at present the case. Hence it happens that a large part of the time of those occupied with descriptive entomology is spent in trying to ascertain the value of the names used by their predecessors; and it by no means unfrequently happens that the efforts of those predecessors have considerably increased instead of diminishing the work to be accomplished by their successors. The practical result of these difficulties is, that an increasing importance is attached to type specimens. This, in my opinion, is a perfectly natural and satisfactory result. Science teaches us to deal with facts as directly as possible, and the actual specimens described by an author afford a certainty as to the species he intended which can be attainable by no other method.

The permanent preservation of valuable and fragile specimens and the associating of them in an inalienable manner with the name given to them, is, however, no easy task. But, in order to accomplish it in the case of the fragile insects here described, I have devised a method of covering and hermetically sealing the type specimens, which will, I believe, accomplish their complete protection from all destroying agencies except fire and rude physical catastrophes. Nearly all the smaller species here described, as well as a considerable proportion of the larger species, I have preserved by this method; and, rendered bold by the valuable results it promises, I have ventured to describe even minute species where I had but a single example of it from which to draw up my description. I have taken some steps to test the efficacy of this mode of preservation, and hope soon to be able to publish a description of the method.

As regards the terms "South America" and "tropical America," constantly used in this paper, I should explain that I mean all the parts of the New World south of the United States of North America, including the West India Islands, but excluding Chili and Patagonia. The fauna of Chili is known to be very different from that of the countries on the eastern side of the Andes, and of the *Staphylinidæ* of Patagonia I know absolutely nothing.

The 487 species here enumerated are divided among

the ordinarily accepted sub-families of *Staphylinidæ* in the following manner:—

Aleocharini	44 species.
Tachyporini	18 ,,
Quediini	9 ,,
Staphylinini	93 ,,
Xantholinini	33 ,,
Pæderini	126 ,,
Pinophilini	66 ,,
Stenini	28 ,,
Oxytelini	27 ,,
Omalini	1 ,,
Piestini	31 ,,
Platyprosopus	10 ,,
Turellus	1 ,,
Total	487 species.

It would be premature for me to attempt to draw any important generalizations as to the geographical distribution of the different groups, for our knowledge of tropical *Staphylinidæ* is yet far too meagre to justify this; but on comparing the number of species contained in each sub-family with the number of species representing the same sub-family in the European fauna, one or two contrasts are so striking that they may be mentioned; they are the great comparative predominance of *Pinophilini* and *Piestini* in the Amazons, and, on the other hand, the diminished number of species of *Aleocharini* and *Omalini*. This latter fact cannot, however, be accepted as more than a negative temporary conclusion; and all I think we can at present say as the result of a comparison of this sort is that the groups *Piestini* and *Pinophilini*, which are barely represented by two or three species in Europe, are richly represented in the Amazons. In this respect the fauna of the Amazons will, I believe, be found to be similar to that of the other warmer parts of the world.

As regards genera, I have referred the Amazonian insects here dealt with to eighty different genera, of which I have established twelve as new, but of this part of my work I am unable to speak with any feeling of satisfaction. My main object in commencing this work was to describe the *species* of Amazonian *Staphylinidæ*, and I have only concerned myself with genera, because questions of nomenclature compelled me so to do. When a man describes a new species, the best thing he can do for the assistance of others is to mention what previously described species it is most nearly allied to. The system at present in vogue permits him, however, to avoid doing this by mentioning a genus to which he supposes it to belong; and thus it happens that when dealing with such an enormous mass of species as exist in the *Insecta*, the very greatest confusion arises. We are not only practically at extreme variance with our predecessors as to what constitutes a genus, but the very greatest discrepancies of opinion prevail among presently active students on this point. I should therefore have preferred in this paper never to have used the word genus at all, and not to have concerned myself with the question of genera, for I am quite convinced that when dealing with a limited geographical fauna the student is not in a position to decide as to questions of genera; and this, I hold, would still be the case, even if an agreement as to what constitutes a genus prevailed among naturalists. The system at present in vogue, however, has not allowed me to do this; and in accordance with the usual custom of naturalists I have had to constantly use the word genus, and to make use of the generic system as the basis of my dealing with species. I have therefore adopted the plan of making as few new generic names as possible. Some farther observations on this point will be found among the remarks prefixed to the descriptions of the species of *Sunius* in this paper.

In examining these insects for description I have paid a good deal of attention to the sexual characters, and have ascertained in a great number of species not only what are the secondary sexual characters of the males, but also what is the actual structure of the ædeagus. It has long been known that the secondary male characters afford a most valuable aid to the distinction of the species of insects of various orders, and it has also been ascertained in several groups of *Coleoptera* that the ædeagus itself exhibits very remarkable differences of structure even in the case of closely allied species. After making an examination of the ædeagus in a large number of the species here described, I am led to think that the investigation of the structure and variations of this organ in the *Coleoptera* (and no doubt in other insects), would lead to highly important biological conclusions. I am able to state that in one group of the *Staphylinidæ*, viz., the *Piestini*, the ædeagus is excessively small, and varies but little from species to species; while in other groups it becomes a large complex structure, varying greatly from species to species. This is the case in many *Pæderini*, *Aleocharini* and *Pinophilini*. The variations of this structure in certain groups are, however, so great, that they must be examined in a very careful manner, from species to species, before any trustworthy generalizations can be established. Mr. Darwin, in his work on natural selection, has attempted to explain the meaning and importance of the extraordinary secondary sexual characters which are so striking in some species of insects; but I am strongly of opinion that an inquiry into the importance of secondary sexual characters must be preceded by a thorough investigation of the primary sexual characters, and after that is gained I think it probable we shall be better able to deal with the secondary characters. In the case of the vegetable kingdom, Mr. Darwin has himself shown, in a manner that has delighted all naturalists, how important and radical is the connection between the actual organs of reproduction and the accessory parts of the inflorescence; and I think it highly probable that a similar course of inquiry, if carried out with insects, would make us acquainted with a direct connection between the primary and accessory sexual peculiarities. The difficulties in making the observations and dissections necessary in the prosecution of such researches is, however, very great in the case of organisms so small in size, and so complicated in structure as are the great majority of the *Insecta*.

Besides the enumeration and description of species with which this paper is chiefly occupied, there will be found prefixed to each genus some slight observations on distribution, and on structural points, and a few critical remarks.

LIST OF SPECIES OF AMAZONIAN STAPHYLINIDÆ.

ALEOCHARINI.

EUDERA.
Eudera cava, n. sp.

FALAGRIA.
Falagria Paræ, n. sp.
„ varicornis, n. sp.
„ curtipennis, n. sp.

PLACUSA.
Placusa confinis, n. sp.

EPIPEDA.
Epipeda cava, n. sp.
„ rufa, n. sp.

DIESTOTA.
Diestota sperata, n. sp.

BRACHIDA.
Brachida Batesi, n. sp.
„ Reyi, n. sp.

MYRMIGASTER (n. gen.).
Myrmigaster singularis, n. sp.

MYRMEDONIA.
Myrmedonia scabripennis, n. sp.
„ pollens, n. sp.
„ Batesi, n. sp.
„ spinifer, n. sp.
„ fortunata, n. sp.
„ nitidula, n. sp.

ALEOCHARINI—*continued.*

CALODERA.

Calodera synthota, n. sp.

HOMALOTA.

Homalota capta, n. sp.
,, tenax, n. sp.
,, brevis, n. sp.
,, gilva, n. sp.
,, Traili, n. sp.
,, culpa, n. sp.

TACHYUSA.

Tachyusa picticornis, n. sp.
,, extranea, n. sp.

OXYPODA.

Oxypoda aliena, n. sp.

ALEOCHARA.

Aleochara prisca, n. sp.
,, verecunda, n. sp.
,, auricoma, n. sp.
,, mundana, n. sp.

GYROPHÆNA.

Gyrophæna pumila, n. sp.
,, parvula, n. sp.
,, parca, n. sp.
,, lævis, n. sp.
,, juncta, n. sp.
,, convexa, n. sp.
,, sparsa, n. sp.
,, quassa, n. sp.
,, tridens, n. sp.
,, boops, n. sp.
,, debilis, n. sp.

DEINOPSIS.

Deinopsis Matthewsi, n. sp.
,, longicornis, n. sp.

TACHYPORINI.

COPROPORUS.

Coproporus rotundatus, n. sp.
,, similis, n. sp.
,, obesus, n. sp.
,, retrusus, n. sp.
,, curtus, n. sp.
,, politus, n. sp.
,, brevis, n. sp.
,, ignavus, n. sp.
,, inclusus, n. sp.
,, cognatus, n. sp.
,, conformis, n. sp.
,, rufescens, n. sp.
,, tinctus, n. sp.
,, distans, n. sp.
,, duplex, n. sp.
,, scutellatus, n. sp.

TACHYPORINI—*continued.*

CONURUS.

Conurus latus, n. sp.
,, setosus, n. sp.

QUEDIINI.

TANYGNATHUS.

Tanygnathus longicornis, n. sp.
,, nasutus, n. sp.
,, flavicollis, n. sp.

ACYLOPHORUS.

Acylophorus punctiventris, n. sp.
,, angusticeps, n. sp.
,, acuminatus, n. sp.
,, iridescens, n. sp.

QUEDIUS.

Quedius clypealis, n. sp.

CORDYLASPIS.

Staphylinus pilosus, Fab.

PLATYPROSOPUS.

Platyprosopus major, n. sp.
,, laticeps, n. sp.
,, parallelus, n. sp.
,, puncticeps, n. sp.
,, rectus, n. sp.
,, minor, n. sp.
,, rufescens, n. sp.
,, opacifrons, n. sp.
,, frontalis, n. sp.
,, similis, n. sp.

STAPHYLININI.

BRACHYDIRUS.

Brachydirus maculiceps, n. sp.
,, antennatus, n. sp.
,, styloceros, n. sp.
,, cribricollis, n. sp.
,, simplex, n. sp.
,, amazonicus, n. sp.
,, Batesi, n. sp.
,, longipes, n. sp.
,, æneiceps, n. sp.

PLOCIOPTERUS.

Plociopterus tricolor, n. sp.
,, fungi, n. sp.
,, nigripes, n. sp.
,, affinis, n. sp.
,, dimidiatus, n. sp.
,, lætus, n. sp.
,, ventralis, n. sp.
,, Traili, n. sp.
,, virgineus, n. sp.
,, mirandus, n. sp.

STAPHYLININI—*continued.*

XANTHOPYGUS.
Staphylinus sapphirinus, Er.
Xanthopygus Solskyi, n. sp.
„ cyanipennis, n. sp.
„ apicalis, n. sp.
„ violaceus, n. sp.
„ depressus, n. sp.
„ nigripes, n. sp.
Staphylinus xanthopygus, Nord.
Xanthopygus cognatus, n. sp.
Philonthus analis, Er.
Staphylinus bicolor, Lap.

PHILOTHALPUS.
Philothalpus luteipes, n. sp.
„ latus, n. sp.
„ incongruus, n. sp.

GASTRISUS (n. gen.).
Gastrisus obsoletus, n. sp.
„ laevigatus, n. sp.
„ punctatus, n. sp.

EUGASTUS (n. gen.).
Eugastus bicolor, n. sp.
„ mundus, n. sp.

ISANOPUS (n. gen.).
Isanopus tenuicornis, n. sp.

TRIGONOPSELAPHUS.
Trigonopselaphus opacipennis, n. sp.
„ mutator, n. sp.
„ violaceus, n. sp.
„ venustus, n. sp.

GLENUS.
Glenus Kraatzi, n. sp.
„ Batesi, n. sp.
„ amazonicus, n. sp.
„ vestitus, n. sp.

LEISTOTROPHUS.
Staphylinus versicolor, Grav.

STAPHYLINUS.
Staphylinus subcyaneus, n. sp.
„ parviceps, n. sp.
„ ochropygus, Nord.
„ gratiosus, n. sp.
„ gratus, n. sp.
„ amazonicus, n. sp.
„ antiquus, Nord.
„ priscus, n. sp.
„ vetustus, n. sp.

BELONUCHUS.
Staphylinus haemorrhoidalis, Fab.
Philouthus xanthopterus, Nord.

STAPHYLININI—*continued.*
Belonuchus Batesi, n. sp.
„ grandiceps, n. sp.
„ decipiens, n. sp.
Staphylinus formosus, Grav.
Belonuchus clypeatus, n. sp.
„ holisoides, n. sp.
„ aequalis, n. sp.
„ impressifrons, n. sp.
„ armatus, n. sp.
„ setiger, n. sp.

PHILONTHUS.
Philonthus amazonicus, n. sp.
„ corallipennis, n. sp.
„ deletus, n. sp.
„ muticus, n. sp.
„ gracillimus, n. sp.
„ aeneiceps, n. sp.
„ cognatus, n. sp.
„ Traili, n. sp.
„ capitalis, n. sp.
„ lustrator, n. sp.
„ aeneicollis, n. sp.
„ palpalis, n. sp.
„ aberrans, n. sp.
„ conformis, n. sp.
„ propinquus, n. sp.
„ regillus, n. sp.
„ abactus, n. sp.
„ longipes, n. sp.
„ serraticornis, n. sp.

HOLISUS.
Holisus depressus, n. sp.
„ picipes, n. sp.
„ excavatus, n. sp.
„ umbra, n. sp.
„ discedens, n. sp.

XANTHOLININI.

DIOCHUS.
Diochus longicornis, n. sp.
„ vicinus, n. sp.
„ tarsalis, n. sp.
„ flavicans, n. sp.

STERCULIA.
Sterculia amazonica, n. sp.
„ pauloensis, n. sp.
„ discolor, n. sp.
„ funebris, n. sp.
„ fimetaria, n. sp.
„ clavicornis, n. sp.
„ minor, n. sp.

AGRODES.
Agrodes conicicollis, n. sp.
„ longiceps, n. sp.

XANTHOLININI—*continued.*

TESBA (n. gen.).
Tesba gigas, n. sp.
„ laticornis, n. sp.

LINIDIUS (n. gen.).
Linidius recticollis, n. sp.
„ tenuipes, n. sp.
„ extremus, n. sp.

XANTHOLINUS.
Staphylinus rutilus, Perty.
Eulissus Mannerheimii, Lap.
Xantholinus bicolor, n. sp.
„ anticus, n. sp.
„ pygialis, n. sp.
„ temporalis, n. sp.
„ ænciceps, n. sp.
„ Batesi, n. sp.
„ amazonicus, n. sp.
„ attenuatus, Er.

LEPTACINUS.
Leptacinus nitidus, n. sp.

LITHOCHARODES (n. gen.).
Lithocharodes fuscipennis, n. sp.

METOPONCUS.
Leptacinus filarius, Er.
Metoponcus basiventris, n. sp.
„ holisoides, n. sp.

PÆDERINI.
OPHITES.
Ophites stilicoides, n. sp.

SCOPÆODES (n. gen.).
Scopæodes gracilis, n. sp.
„ fusciceps, n. sp.

CRYPTOBIUM.
Cryptobium gigas, n. sp.
„ plagipenne, n. sp.
„ opacum, n. sp.
„ opacifrons, n. sp.
„ longiceps, n. sp.
„ ruficorne, n. sp.
„ subfractum, n. sp.
„ longicorne, n. sp.
„ scutigerum, n. sp.
„ alternans, n. sp.
„ punctipenne, n. sp.
„ scrobiculatum, n. sp.
„ fuscipenne, n. sp.
„ angustum, n. sp.
„ cylindricum, n. sp.
„ laticolle, n. sp.

PÆDERINI—*continued.*
Cryptobium angustifrons, n. sp.
„ alienum, n. sp.
„ triste, n. sp.
„ Traili, n. sp.

SPHÆRINUM (n. gen.).
Sphærinum opacum, n. sp.
„ depressifrons, n. sp.
„ carinifrons, n. sp.
„ elongatum, n. sp.
„ carinicolle, n. sp.
„ pallidum, n. sp.

LATHROBIUM.
Lathrobium macrocephalum, n. sp.
„ opalescens, n. sp.
„ decisum, n. sp.
„ puncticeps, n. sp.
„ parallelum, n. sp.
„ mendax, n. sp.
„ certum, n. sp.
„ rufulum, n. sp.
„ proximum, n. sp.
„ amazonicum, n. sp.
„ tardum, n. sp.
„ tenuicorne, n. sp.
„ Batesi, n. sp.
„ minor, n. sp.
„ simplex, n. sp.
„ chloroticum, n. sp.
„ necatum, n. sp.
„ deletum, n. sp.
„ integrum, n. sp.
„ pictum, n. sp.
„ hilare, n. sp.
„ nanum, n. sp.
„ glabrum, n. sp.
„ politum, n. sp.
„ pumilum, n. sp.

DOLICAON.
Dolicaon distans, n. sp.

SCOPÆUS.
Scopæus tarsalis, n. sp.
„ ornatus, n. sp.
„ pauper, n. sp.
„ chloroticus, n. sp.
„ distans, n. sp.
„ laxus, n. sp.
„ lævis, n. sp.

LITHOCHARIS.
Lithocharis latro, n. sp.
„ simplex, n. sp.
„ condita, n. sp.
„ diffinis, n. sp.
„ comes, n. sp.
„ sobrina, n. sp.

PÆDERINI—continued.

Lithocharis crassula, n. sp.
„ vestita, n. sp.
„ integra, n. sp.
„ compressa, n. sp.
„ discedens, n. sp.
„ convexa, n. sp.
„ oculata, n. sp.
„ quadrata, n. sp.
„ egena, n. sp.
„ humilis, n. sp.
„ ardua, n. sp.
„ munda, n. sp.
„ polita, n. sp.
„ germana, n. sp.
„ pagana, n. sp.
„ picta, n. sp.

STILICUS.

Stilicus amazonicus, n. sp.
„ punctatus, n. sp.

MONISTA (n. gen.)

Monista certa, n. sp.
„ longula, n. sp.
„ divisa, n. sp.

ECHIASTER.

Echiaster boops, n. sp.
„ fumatus, n. sp.
„ signatus, n. sp.
„ carinatus, n. sp.
„ latifrons, n. sp.
„ mamillatus, n. sp.
„ muticus, n. sp.
„ tibialis, n. sp.
„ Batesi, n. sp.
„ scissus, n. sp.

LINDUS (n. gen.).

Lindus religans, n. sp.

PÆDERUS.

Pæderus solidus, n. sp.
„ tridens, n. sp.
„ lingualis, n. sp.
„ mutans, n. sp.
„ protensus, n. sp.
„ amazonicus, n. sp.
„ punctiger, n. sp.

SUNIUS.

Sunius amicus, n. sp.
„ vittatus, n. sp.
„ serpens, n. sp.
„ ventralis, n. sp.
„ strictus, n. sp.
„ marginatus, n. sp.
„ brevis, n. sp.
„ modestus, n. sp.

PÆDERINI—continued.

Sunius crassus, n. sp.
„ pictus, n. sp.
„ confinis, n. sp.
„ catena, n. sp.
„ peltatus, n. sp.
„ palpalis, n. sp.
„ bidens, n. sp.
„ bispinus, n. sp.
„ spinifer, n. sp.
„ celatus, n. sp.
„ insignis, n. sp.

PINOPHILINI.

TÆNODEMA.

Tænodema plana, n. sp.
„ lævis, n. sp.
„ recta, n. sp.
„ lenta, n. sp.
„ dubia, n. sp.
„ quadrata, n. sp.
„ tarsalis, n. sp.
„ bella, n. sp.
„ cinerea, n. sp.
„ vicina, n. sp.
„ similis, n. sp.
„ rudis, n. sp.
„ filum, n. sp.
„ producta, n. sp.
„ laticornis, n. sp.
„ serpens, n. sp.
„ tecta, n. sp.
„ lurida, n. sp.

PINOPHILUS.

Pinophilus dux, n. sp.
„ ater, n. sp.
„ rectus, n. sp.
„ æqualis, n. sp.
„ minus, n. sp.
„ modestus, n. sp.
„ tenuis, n. sp.
„ distans, n. sp.
„ incultus, n. sp.
„ proximus, n. sp.
„ angustus, n. sp.
„ oblatus, n. sp.
„ extremus, n. sp.
„ sulcatus, n. sp.
„ duplex, n. sp.
„ laxus, n. sp.
„ aberrans, n. sp.
„ bicolor, n. sp.
„ Batesi, n. sp.
„ debilis, n. sp.
„ minor, n. sp.
„ affinis, n. sp.
„ egens, n. sp.
„ abax, n. sp.

PINOPHILINI—*continued.*

ŒDODACTYLUS.
Œdodactylus errans, n. sp.
„ anceps, n. sp.

ŒDICHIRUS.
Œdichirus optatus, n. sp.

PALAMINUS.
Palaminus simplex, n. sp.
„ longicornis, n. sp.
„ modestus, n. sp.
„ crassus, n. sp.
„ robustus, n. sp.
„ breviceps, n. sp.
„ discretus, n. sp.
„ sinuatus, n. sp.
„ apicalis, n. sp.
„ fragilis, n. sp.
„ niger, n. sp.
„ anceps, n. sp.
„ sobrinus, n. sp.
„ puncticeps, n. sp.
„ parcus, n. sp.
„ pellax, n. sp.
„ fuscipes, n. sp.
„ stipes, n. sp.
„ sellatus, n. sp.
„ gracilis, n. sp.
„ distans, n. sp.

STENINI.

STENÆSTHETUS.
Stenæsthetus illatus, n. sp.

STENUS.
Stenus inspector, n. sp.
„ obductus, n. sp.
„ tinctus, n. sp.
„ cognatus, n. sp.
„ vacillator, n. sp.
„ cursitor, n. sp.
„ fallax, n. sp.
„ simulator, n. sp.
„ certatus, n. sp.
„ Traili, n. sp.
„ pedator, n. sp.
„ ventralis, n. sp.
„ extensus, n. sp.
„ genalis, n. sp.
„ Paræ, n. sp.
„ nigricans, n. sp.
„ excisus, n. sp.
„ laticeps, n. sp.
„ tricolor, n. sp.
„ heres, n. sp.
„ cevictus, n. sp.
„ Batesi, n. sp.

STENINI—*continued.*
Stenus collaris, n. sp.
„ parviceps, n. sp.
„ proximus, n. sp.

MEGALOPS.
Megalops spinosus, n. sp.
„ impressus, n. sp.

OXYTELINI.

OSORIUS.
Osorius stipes, n. sp.
„ nitens, n. sp.
„ simplex, n. sp.
„ integer, n. sp.
„ solidus, n. sp.
„ affinis, n. sp.
„ oculatus, n. sp.

HOLOTROCHUS.
Holotrochus durus, n. sp.
„ syntheticus, n. sp.
„ pubescens, n. sp.
„ subtilis, n. sp.
„ clavipes, n. sp.
„ Fauveli, n. sp.

BLEDIUS.
Bledius albidus, n. sp.
„ rarus, n. sp.
„ addendus, n. sp.
„ simplex, n. sp.
„ muticus, n. sp.
„ similis, n. sp.
„ modestus, n. sp.

TROGOPHLŒUS.
Trogophlœus mundus, n. sp.
„ breviceps, n. sp.
„ latifrons, n. sp.
„ hilaris, n. sp.
„ vicinus, n. sp.

APOCELLUS.
Apocellus planus, n. sp.
„ lævis, n. sp.

OMALINI.

OMALIUM.
Omalium nanum, n. sp.

PIESTINI.

PIESTUS.
Piestus validus, n. sp.
„ bicornis, Oliv.
„ spinosus, Fab.
„ frontalis, n. sp.

PIESTINI—*continued.*
Piestus rectus, n. sp.
„ minutus, Er.
„ pygmæus, Lap.
„ sulcatus, Grav.
„ rugosus, n. sp.
„ aper, n. sp.

HYPOTELUS.
Hypotelus micans, n. sp.

ISOMALUS.
Isomalus agilis, n. sp.
„ dubius, n. sp.
„ tenuis, Fauv.

LISPINUS.
Lispinus striola, Er.
„ catena, n. sp.
„ apicalis, n. sp.
, terminalis, n. sp.

PIESTINI—*continued.*
Lispinus punctatus, n. sp.
„ cognatus, n. sp.
„ modestus, n. sp.
„ planus, n. sp.
„ depressus, n. sp.
„ simplex, n. sp.
„ lætus, n. sp.

THORAXOPHORUS.
Thoraxophorus opacus, n. sp.
„ crassus, n. sp.

LEPTOCHIRUS.
Leptochirus fontensis, n. sp.
„ brunneoniger, Perty.
„ latro, n. sp.
„ maxillosus, Fab.

TURELLUS (n. gen.).
Turellus Batesi, n. sp.

EUDERA.

This genus was established by Fauvel (Notices Entomologiques, 4me part. p. 8), for a small beetle from Chili: and at the same time this savant established another genus (op. cit. p. 10) with the name *Ophioglossa*, for a closely allied insect from the same country. Some differences in the trophi (of which the most important is stated to be the labial palpi bi-articulate in *Eudera*, tri-articulate in *Ophioglossa*), and a slightly longer basal joint of the hind tarsus in *Ophioglossa*, are the only characteristics given to distinguish the two genera; moreover the hind tarsi are figured by the author, and on measuring with compasses the length of the basal joint in the two figures, I find it to be exactly the same: the distinction between the two genera rests therefore entirely on the trophi; and very unsatisfactory such a distinction is in the case of two such minute insects. I feel considerable doubt, after examining a specimen of the insect I here describe under the name of *Eudera cava*, as to whether its labial palpi are bi-, or tri-articulate; but as the *Eudera sculptilis* is known to me, and I am able to say that *E. cava* is certainly closely allied thereto, while I do not know the genus *Ophioglossa*, I have chosen the former name for the generic appellation of my new species. The species I here describe is very remarkable by reason of the extremely large and deep transverse impressions on the basal segments of the hind body; this character distinguishes it readily from its allies, viz., *E. sculptilis* and the species of *Falagria* and *Autalia*.

It may not be amiss to remark here, that although M. Rey, in the recently published parts of the "Histoire Naturelle des Coléoptères de France," has placed the two genera *Autalia* and *Falagria* in different primary divisions of the *Aleocharidæ*, still the two genera are really allied, as the *Eudera sculptilis* and *cava* undoubtedly indicate.

1. *Eudera cava*, n. sp. Rufescens, antennis basi excepto, capite, elytris, pectoreque obscurioribus; thorace elytrisque dense subtilissime punctatis, abdomine fere impunctato, segmentis 2—4 basi profunde transversim impressis; thorace fortiter transverso, anterius obsolete impresso, basi transversim bifoveolato. Long. corp. 1⅛ lin.

Antennæ short and stout, the basal joints reddish, the others obscure in colour; 3rd joint stout, a little shorter than 2nd; joints 4—10 similar to one another in length, and each distinctly broader than its predecessor, the 4th not so long as broad, the 10th strongly transverse; 11th joint pointed, as broad as, and more than twice as long as 10th. Head nearly black, short and broad, with a narrow and remarkably abrupt neck, with a transverse impression in front at the insertion of the antennæ, the surface very finely, scarcely visibly, punctured. Thorax obscure reddish, broader than the head, but narrower than the elytra, strongly transverse, the front angles rounded and depressed, the hind ones rectangular; on the middle, in front, is a small indistinct impression, and in the middle, in front of the base, is a kind of transverse impressed line, which is interrupted in the middle, and in certain lights appears to consist on each side of two or three very minute foveæ placed extremely close to one another; the surface is very finely and closely punctured and pubescent. Elytra short and broad, longer than the thorax, the suture depressed at the base behind the scutellum; their colour obscure castaneous, their punctuation very fine, and not so dense as that of the thorax. Hind body broad, above shining and flat; the 2nd, 3rd and 4th segments each with a peculiar large deep transverse impression at the base; at each front angle of these impressions there is a kind of tubercle or projection. Legs rather short; middle coxæ widely separated.

Pará; eight individuals taken two or three years ago; I notice no indications of sexual distinctions among them.

Obs.—A closely allied but distinct species occurs at Rio de Janeiro.

FALAGRIA.

The species of this widely distributed genus appear to be more numerous in tropical America than elsewhere; 23 species of the genus were described by Erichson, and of this number no less than 14 were from the quarter of the globe above mentioned. I here describe three new species, but I have no doubt numerous others are to be found in the Amazon valley. These three species are very dissimilar *inter se;* the *F. Paræ* appears to be rather closely allied to the Eastern *F. flavocincta*, Kr., and *F. forea*, Sharp; *F. varicornis* is remarkable not only from the great development and elegant colour of its antennæ, but also from the point of insertion of these organs, this appearing to be actually nearer to the vertex than to the front margin of the head. *F. curtipennis* is a very peculiar species; its abbreviated elytra and breast, together with the slender and elongate limbs, give it a peculiar facies; and moreover the mesosternum shows no trace whatever of that division into distinct plates, which is so conspicuous in *F. obscura*, and others of the genus; it is, therefore, very probable that the species may ultimately be considered to belong to a distinct genus.

1. *Falagria Paræ*, n. sp. Rufescens, antennis medio infuscatis, basi cum pedibus testaceis, femoribus basi excepto fuscis, elytris abdominēque fuscis, illis humeris, hoc basi testaceis; capite thoraceque lævis, nitidis, hoc profunde canaliculato; abdomine dense subtilissime punctulato. Long. corp. 1⅛ lin.

Antennæ rather stout, very nearly half a line in length; the two or three basal joints pale reddish, the following ones infuscate; the 10th and 11th again reddish; 3rd joint shorter than 2nd; 4—10 differing very little from one another in length, each a little stouter than its predecessor; the 10th about as long as broad; 11th scarcely broader than, but twice as long as the 10th, obtusely pointed. Head obscure reddish, shining, impunctate, with a transverse impression between the points of insertion of the antennæ, its breadth about equal to the thorax. Thorax reddish, impunctate, with a deep channel along the middle, rather longer than broad, much narrowed towards the base. Elytra yellowish at the base, elsewhere infuscate, much broader than the thorax, but scarcely so long, with a small deep fovea behind the scutellum, extremely finely—almost imperceptibly—punctured. Hind body slender, but distinctly narrowed at the base; blackish except at the base, which is yellowish, the segments very finely punctured. Legs yellow, the femora infuscate, except at the base; the front ones however almost yellow.

Pará; a single individual taken two or three years ago by Mr. Smith.

Obs.—This species is smaller and more slenderly formed than our European *F. sulcatula*, and the antennæ are similar in structure to those of *F. sulcatula*, but are not quite so much incrassate at the apex.

2. *Falagria varicornis*, n. sp. Rufo-brunnea, nitida, fere lævis, antennis articulo ultimo, femoribus posticis basi, tarsis omnibus, abdomineque basi pallidis, hoc segmentis 3—6 nigris; thorace subcordato, angulis posticis minutis, prominulis, late profundeque canaliculato; elytris hoc longioribus. Long. 1¼ lin.

Of an obscure reddish colour, with the elytra rather darker; the hind body, with the exception of its pale base, nearly black; the legs and antennæ variegated. The antennæ are very long and reach beyond the extremity of the elytra, they are reddish at the base; the joints from the middle to the tenth get gradually darker, but the last joint is very pale, nearly white; this joint is elongate and pointed, about the length of the three preceding together. The head is broader than the thorax, reddish, almost impunctate. The thorax is reddish, its length about as much as its greatest width. The posterior angles are very minute and pointed, and directed outwards; it has a very broad and very deep longitudinal channel. The scutellum is delicately margined and furnished with two or three indistinct raised lines. The elytra are darker in colour than the head and thorax, very convex, depressed along the suture and round the scutellum, shining and almost glabrous. The hind body has the two basal segments testaceous, the rest nearly black; the very extremity obscurely paler; it is narrowed at the base, impunctate on the upper side, and only obscurely furnished with hairs. The legs are very long and slender, they are reddish-yellow, the anterior with the tarsi paler, the middle ones have the base of the tibiæ darker, the apex as well as the tarsi pale; the hind legs have the base of the femora pale, the rest as well as the tibiæ pitchy; but the extremity of the tibiæ and the tarsi are pale.

Of this most elegant species I have seen but one specimen, from Ega.

3. *Falagria curtipennis*, n. sp. Elongata, rufo-testacea, nitida, fere lævis, abdomine subdilatato, segmento quinto piceo. Long. 1½ lin.

This species has a peculiar aspect for this genus, owing to its elongate form, and its hind body being broader than the anterior parts. It is shining, of a yellowish colour, with the fifth segment of the hind body darker; its upper surface is smooth, shining and impunctate; the under surface of the hind body is furnished towards the extremity with numerous hairs. The antennæ are yellow, and reach quite to the extremity of the elytra; they are stout for this genus, and considerably thickened towards the apex; the 2nd, 3rd and 4th joints are slender and elongate, the 3rd a little longer than the 2nd, much longer than the 4th; from the 5th to the 10th each joint is a little broader than the preceding one, the 10th being about as long as broad; the 11th joint is pointed, about as broad as the 10th, but twice as long. The head is quite smooth and shining, quite as broad as the thorax; the thorax is rather longer than its greatest breadth, convex, smooth and shining, with two or three setæ towards the sides; it is subcordate, with the anterior and posterior angles rounded; it is delicately margined at the sides and behind, without channel or fovea. The elytra are not so long as the thorax, and scarcely attain the width of its broadest part; they are just a little darker in colour, impunctate and shining, with a most delicate and sparing pubescence. The hind body is rounded at the sides; it is elongate and considerably broader in the middle than at the base and apex. The legs are yellowish and elongate, the tarsi very slender.

Several specimens from Tapajos.

Obs.—A very closely allied species to this one occurs at Monte Video.

PLACUSA.

No species with this generic name has yet been indicated as inhabiting the New World; and the species I here describe does not accord very closely in all its characters with our European species, as will be seen on comparing the observations I have made at the conclusion of the description of *P. confinis* with the generic characters of *Placusa*, as indicated by Rey (Hist. Nat. Col. Brev. Al. Bol. p. 103). I have not ascertained whether the labial palpi of this minute insect are two- or three-jointed.

1. *Placusa confinis*, n. sp. Angustula, subopaca, nigricans, antennarum articulo primo, pedibus elytrisque sordide testaceis, his basi lateribusque fuscis; supra crebre subtiliterque punctata. Long. corp. ⅞ lin.

Antennæ short, rather slender, blackish, with the base indistinctly paler; 3rd joint small, much shorter than 2nd; 4th and 5th smaller than those following; 6—10 differing but little from one another, transverse; 11th small. Head small, a good deal narrower than the thorax, blackish, finely and not densely punctured. Thorax small, rather strongly transverse, the base rounded, and not perceptibly sinuate, the hind angles very indistinct, the sides distinctly narrowed towards the front; it is blackish in colour, finely and not densely punctured, so as to be a little shining. Elytra short but a little longer than the thorax, blackish at the base and sides, shading into yellow towards the extremity, rather finely and closely punctured. Hind body narrow, pointed, all the segments finely and closely punctured, but still slightly shining. Legs yellow.

In the male the hind margin of the dorsal plate of the 7th segment of the hind body projects a little on each side of the middle, so as to form two very short, obtuse, distant projections, which are not long enough to be called teeth.

A pair of this species, ♂ and ♀, were found at Lagos, on the 5th January, 1875, by Dr. Trail.

Obs.—This species, though it has quite the facies of our European *Placusæ*, is narrower than any one of them I am acquainted with. The middle coxæ are quite contiguous, and the middle portion of the front of the metasternum is less acuminate than in the European species; the basal joint of the 4-jointed middle tarsi is distinctly longer than the 2nd, but not nearly so long as the 2nd and 3rd together; and the basal joint of the hind tarsus, though more elongate than that of the middle ones, is not so long as the 2nd and 3rd together.

EPIPEDA.

This genus has recently been established and named by Rey, to express the *Aleochara plana*, Gyll., and *Homalota arcana*, Er. The first of these species (the second I do not know) certainly cannot be associated with the *Homalotæ*, on account of the 4-jointed intermediate tarsi; and it appears to me probable that Rey is correct in placing *Epipeda* near *Placusa*. The two species of the genus here described, depart somewhat in their structure from the *Aleochara plana*, and apparently approximate to *Diestota*. Indeed the relationship of *Diestota* with *Epipeda* seems to me to be probably (for I do not know the *D. Mayeti*) much closer than is suggested by Rey, who places the two genera in different "rameaux," on account of the separation of the middle coxæ in the one (*Diestota*) and their contiguity in the other (*Epipeda*); it is precisely in this character that the two species, here described as *Epipeda*, depart from the European species of the genus, and appear to exhibit the connecting links with *Diestota*. I may add that Dr. Trail also brought back an insect which I can scarcely class with either of the two genera, for the middle coxæ are widely separated, while the genæ are very strongly margined. I have unfortunately not been able to describe this interesting species, as the only exponent of it I have received has lost its elytra.

1. *Epipeda cava*, n. sp. Linearis, subdepressa, opaca, nigro-fusca, antennarum articulo primo pedibusque testaceis; prothorace subquadrato, medio late, profundeque impresso; abdomine segmentis 2—4 crebre subtiliter punctatis, 5 et 6 fere impunctatis. Long. 1 lin.

Allied to *Homolata plana*, but smaller, and with much longer antennæ. These are longer than head and thorax, pitchy in colour, with the base paler; they are moderately stout, and scarcely thickened towards the apex; 3rd joint more slender than the 2nd, and scarcely so long; 4—10 differing but little from one another; 4th longer than broad, 10th scarcely so long as it is broad; last joint elongate, quite twice as long as the two preceding together. Head a little narrower than the thorax, formed as in *H. plana*, the front a little depressed; it is of a dark colour, quite dull, punctuation quite indistinct. Thorax about a third broader than long, shaped much as in *H. plana*, dull, its punctuation very dense and indistinct, of a dark fuscous colour in the middle, with a very broad and deep impression, occupying the whole of the middle portion of the thorax. Elytra but little longer than the thorax, of the same colour, and with similar punctuation. Hind body shaped much as in *H. plana*, segments 2—4 rather closely and finely punctured, 5th sparingly punctured, 6th nearly impunctate. Legs yellowish.

St. Paulo (Amazons); one specimen. I think it is a male; if so, it has no well marked peculiarity as such, unless the very large impression of the thorax be characteristic of this sex, as is quite probable.

Obs.—Though much smaller than the European *Epipeda plana*, this species appears to be structurally closely allied thereto; the middle coxæ are however slightly more widely separated, so that the meso- and meta-sternal processes are rather less slender and acuminate in *E. cava*.

2. *Epipeda rufa*, n. sp. Angustula, depressa, dense punctata, rufescens, antennis extrorsum, abdomine pectoreque infuscatis, elytris pedibusque testaceis; prothorace subquadrato, angulis posterioribus obtuse rectis. Long. corp. ⅞ lin.

Antennæ moderately long, not stout, scarcely at all thickened towards the extremity, the two or three basal joints reddish, the rest darker; 2nd joint about as stout as the 1st; 3rd shorter than 2nd; 4th smaller than the others; 5—10 differing but little from one another; 11th moderately long. Head rather narrow, a good deal narrower than the thorax, distinctly narrowed behind the eyes, infuscate reddish, not at all shining, closely and finely punctured. Thorax slightly narrower than the elytra, almost as long as broad, somewhat straight at the sides, being only a little narrowed behind, and but little rounded near the front angles, the hind angles not at all rounded but distinctly obtuse; dull reddish in colour, finely and closely punctured, quite dull, longitudinally depressed along the middle. Elytra rather elongate, a good deal longer than the thorax, yellow, closely and finely punctured. Hind body parallel, all the segments closely, finely and evenly punctured, infuscate reddish, with the extremity paler. Legs, including the coxæ, clear yellow; metasternum infuscate at the sides, reddish in the middle.

Ega; two individuals in which I see no sexual characters.

Obs.—In this species the middle coxæ are quite separated, the mesosternal process is elongate, truncate at the extremity and meets the middle process of the metasternum. The middle tarsi are 4-, the hinder 5-jointed. The genæ are not margined.

DIESTOTA.

This genus has been recently established by Rey, for a minute beetle found at Cette, in flowers of *Cistus:* and it is moreover considered by the French author to form of itself a distinct rameau of his branch *Bolitocharaires*. I am not acquainted with this French insect, and I cannot therefore speak in an unhesitating manner as to the close affinity therewith of the species here described ; but so far as I can judge the two form really part of one genus. The relationship of *Diestota* with *Epipeda* seems to me to be closer than supposed by M. Rey ; and the two species of the latter genus here described, indicate this affinity in a still more certain manner than does the European *Aleochara plana* (*Epipeda*, Rey).

Rey describes *Diestota Mayeti* as possessing a "sensible" margin to the genæ, but I cannot see any trace of such raised margin in *D. sperata*. I may also add, that trusting to my memory, I believe the insect described by me (Trans. Ent. Soc. Lond. 1869, p. 166) as *Homalota cribriceps*, will be found to belong to this genus; and further, that I think it highly probable that *Diestota* will prove to be synonymous with *Cænonica*, Kr.

1. *Diestota sperata*, n. sp. Parallela, subdepressa, fusca, prothorace dilutiore ; abdomine pedibusque testaceis, illo cingulo lato ante apicem fusco ; prothorace valde transverso, basi impresso ; abdomine subtiliter, basi crebre, apice parce punctato. Long. corp. 1 lin.

Antennæ rather slender, moderately long, only slightly thickened towards the apex; the three basal joints elongate, 3rd almost as long as 2nd ; 4th distinctly smaller than 5th, about as long as broad ; 5—10 each of about the same length ; the 5th about as long as broad ; the 10th distinctly transverse ; 11th joint elongate, longer than the two preceding together ; their colour is smoky black, with the base a little paler. Head narrower than the thorax, short and broad, but much narrowed behind, closely and finely punctured, not shining, fuscous in colour, with the parts of the mouth obscure yellow. Thorax a little narrower than the elytra, strongly transverse, nearly twice as broad as long: the sides sinuate and narrowed behind; the hind angles obtuse but not rounded, with a transverse impression at the base in the middle, infuscate red, very closely and very finely punctured, quite dull. Elytra a good deal longer than the thorax, very closely and finely punctured, quite dull, in colour darker than the thorax, but similar to the head; their hind margin a little sinuate at the outer angles. Hind body parallel, obscure yellowish in colour; the 5th and 6th segments infuscate, the extremity not so pale as the base; the basal segments are closely and finely punctured, the apical ones sparingly punctured so that they are less opaque than the rest of the upper surface. Legs, including the coxæ, clear yellow.

In the male the hind margin of the dorsal plate of the 7th segment of the hind body terminates in six slender teeth; the four middle ones are equidistant, and each one of them is a little thickened at the extremity, the outside one (on each side) is slightly longer than the middle ones, and is pointed, and is separated from the middle ones by a broader space than divides them from one another.

Rio Purus, Amazons; six specimens found by Dr. Trail on the 13th October, 1874.

Obs.—This species appears to vary somewhat in colour and size; and the above description is made from one of the largest and most brightly coloured individuals.

Brachida.

This genus has recently been established by M. Rey, in the "Histoire Naturelle des Coléoptères de France,"* (*Brevipennes, Aléochariens*), for the European *Homalota notha*. The 4-jointed intermediate tarsi, and the structure of the meso- and meta-sterna, fully justify this course, and prove that the relationship of *H. notha* to other *Homalotæ* is only remote. I here describe two species which must be ascribed to Rey's genus *Brachida*, and I have other allied South American species in my collection.

---

* It is to be regretted that the volumes of this work are not numbered; four or five different parts each with separate pagination, and indices, bear the above title (*Brevipennes, Aléochariens*), and detailed reference to the work is not easy. It is true that one might use the year of publication for the purpose, but this is sometimes erroneously indicated on the title page, and sometimes more than one part has been published in the same year.

1. **Brachida Batesi**, n. sp. Convexa, ferruginea, antennis medio, capite, prothorace, elytrorumque angulo externo fuscis; abdomine supra subglabro. Long. 1⅓ lin. (abdomine haud extenso).

Mas: elytrorum angulo externo apicali tuberculo parvo instructo, abdomine segmentis dorsalibus 6 et 7 crebro granulato-asperatis.

Allied to *Homalota notha*, but larger and brighter in colour, and even more like a *Gyrophæna*, by its deflexed head, and curled-up hind body; the colouration is also very much that of a *Gyrophæna*. The antennæ are rather short, considerably thickened towards the apex; the 3rd joint is more short and more slender than the 2nd; the 4th joint is but little stouter than the 3rd, and not quite so long—from this to the 10th each joint is stouter than the preceding one; 5th longer than broad; 10th evidently transverse; 11th joint rather thick, nearly twice as long as the 10th, and rather lighter in colour. Head deflexed, narrower than the thorax, pitchy, very finely and indistinctly punctured. Thorax very short, shaped like that of a *Gyrophæna*, rather narrower than the elytra, more than twice as broad as long, all the angles much rounded; of a pitchy ferruginous colour, almost impunctate. Elytra longer than the thorax, of a tawny colour, a little darker at the external angle, rather closely covered with fine elevated punctures, which are not so distinct about the scutellum. Hind body yellowish, a little darker towards the apex; rather rounded at the sides, narrower at the extremity; shining above, and nearly impunctate, but towards the extremity the segments are furnished with some very fine elevated points; it is clothed on the underside with a dense, fine, erect pubescence. The legs are yellow.

In the male the elytra are provided, near the outer corner, with an elevated tubercle, and the 6th and 7th segments of the hind body are furnished on the upper side with numerous granulations, which are coarser on the 7th than on the 6th segment.

Tapajos; two male individuals.

2. **Brachida Reyi**, n. sp. Rufescens, antennis medio, capite, elytris, abdomineque ante apicem infuscatis; thorace omnium subtilissime punctato, tenuissimeque pubescente; elytris subtiliter punctatis, dense breviterque

pubescentibus; abdomine subtilissime punctato. Long. corp. (capite abdomineque extensis) 1½ lin.

Antennæ slender, scarcely thickened towards the extremity, yellowish, with joints 5—10 nearly black, each joint longer than broad. Head infuscate, very indistinctly punctured, but with a distinct pubescence. Thorax obscure reddish, scarcely visibly punctured, but with a delicate, distinct pubescence; the front angles depressed and rounded, the sides scarcely curved; the hind angles obtuse, the base rounded, very slightly emarginate in front of the scutellum. Elytra a good deal longer, and slightly broader than the thorax, infuscate-red, with the shoulders paler, their punctuation close and fine, but more visible than on the other parts of the surface; the extremity only obsoletely sinuate near the outer angle. Hind body much narrowed towards the extremity, reddish, with the 4th, 5th and 6th segments infuscate, the apex yellowish; the dorsal segments very finely punctured and pubescent. Legs yellowish, rather long and slender.

The male shows no marked external sexual characters; the hind margin of the dorsal plate of the 8th segment is almost truncate, being only very slightly emarginate. In the female the hind margin of the dorsal plate of the 8th segment is deeply emarginate; and the same plate shows, at the extreme base on each side, a deep transverse fovea of which there is no trace in the male.

Pará; several specimens taken two or three years ago.

Obs.—A variety occurs in the same locality; it is almost unicolorous reddish, except the middle of the antennæ.

Myrmigaster, n. gen.

Tarsi anteriores (4-?) intermedii et posteriores 5-articulati.

Prothorax lateribus rotundatis, postice latioribus.

Prosternum carinatum, anterius acute elevatum.

Elytra lateribus haud carinatis.

Abdomen conicum, basi fortiter constrictum, segmentis dorsalibus transversim convexis.

Corpus anterius latius, partibus anterioribus pilis brevibus crassis instructis. Antennæ sat crassæ, evidenter pilosellæ. Thorax coleoptera latitudine excedens, basi utrinque emarginatus, angulis posterioribus acutis retrorsum spectantibus, lateribus rotundatis, angulis anterioribus deflexis, caput amplectentibus. Scutellum transversum, apice conspicuo. Elytra prothorace longiora, apice truncata. Mesosternum inter coxas intermedias longe productum, parte productâ elevatâ. Abdomen convexum, conico-cylindricum, lateribus marginatum, basi fortiter constrictum.

The curious insect for which I propose this generic name is one of the most remarkable of the *Staphylinidæ* found by Mr. Bates. Unfortunately I have seen but a single individual, so that I am able to give its characters in only an incomplete manner. The head is small, deflexed, and much embraced by the angles of the thorax, so that I have been quite unable to see the parts of the mouth; and although I have made a careful examination with the compound microscope, I do not feel quite sure that the front tarsi may not have a minute basal joint. I cannot pronounce on its exact position a confident opinion, but I believe it will ultimately be found allied to *Dinarda*, possibly making an approach from that genus to the wonderful *Corotoca* of Schiödte.

Rey has recently considered the genus *Dinarda* as forming of itself one of the eight primary divisions of the *Aleocharidæ*; this isolated position he assigns to it in consequence of the elytra being compressed and carinated at the sides, which is the case so far as he knows with no other *Aleocharidæ*. The *Myrmigaster singularis* shows, however, no trace of this peculiarity; I myself consider this character to be quite insufficient of itself to justify the prominent isolation given to the two species of *Dinarda* by the learned Frenchman, who actually makes, of the two species of *Dinarda*, the first branch (branche, Dinardaires) of the *Aleocharidæ*.

1. *Myrmigaster singularis*, n. sp. Rufo-picea, subnitida, antennis pedibusque testaceis, obsolete punctulata; thorace fortiter transverso; elytris, hoc ter longioribus, apicem versus subattenuatis. Long. 1½ lin.

Antennæ formed as in the genus *Dinarda*, but much more slender, and with their exserted setæ longer; the joints are closely packed, so that the divisions between them are not striking; the 1st joint stouter and longer than the 2nd; the 2nd and 3rd small, the 2nd being the longer of the two; from the 4th to the 10th each joint is but slightly stouter than the foregoing one; the 10th about as long as broad; the terminal joint not broader than the 10th, pointed, about as long as the two preceding ones together. The head and thorax are of a pitchy or a pitchy-red colour, covered, as are also the elytra, with short, stout, erect setæ. Thorax a little broader than the elytra, more than twice as broad as long, much narrower in front than behind, its middle part a little elevated, the sides deflexed, the middle of the base a little produced in front of the scutellum, the posterior angles acute and projecting backwards. The elytra are rather lighter in colour than the thorax, and about one-third longer; they are narrower at the apex and the base; under the microscope the sculpture of the thorax and elytra is seen to consist of small round smooth spots, between which the surface is coriaceous. The abdomen is much narrower than the thorax; it is nearly impunctate, and very finely pubescent, but the basal segments are also furnished with setæ finer than those of the thorax and elytra: it is very convex above and below, and the first visible segment is much narrowed all round from its apex to its base.

Ega; a single individual, which I suspect to be a male.

MYRMEDONIA.

Of this widely distributed genus, nine or ten species have been already described from tropical America: to this number I now add other six species. Of these six species, the first five are pretty closely allied to one another, while the sixth (viz., *M. nitidula*) is very distinct. Rey has recently divided the genus *Myrmedonia* into a number of different genera, distributed among two distinct branches, *Myrmédoniates*, and *Myrméciates*. If this arrangement were adopted, then *M. scabripennis*, *M. pollens*, *M. Batesi*, *M. spinifer*, and *M. fortunata*, would belong to the branch *Myrmédoniates*, and to Rey's genus *Zyras*, or more probably to a distinct new genus to be placed at the head of the branch: while *M. nitidula* would have to form a distinct branch intermediate between the *Myrmédoniates* and *Myrméciates*. I do not, however, adopt this classification; for while I thoroughly appreciate the great addition M. Rey's labours have made to our knowledge, I am quite convinced that the attempt to found new and complicated classifications on the insects of a single country must prove abortive; and probably worse than useless when applied to the insects of the world at large. Without therefore adopting Rey's names, I prefer to point out that the first five species here described are allied to our *Myrmedonia Haworthi*, but are remarkable by the great development of the spurs at the extremities of the tibiae. *M. nitidula*, on the other hand, is very distinct on account of the basal joint of the intermediate tarsus, which is only as long as the 2nd joint, while the basal joint of the hind tarsus is only a little longer than the following one; and the spurs at the extremity of the tibiae are much less developed. It is only when our knowledge of these insects is much more advanced than it is at present, that we shall be able to point out with something like certainty what we may hope will prove stable generric characters.

1. *Myrmedonia scabripennis*, n. sp. Piceo-nigra, nitidula, antennarum basi, pedibusque pallidis, femoribus subtus piceis; prothorace subquadrato, inæquali, fortiter punctato; elytris dense scabrosis, opacis; abdomine supra concavo, lævigato, subtus crebre subtiliter punctulato. Long. 4¼ lin.

Mas: abdomine segmento secundo dorsali spinis elongatis tribus armato; seg. 6° utrinque obsolete longitudinaliter plicato; 7° asperato, medio obtuse angulatim elevato.

Fem. latet.

Antennæ longer than the head and thorax; slender, scarcely thickened towards the apex, pitchy, paler at the base; 3rd joint elongate, quite twice as long as the 2nd; 4th shorter than 3rd, but considerably longer than the 5th; 10th about as long as broad; 11th pointed, twice as long as the 10th. Head black, shining, strongly punctured, with a narrow smooth space in the middle. Thorax about as long as broad, scarcely narrowed behind, the base rounded, the hind angles very obtuse, the front ones deflexed and rounded; very strongly and closely, but not deeply, punctured, with a transverse impression in front of the scutellum, and two irregular, not very distinct, smooth spaces on each side near the front (placed one behind the other); on each side it is broadly impressed at the sides behind. The elytra are about as long as the thorax, and nearly twice as broad; more than twice as broad as the head; black, not shining, densely scabrous. Hind body pitchy, shining, narrowed behind; the margins large, and much turned upwards; impunctate, and shining above;

very convex on the under surface, and there closely and
distinctly punctured. The legs are pale yellow, the
femora pitchy on their hind margin; the hind tarsi long
and slender; their basal joint, though not quite so long as
the other four together, is longer than the three following
taken together.

In the male the second segment of the hind body has
on the upper side three long spines, reaching about to the
extremity of the next segment, the middle one not quite
so long as the side ones; the third segment has its hind
margin a little emarginate; the sixth segment has an elevation near each side extending from the base to the
extremity, between these it is a little impressed, and
immediately in front of the hind margin is furnished with
four or five small asperities; the seventh segment is
covered above with similar asperities, and is longitudinally
elevated or swollen in the middle.

Ega; one male.

2. *Myrmedonia pollens*, n. sp. Picea, antennis pedibusque pallidioribus, thorace elytrisque opacis creberrime
ruguloso-punctatis, illo subquadrato, canaliculato; abdomine
supra nitidulo obsolete punctulato, subtus crebre punctato.
Long. 4¼ lin.

The single specimen before me of this species is clearly
immature, so that I shall not give details of its colouring
further than to say, that it is of a pitchy colour, with the
basal portion of the antennae and the legs (more particularly the femora) paler. Antennae with the 3rd joint
about twice as long as the 2nd, and one and a-half times
the length of the 4th; 5th much shorter than the 4th,
longer than broad; 6—10 differing but little in length;
the 10th scarcely so long as broad; last joint pointed,
about twice as long as the 10th. The head is closely
and rather coarsely punctured, with a smooth shining
space in the middle. The thorax is about as long as
broad, with a distinct longitudinal channel; scarcely narrowed behind; the sides nearly straight from the front
to the hind angles; it is moderately finely and very
closely punctured, so that it is not at all shining. The
elytra are about as long as the thorax and one and a-half
times its width, their punctuation extremely close, and
finer than that of the thorax, quite dull. The hind body
is shining above, and obsoletely but distinctly punctured,

the punctuation being less evident on the apical segments. Beneath it is closely and distinctly punctured. The basal joint of the posterior tarsus about as long as the three following together.

In the female the 6th segment above has a longitudinal elevation on each side; the 7th segment is covered with granulations not much elevated, but with their points directed backwards; these granulations are larger towards the extremity of the segment, and the hind margin is slightly emarginate.

Ega; one specimen, which I consider to be a female.

Obs.—Though rather closely allied to *M. scabripennis*, *M. pollens* will be very readily distinguished therefrom by the denser and finer punctuation of the thorax and elytra, and by the less uneven surface of the thorax.

3. *Myrmedonia Batesi*, n. sp. Nigro-picea, nitidula, elytris opacis, brunneis, apicem versus late infuscatis, antennarum basi pedibusque pallidis; prothorace subquadrato, inæquali, fortiter punctato; elytris dense subtiliter scabrosis; abdomine supra lævigato, subtus crebre subtiliter punctulato. Long. 4 lin.

Antennæ moderately long and slender, pitchy, paler at the base; 3rd joint twice as long as the 2nd; 4th shorter than the 3rd, twice as long as the 5th; 5—10 each a little shorter and a little broader than the preceding one, 5th about as long as broad, 10th with the breadth greater than the length; 11th pointed, quite twice as long as the 10th. Head black and shining; front punctured on each side, smooth in the middle. Thorax black, its breadth one and a third times its length, scarcely narrowed behind; the base rounded; the posterior angles extremely obtuse, the anterior rounded and deflexed; its upper surface strongly punctured and uneven, with an impression in front of the scutellum, and with an indistinct longitudinal channel, and a large ill-defined impression on each side, reaching from the posterior angles nearly to the front. Elytra about as long as the thorax, reddish, but all the hind part smoky black, densely and finely scabrous, punctate, more strongly so near the suture behind. Hind body pitchy, with the margins of the segments paler, narrowed behind, and with the basal segment a little constricted, smooth, shining and impunctate above, finely and closely punctured below; the side margins of the segments much

elevated. Legs yellowish; the basal joint of the hind
tarsus about as long as the three succeeding together.

In the female the upper side of the sixth segment has
an indistinct longitudinal elevation (or two) on each side,
and also some scattered fine asperities, with their points
directed backwards; the 7th segment is covered above
with similar, but larger and more numerous, asperities,
and its hind margin is irregularly serrate; the teeth six or
eight in number; the lateral margins of these two segments are also indistinctly serrate.

Ega; one female individual.

4. *Myrmedonia spinifer*, n. sp. Rufo-testacea, nitidula,
capite, elytrorum maculâ externâ, abdomineque medio
nigris, antennis apicem versus infuscatis; prothorace subquadrato, basin versus subangustato, ante scutellum foveolato, fortiter punctato, punctis leviter impressis; elytris
dense scabrosis; abdomine supra lævigato, subtus crebre
subtilissime punctulato. Long. 3½ lin.

Mas: abdomine segmento 2° dorsali, spinâ elongatâ,
tenui armato, marginibus lateralibus angulo acuminato,
sexto utrinque longitudinaliter plicato ante apicem granulis
duobus instructo, 7° dense granulato, medio tuberculo majori
armato, margine apicali obscure serrato, medio leviter emarginato.

Fem. abdomine simplice.

Antennæ rather long and stout, reddish-yellow, infuscated towards the apex; 3rd joint quite twice as long as
the 2nd, one and a-half times the length of the 4th; from
5—10 each joint is a little shorter than the preceding,
the 5th more than half the length of the 4th, much longer
than broad, the 10th about as long as broad; the 11th
joint pointed, quite twice as long as the 10th. Head
black and shining, strongly punctured, but with a smooth
space in the middle. Thorax shining, yellowish, about
as long as broad, the sides a little dilated in front; the
posterior angles extremely obtuse, with a fovea in front
of the scutellum, and an indistict impression on each side
behind, covered with large but little-impressed punctures;
these punctures nearly disappear in front, and are quite
absent at the anterior angles. The elytra are about as
long as the thorax and nearly twice as broad, reddish-
yellow, with a black spot on the external margin behind,
densely covered with coarse asperities or elevations. The
hind body is yellowish, marked much with black in the

middle; it is smooth and shining and impunctate, with lateral margins broad and turned outwards as well as upwards. The legs are yellowish; the basal joint of the posterior tarsus about as long as the three following together. In the male the 2nd segment is armed in the middle with a long sharp spine, about as long as the segment itself; the lateral margins of the segments are detached and acuminated behind, but not produced into spines (as they are in *M. scabripennis*); the 6th segment is furnished with a longitudinal plica on each side, and has two tubercles a little before the hind margin. The 7th segment is thickly covered with rather fine granulations, and has a larger tubercle on the middle; its hind margin is obscurely serrated and has a shallow notch in the middle. In the female the 2nd segment is simple and the tubercle of the 7th segment is absent.

Ega; three individuals,—two male, one female.

5. *Myrmedonia fortunata*, n. sp. Rufo-testacea, nitidula, capite nigro, antennis apicem versus, pectore, elytrorum lateribus, abdominisque segmento 5° infuscatis; capite prothoraceque subquadrato crebre fortiter punctatis; elytris hoc fere brevioribus, crebre minus fortiter punctatis; abdomine supra lævigato, subtus sat crebre punctato. Long. $3\frac{1}{4}$ lin.

Antennæ rather long, moderately stout, the four basal joints yellowish, the rest infuscated; 3rd joint twice as long as the 2nd; 4th intermediate in length between the 2nd and 3rd; 5th much shorter than the 4th, rather longer than broad; 5—10 differing but little in length, each slightly stouter than its predecessor; 11th rather stouter than the 10th, but not quite so long. Head blackish, shining, very coarsely punctured. Thorax yellowish, about as long as broad, slightly narrowed behind, very coarsely punctured, obsoletely depressed on each side the disc. The elytra yellowish, infuscate about the apical angle, scarcely so long as the thorax but much broader, closely and moderately finely punctured. The hind body is yellowish, the segments irregularly marked with a pitchy colour; before the apex nearly entirely pitchy; smooth, shining and impunctate above; *distinctly* and *moderately closely* punctured beneath. Legs pale yellow, with the basal joint of the hind tarsus rather longer than the three next together.

In this species the sculpture of the elytra consists of true impressions closely packed, so that the interstices appear like rough irregularly waved lines.

Ega; one female individual.

6. *Myrmedonia nitidula*, n. sp. Castanea, parce subtiliter pubescens, parce obsoleteque punctulata; thorace convexo, transversim subquadrato; elytris hujus longitudinis, abdomine magno, supra omnino lævigato. Long. 2 lin.

Mas (?): abdomine segmento sexto dorsali ante apicem lineis quatuor minus elevatis instructo, inter lineas subtilissime longitudinaliter striguloso; 7° ante apicem granulato, apice serrato.

A shining unicolorous species, with the head and thorax small in proportion to the hind body. Antennæ moderately long, a little thickened towards the extremity, their pubescence well marked; 3rd joint a little longer and rather more slender than the 2nd; 4th joint small, about half the length of the 3rd, and rather smaller than the 5th; 5—10 each is a little shorter and broader than its predecessor, each much narrowed to its base, 5th considerably longer than broad, 10th not so long as broad; 11th joint pointed, fully as stout as the 10th and twice as long. The head is but little narrower than the thorax, shining and impunctate. The thorax is convex, rather broader than long, not at all narrowed behind, the sides being straight, with the anterior and posterior angles rounded; it has a few elevated punctures, from each of which springs a fine short hair. The elytra are about as long as, but much broader than the thorax, a little narrowed at the shoulders; sparingly punctured, the points consisting of very slight elevations, from each of which rises a fine but rather long hair. The hind body is entirely smooth, shining, and impunctate above; it has the margin much developed, and directed outwards as well as upwards; beneath it is finely and sparingly punctured and very delicately pubescent. The legs are yellowish, the tibiæ much ciliated, the basal joint of the hind tarsi not greatly longer than the second.

The male has on the upper side of the 6th segment four slight raised lines, more distinct at their termination; a little before the extremity of the segment, between these lines, are numerous very fine, closely packed, indistinct lines; the hind margin of the 7th segment is serrate, and in front of the margin are some longitudinal granulations.

Ega; a single individual, which I believe to be a male, though I have not seen the ædeagus.

CALODERA.

The insect to which I have given this generic name is one which, in relation to our European species, must be considered very anomalous. The tarsi are all 5-jointed; the head is distinctly constricted behind; the antennæ are elongate, and the basal joint of the hind tarsi, though moderately elongate, is not equal to the two following together; the middle coxæ are rather widely separated, the suture between the meta- and meso-sterna being near the front of the coxæ, and nearly straight. The insect, therefore, is intermediate between Rey's two rameaux of the *Aléocharaires*, which he names *Phlœoparates* and *Caloderates*. I have made use of the generic name *Calodera* rather than *Phlœopora*, because the latter is used only in a restricted sense—that is, for a few species closely resembling one another—while *Calodera* is a very elastic and general term, applied to a number of insects with very different facies, and varying much in structure.

1. *Calodera syntheta*, n. sp. Capite thoraceque fuscis, elytris fusco-testaceis, basi dilutiore, abdomine medio apiceque nigris, pedibus antennisque testaceis, his apicem versus infuscatis; angustula, subtiliter punctata, abdomine fere lævigato, antennis elongatis, prothorace basin versus angustato. Long. corp. 1½ lin.

Antennæ elongate, rather slender, but a little incrassate towards the extremity, the three or four basal joints pale yellow, the following ones darker, and the three or four apical ones strongly infuscate; 3rd joint slightly longer than 2nd; 4th slender and rather long; 10th about as long as broad; 11th joint elongate, about as long as the three preceding joints together, acuminate, its apex rather paler in colour than its base. Head about as broad as the thorax, narrowed behind, with a moderately broad neck; the eyes large, the genæ not margined; blackish in colour, finely punctured, a little shining. Thorax not quite so long as broad, narrower than the elytra, the sides much rounded at the front angles, narrowed and sinuate towards the base, which is rounded; the lateral and basal margins distinct, the hind angles depressed, but a little prominent, and rather obtuse; the colour is deeply infuscate-yellow; the surface indistinctly and not densely punctured, without channel or fovea. Scutellum large; elytra rather longer than the thorax, yellow at the base, infuscate towards the apex, finely and not densely punctured, their hind margin almost straight. Hind body rather slender, pale yellow at the base, with the 5th and 6th segments black, the 7th reddish, the apex again dark, the basal segments finely, indistinctly and not closely punctured, the apical ones impunctate. Legs pale yellow.

Garrno; three individuals found in fungus on the 11th November, 1874, by Dr. Trail. I see no characters from which to infer their sex.

HOMALOTA.

It would not be easy at the present moment to say what the above word represents: to Erichson and Kraatz it represented a vast number of species of minute *Staphylinidæ*; to C. J. Thomson a single species, viz. *Aleochara plana*, Gyll.; while to Rey it appears to have two different and yet simultaneous meanings; first, as representing about forty-five species of French *Aleocharidæ*, and the second, with the affix "*vera*," as representing only ten of those forty-five species.

As to the species here described, I use the word in the general sense of Kraatz and Erichson; for the first five of the six species here described would no doubt have been referred by those authors to the genus *Homalota*; the other species, however, *H. culpa*, is anomalous, inasmuch as the basal joint of the hind tarsus is more elongate than is supposed to be the case with *Homalota*: but, as in the present state of confusion as to the nomenclature of the groups of species of *Aleocharidæ*, I decline to be responsible for a new generic name, I have had no option but to refer this as well as the other species to the genus *Homalota*.

1. *Homalota capta*, n. sp. Nigra, sat nitida, pedibus testaceis, antennarum basi fusco-testacco; antennis sat brevibus; capite fere transverso, vertice leviter impresso; prothorace valde transverso, dorso minus distincte cum elytris, crebre subtiliter punctatis, his illo longioribus; abdomine basi subtiliter crebre punctato, apice fere lævigato. Long. corp. ⅞ lin.

Antennæ rather short, and not stout, black, with the base paler, a little thickened towards the extremity; 1st joint thick, and much longer than 2nd; 3rd rather short and stout, shorter than 2nd; 4th to 10th scarcely different from one another in length, but the 4th not more than half as broad as the 10th; this latter as well as the preceding joints strongly transverse; 11th joint moderately long, quite as long as the two preceding together. Head short and broad, but distinctly narrower than the thorax, black, very finely punctured, the punctures absent from a longitudinal space on the middle, and the hinder part slightly impressed on the middle. Thorax almost as broad as the elytra, twice as broad as long, very finely and moderately closely punctured, with a slight depression on the disc. Elytra a good deal longer than the thorax, finely and moderately closely punctured. Hind body distinctly narrowed towards the extremity, the basal segments finely and rather closely, the apical ones sparingly punctured. Legs yellow, but a little infuscate.

In the male, the hind margin of the dorsal plate of the 7th segment of the hind body is furnished in the middle with three rather approximate, equidistant, similar spines, and on each side with an extremely long, slender, pointed spine, which is a little incurved.

A single male was found by Dr. Trail at Barreiras de Janarape, Rio Solimoes, on the 9th January, 1875.

Obs.—This species has the middle tarsi 5-jointed; the middle coxæ are distinctly separated, the meso- and metasternal processes both acuminate, and meeting one another about half-way between the coxæ; the genæ appear to me to be very finely margined. It much resembles the European *H. indiscreta*, Sharp, but is decidedly smaller, and has the male characters very different.

2. *Homalota tenax*, n. sp. Fusca, pedibus testaceis, antennarum basi fusco-testaceo, thorace abdomineque segmentis basalibus obscure rufescentibus; antennis crassiusculis; thorace transverso, elytris breviore; abdomine subparallelo, segmentis basalibus sat crebre, apicalibus parce punctatis. Long. corp. 1⅛ lin.

Antennæ stout, moderately long, black, the basal joint obscurely pale; 3rd joint stout, nearly as long as 2nd; 4th smaller than the others and scarcely so long as broad; 5—10 differing but little from one another, each distinctly transverse; 11th joint rather elongate, fully twice as long as the 10th. Head short and broad, finely and not densely punctured, nearly black. Thorax a little narrower than the elytra, broader than long, slightly rounded at the sides and a little narrowed towards the front, the base much rounded, the hind angles quite indistinct; it is rather paler in colour than the head and elytra, finely and not closely punctured. The elytra are distinctly longer than the thorax, but their punctuation is similar to it; their hind margin is not sinuate. Hind body with the basal segments distinctly and not densely, the apical one sparingly punctured; the three basal segments reddish, the others darker. Legs yellow.

In the male, the hind margin of the dorsal plate of the 7th segment of the hind body has in the middle a long and broad projection which is divided in the middle, each side being rounded, while at each side the hind margin terminates in a long, slender, pointed and incurved spine, which projects fully as far backwards as the central process.

Barreiras de Janarape, 9th January, 1875, a single male brought back by Dr. Trail.

Obs.—Though much smaller, this species is somewhat similar to *H. fungicola*, particularly in the structure of its antennæ. The genæ are very finely margined, the middle coxæ distinctly separated, the mesosternal process much produced between them.

3. *Homalota brevis*, n. sp. Opaca, breviuscula, testacea, nigro-cingulata, antennis extrorsum infuscatis; oculis magnis; prothorace valde transverso; abdomine crebre punctato. Long. corp. ¾ lin.

Antennæ stout, the three or four basal joints pale yellow, the others infuscate; 3rd joint more slender and rather shorter than 2nd; 4—10 about equal to one another in length, each a little broader than its predecessor, the 4th about as long as broad, the penultimate ones transverse; 11th broad, quite as long as the two preceding together. Head broad and short, but a good deal narrower than the thorax, the eyes occupying nearly all the sides; blackish in colour, finely punctured. Thorax distinctly narrower than the elytra, strongly transverse, the base much rounded, but a little truncate in front of the scutellum, the hind angles indistinct; it is yellowish in colour, finely punctured, without channel or impression. Elytra broad, longer than the thorax; black with a violet tinge, closely and finely punctured, their hind margin not at all sinuate at the outer angles. Hind body broad and short; the three or four basal segments yellow; the following three abruptly black, the extremity infuscate-yellow; the segments are finely and rather closely punctured, the apical ones not so densely, however, as the basal ones. The legs, including the coxæ, are pale yellow.

Garrao; five individuals found in fungus by Dr. Trail on the 11th November, 1874. I see no indications of external sexual characters.

Obs.—This species is similar in size and form to our European *H. celata*, but the colouration is rather that of *Gyrophænæ*. The genæ are finely margined; the middle coxæ quite separated; the metasternum is much produced between them, and is separated only by a narrow space from the extremity of the mesosternal process, the extremities of both processes being rounded; the middle tarsi are 5-jointed, the basal joint a little longer than the 2nd; 2—4 about equal to one another; hind tarsi with the 2nd joint a little shorter than the 1st, and the 3rd a little shorter than the 2nd, slightly longer than the 4th.

4. *Homalata gilva*, n. sp. Testacea, elytris abdomineque cingulo ante apicem nigricantibus; prothorace transverso, basi rotundato; abdomine basi crebre subtiliter, apice parce punctato. Long. corp. ¾ lin.

Antennæ moderately long and stout, yellow; 3rd joint shorter and more slender than 2nd; 4th joint very small; 5th a good deal larger than 4th; 6—10 each slightly shorter and broader than its predecessor, the 6th about as long as broad, the 10th rather strongly transverse; 11th joint large, longer than the two preceding ones together. Head short and broad, a good deal narrower than the thorax; dark yellow, very finely punctured, not shining; the eyes large. Thorax strongly transverse, nearly as broad as the elytra, the sides slightly narrowed behind, the base much rounded; it is yellow in colour, not at all shining, very finely punctured, without channel or fovea. Elytra short, but a little longer than thorax, infuscate; the base paler, finely and closely punctured, quite dull; the hind margin straight. Hind body yellow, with the 5th segment black; the basal segments finely punctured, the sixth nearly impunctate. Legs pale yellow.

In the male the dorsal plate of the 7th segment of the hind body is deeply emarginate on each side, so as to form a large central lobe, and on each side a projecting spine; the central lobe is itself emarginate at the hind margin, and its angles are quite rounded; the lateral spine on each side is pointed, and very nearly straight, and does not extend quite so far backward as the most prominent part of the central lobe.

Ciarrao; a single male, found with *H. brevis*.

Obs. I.—The middle tarsi in this species are 5-jointed, the two basal joints being short, and similar to one another in length; the four basal joints of the hind tarsus also differ but little from one another in length; the middle coxæ are distinctly separated; the apices of the meso- and metasternal processes are rounded and widely separated from one another by a black space.

Obs. II.—This species is smaller and narrower than *H. brevis*, to which it bears a considerable resemblance; the antennæ are rather more slender, and the hind body less densely punctured. The individual described is perhaps immature.

5. *Homalota Traili*, n. sp. Testacea, capite, elytris, pectore, abdomineque cingulo ante apicem infuscatis; antennis fuscis, basi apiceque testaceis; prothorace transverso, medio indistincte impresso; abdomine basi sat crebre, apice parce, punctato. Long. corp. $\frac{7}{8}$ lin.

Antennæ rather stout, only moderately long, the two or three basal joints obscure yellow, the following ones infuscate, the 11th clear yellow; 3rd joint slender at base, but broad at apex, almost equal to 2nd joint in length and breadth; 4th not shorter and only slightly narrower than 5th, hardly so long as broad; 5—10 differing but little from one another, each transverse, the 10th more strongly so than the 5th; 11th joint rather elongate, quite as long as the two preceding together. Head broad and short, the eyes moderately large, infuscate-yellow, finely punctured, with the vertex depressed in the middle. Thorax rather strongly transverse, slightly narrower than the elytra, nearly straight at the sides, with the base rounded, yellowish in colour, very finely punctured and obscurely impressed along the middle. Elytra a little longer than the thorax, finely and closely, but distinctly, punctured. Hind body yellow, with the 5th and 6th segments infuscate, very finely and not densely punctured, the apical segments rather more sparingly so than the basal ones. Legs pale yellow.

In the male the hind margin of the dorsal plate of the 7th segment of the hind body is emarginate in the middle, and on each side of the middle, so as to form four almost equidistant teeth; the middle ones are short prominences, the lateral ones are quite as elongate as the middle ones and more slender and more spine-like.

Rio Purus; a single male found by Dr. Trail (after whom I have named the species), on the 13th Oct. 1874.

Obs.—This species in the structure of its tarsi and sternal processes much resembles *H. gilva*; it is rather larger than that species, has the surface less opaque because less densely punctured; the antennæ differently coloured, with the 4th joint larger, and the male characters different. From *H. brevis*, the less densely punctured upper surface, and the different structure of the intermediate joints of the antennæ, readily distinguish it. It is probable that the individual described is a little immature.

6. *Homalota culpa*, n. sp. Rufo-testacea, antennis apicem versus pallidioribus; capite, elytrorum medio, abdomineque ante apicem, infuscatis; capite, thorace, elytrisque subtiliter punctatis, abdomine impunctato; prothorace transversim subquadrato. Long. corp. 1⅜ lin.

Antennæ stout, moderately long, a good deal thickened towards the extremity, the basal portion yellowish, the apical very pale yellow; 2nd and 3rd joints elongate, about equal to one another; of 4—10 each is distinctly broader, but scarcely shorter than its predecessor, the 4th longer than broad, the 10th rather strongly transverse; the 11th joint rather stout, as long as the two preceding together. Head broad and short, without distinct neck, a good deal narrower than the thorax, finely punctured, infuscate. Thorax about as broad as the elytra, a good deal broader than long, the sides a little rounded in front, distinctly narrowed behind the middle, the hind angles distinct but very obtuse, the sides and base with a very fine but distinct margin; the colour is yellowish, the surface very finely punctured without channel or fovea. Elytra broad and short, a little longer than the thorax; their hind margin not (or scarcely) sinuate at the outer angles, their colour yellowish, but with a very broad, ill-defined, smoky-violet colour across the middle, their punctuation fine and rather dense. Hind body broad, a little rounded at the

sides and narrowed behind, yellowish in colour, with the
6th segment infuscate, shining and without punctuation;
beneath finely and closely punctured and pubescent. Legs
pale yellow. Tibiæ slender and elongate.

Tapajos; two specimens. Perhaps found in company
with *Oxypoda aliena*, which the species much resembles,
though it is a good deal larger.

Obs.—In this species the genæ are immarginate; the
middle coxæ are widely separated; the apex of the metasternum is truncate and separated from the little produced
portion of the mesosternum by a narrow black space. The
anterior tarsi are 4-, the intermediate and posterior
5-jointed; the basal joint of the middle tarsus is about
equal in length to the 2nd joint; the basal joint of the
hind tarsus is elongate and about equal in length to the
two following joints together.

TACHYUSA.

About thirty species, most of them European, are at
present included under this generic name; none of them
are from tropical America, but two are described by
Fauvel, from Chili. The genus has not yet been treated
of by Rey, but we may confidently expect that in the
forthcoming part of his work the name *Tachyusa* will be
used with a very different application to that of the works
of Erichson and Kraatz.

As regards the two species here described I need only
remark that they appear closely allied inter se, and that
they are not nearly allied to any of our European species.

1. *Tachyusa picticornis*, n. sp. Testacea, nitidula, fere
impunctata, antennis articulis 3—6 rufescentibus, 7—11
albidis. Long. corp. 1¾ lin.

Broader than *T. ferialis*. Antennæ long, reaching
about to the end of the elytra, rather stout for this genus,
a little thickened towards the extremity; joints 1 and 2
yellowish; 3rd joint one and a half times the length
of the 2nd; 4th about as long as the 2nd, but stouter;
5—10 each just a little shorter and stouter than the preceding, 5th much longer than broad, 10th about as long
as broad; last joint long and pointed, two and a half times
the length of the tenth; joints 3—6 brownish, those beyond abruptly paler, almost white. Head about as broad
as the thorax, yellowish; the eyes black, large and pro-

minent. Thorax about as long as broad, dilated in front, with the anterior angles rounded; again just a little broader at the posterior angles, which are deflexed and obtuse; yellowish, shining and impunctate. Elytra much broader and a little longer than the thorax, also a little darker in colour. Hind body smooth, shining and impunctate, its sides gently rounded, scarcely any narrower at the base than at the 6th segment. Legs yellow, elongate and slender; posterior tarsi with the three basal joints elongate, each a little shorter than the preceding one; middle coxæ widely separated, mesosternum very little produced between them, and separated from the produced point of the metasternum by a broad space.

Tapajos; one individual.

2. *Tachyusa extranea*, n. sp. Rufo-testacea, capite, elytris abdominisque segmento sexto obscurioribus; subtilissime punctulata, thorace subquadrato, obsolete canaliculato, elytris hujus longitudinis. Long. corp. 1¼ lin.

This species has much the form of a *Homalota*. The antennæ are yellowish, elongate and rather stout, slightly thickened towards the apex; 3rd joint a little longer than the 2nd; 4th shorter than the 2nd; 5—10 differing but little in length, 10th about as long as broad; 11th joint long, rather pointed, more than twice as long as the 10th. Head brownish-yellow, slightly narrower than the thorax, short, smooth, shining and impunctate. Thorax yellowish, with the sides in front dilated, rounded and much deflexed; it is about as long as broad, indistinctly channelled, almost imperceptibly punctured. The elytra are broader than the thorax, about as long, a little darker in colour, and with their punctuation not quite so obsolete. The hind body is but little narrowed at the base; it is yellowish, with the 6th segment a little darker; it is smooth, shining and impunctate. The legs are yellowish; the posterior tarsi slender and moderately long.

Tapajos; two individuals, much mutilated.

Obs.—This species appears to be nearly allied to *T. picticornis*, at any rate in so far as the structure of the sternum is concerned; its tarsi I have been unable to examine. Its smaller size and unicolorous antennæ render it easily distinguishable from *T. picticornis*.

OXYPODA.

Only one species with this generic name is as yet described from tropical America; a number of species have, however, been described from Chili by M. Fauvel. The single species I here describe differs a good deal in its form from our European species of the genus, and resembles rather the Chilian *O. scutellata.* I have also one or two other closely allied Brazilian species in my collection, so that probably these South American species will be distinguished ultimately as a separate genus.

1. *Oxypoda aliena,* n. sp. Latior, omnium subtilissime punctulata, prothorace valde transverso, elytris latiore; abdomine apicem versus attenuato, evidentius pubescente; testacea, elytris abdomineque ante apicem fusco-signatis. Long. corp. 1 lin.

This species is remarkable for its short and very broad form, its prothorax being particularly broad and short. Antennæ yellow, rather long, distinctly thicker towards the apex; joints 2 and 3 about of equal length; 4th joint much shorter than the 3rd; from the 4th to the 10th the joints differ but little in length, each is just a little broader than its predecessor, 4th joint slender, much longer than broad, 10th about as long as broad; last joint long, quite twice as long as the 10th, rather pointed. Head broad, nearly half as wide as the thorax, yellow, shining, and with the finest possible punctuation and pubescence. Thorax three times as broad as long, the sides very gently rounded, just a little narrower at the front than at the hind angles; all the angles, especially the front ones, extremely rounded, without channel or fovea, yellow, and with an almost invisible punctuation and pubescence. Elytra distinctly narrower than the thorax, but about as long, yellow, slightly obliquely darker across the middle with the external angle a little paler, with an extremely fine punctuation and pubescence. Hind body much narrowed at the apex, extremely finely and closely punctured, with a depressed, long and distinct, though extremely fine pubescence; the 5th segment a little darker in colour than the others. Legs yellow; hind tarsi with the joints elongate and slender, the 1st more than twice as long as the second.

Tapajos; four individuals.

ALEOCHARA.

The four species described under this generic name appear to be closely allied in structure to one another and to the European *A. fuscipes*. Allied species appear to be found all over the world, and some species have an extremely wide geographical range. I have, however, seen no specimens of any of the species here described from any other locality than the Amazons, and I have also failed to identify any of the four with species previously described from tropical America; in which part of the world, I may remark, that allied species are no doubt pretty numerous, though as yet scarcely a dozen have been described as purporting to belong to the genus and to the locality mentioned.

1. *Aleochara prisca*, n. sp. Latior, nigra, antennis medio subincrassatis, elytris fuscis, thorace brevioribus; pedibus fusco-testaceis, ano rufo-testaceo, abdomine sat crebre minus fortiter punctato. Long. corp. 4 lin.

Allied to *A. fuscipes*. Antennae black, very stout, with the joints very short; 2nd and 3rd joints slender, the latter the longer; 4th strongly transverse; 5—7 similar to one another, 5th considerably broader than the 4th, after the 7th slightly narrower again to the extremity. Head very much narrower than the thorax, distinctly but not closely punctured. Thorax rather narrowed in front, its breadth about one and a half times its length, moderately closely and distinctly punctured. Elytra considerably shorter than the thorax, not quite black, closely and distinctly punctured. The hind body is broad, not narrowed till the 6th segment, moderately closely and strongly punctured; the punctuation of the basal segments finer than that of the rest; the 7th segment, as well as the hind margin of the 6th, reddish. Legs pitchy, with the anterior tibiae yellowish.

Ega; one specimen, which appears to be a female.

Obs.—This species appears to be very closely allied to the European *A. fuscipes*, but it has the elytra darker and the extremity of the hind body paler; the 4th joint of the antennae is more transverse; the head has an impunctate space on the disc, and the punctuation of the hind body is rather finer and closer.

2. *Aleochara verecunda*, n. sp. Nigra, antennis basi,

pedibusque testaceis, clytris thorace brevioribus anoque obscure rufis; thorace crebre minus subtiliter punctato; abdomine apicem versus subattenuato. Long. corp. 3½ lin.

Allied to *A. fuscipes*, and almost equal in size thereto, but rather more slender, and with the red of the extremity of the hind body more distinct. The antennæ are short and stout, the three first joints yellow; the 3rd a little longer than the 2nd; the 4th strongly transverse, but much narrower than the 5th; the rest strongly transverse, slightly narrower after the 7th to the extremity; last joint quite twice as long as the preceding one. Palpi obscurely yellowish. Head narrow, scarcely half so wide as the thorax, rather coarsely but not closely punctured. Thorax narrowed to the front, its breadth quite one and a half times its length, closely and strongly punctured. Elytra considerably shorter than the thorax, reddish, closely and rather strongly punctured. Hind body slightly narrowed to the extremity; 2nd and 3rd segments sparingly and finely punctured, the rest coarsely but not closely; the 7th segment and the hind portion of the 6th reddish. Legs dull yellowish.

Tapajos; one specimen, which is, I have no doubt, a female.

Obs.—Though closely allied to *A. prisca*, this species may be readily distinguished from it by the pale basal joints of the antennæ.

3. *Aleochara auricoma*, n. sp. Rufo-testacea, fulvo-pubescens, capite nigro, antennis abdominisque segmentis 2—5 nigricantibus; clytris thorace brevioribus; abdomine segmentis 3—7 sat crebre punctatis. Long. corp. 4 lin.

Antennæ blackish, with the apical joint blackish-yellow, stout; 3rd joint one and a half times the length of the 2nd; 4th scarcely so long as broad; 5—10 strongly transverse; last joint elongate, rounded at the extremity, more than twice as long as the tenth. The head is black, rather strongly but not closely punctured. The thorax is yellow, ample, its breadth one and a third times its length; its base and hind angles much rounded, narrower in front than behind; it is moderately, finely and closely punctured, without channel or fovea. The elytra are not more than two-thirds the length of the thorax, with a similar colour and punctuation to it. The hind body is pitchy-red at the extreme base, the segments becoming darker till the 5th, which is quite black; the 6th and 7th bright-orange; segment 2 almost impunctate; the 3rd rather sparingly and moderately finely, the 4th—7th evenly and distinctly, but not densely, punctured. Legs yellow.

Ega; two specimens, one of which I have ascertained by dissection to be a male, while the other I suppose to be a female. The male carries no external indication of its sex; the dorsal and ventral plates of the 7th segment are both truncate, with the angles rounded, and without visible crenulations.

4. *Aleochara mundana*, n. sp. Ferruginea, capite abdomineque (apice excepto) nigricantibus; crebre sat fortiter punctata; elytris thorace brevioribus. Long. corp. 2½— 2¾ lin.

Similar in build to *A. tristis*, but rather larger, very differently coloured. Antennæ tawny, the 3rd joint a little longer than the second, 4th short and transverse, 5th considerably broader than the 4th; after this the joints become no broader, each markedly transverse; the last joint twice as long as the preceding. Palpi tawny. Head black, half the width of the thorax, moderately distinctly punctured, with a well-marked yellow pubescence. Thorax tawny, nearly twice as broad as long, a little narrowed in front, rather closely and finely punctured, with a yellow pubescence. Elytra tawny, very short, considerably shorter than the thorax, rather closely and finely punctured. Hind body tawny black at the base, darker till the 5th segment, which is quite black; the 6th segment (except the base) and the 7th orange-coloured; rather closely punctured, the basal segments more finely than the apical ones; scarcely narrowed till the 6th segment. The legs are yellow.

This is probably a very common species in the Amazon district, extending from Pará to Ega. I have five specimens before me, coming from Pará, Tapajos and Ega; one of them bears a ticket,—" in dung."

Obs.—This species, though very closely allied to *A. auricoma*, is easily distinguished therefrom; it is a little smaller, and its colours are not quite so brightly contrasted; the antennæ are paler and less stout than in *A. auricoma*, and the punctuation of the upper surface is a little finer and closer. There are no external marks of the sexes to be seen.

GYROPHÆNA.

Eleven species referred by me to this genus are here described: one of these, viz., *G. pumila*, is among the smallest of known *Staphylinidæ*; two others, *G. boops* and *G. debilis*, depart widely from the European species of the genus, inasmuch as they have the eyes of unusual size, far surpassing in this particular any other *Aleocharidæ* known to me. Erichson has already described, from Brazil and North America, one or two species resembling them in this respect, and the European species differ somewhat from one another in the size of the eyes and the form of the head, so that I do not consider that it is at present advisable to make a distinct generic name for these insects.

The genus, as at present understood, is probably distributed over nearly all quarters of the world, but the extra-European species as yet described are not very numerous; only three, in fact, have yet been described from tropical America; though, judging from the number here described, as well as from numerous other species in my collection, it is pretty certain that these insects will prove to be very rich in species in South America.

1. *Gyrophæna pumila*, n. sp. Fusca, nitidula, fere impunctata, antennis pedibusque testaceis, prothorace valde transverso, basi et lateribus rotundatis. Long. corp. $\frac{1}{2}$ lin.

Antennæ short and stout, yellow; 2nd joint stout; 3rd joint small, much shorter and thinner than 2nd, the basal much narrower than the apical portion; 4th joint much smaller than the following ones; 5—10 nearly equal to one another in breadth, the 10th, however, a little broader and longer than the 5th, each of them transverse; 11th joint stout, obtuse. Head small, a good deal narrower than the thorax; the eyes small, the surface smooth and shining. Thorax very transverse, a little narrower than the elytra, more than twice as broad as long; the base greatly rounded, the surface smooth and shining. Elytra short, a little longer than the thorax, with a few indistinct and distant punctures. Hind body impunctate. Legs yellow.

Rio Purus, 24th September, 1874, Dr. Trail; a considerable number of examples were found, but most of them have come to pieces in the spirit in which they were preserved.

Obs.—This minute insect is smaller than the European *G. boleti*, and in size scarcely equals our smallest *Oligotæ*. I find it not easy to see with distinctness the form of the hind margin of 7th segment, but I believe the dorsal plate has the hind margin a little obtusely prominent in the middle, and has a curved spine on each side.

2. *Gyrophæna parvula*, n. sp. Nigro-fusca, nitidula, fere lævis, elytris parce punctatis, antennis pedibusque testaceis; prothorace valde transverso, basi et lateribus rotundatis. Long. corp. $\frac{3}{8}$ lin.

This species is extremely similar to *G. pumila*, but the elytra are more distinctly, and not quite so sparingly strewed with punctures, which on examination with a high power appear to me to be fine elevated granules; besides this, it is larger and broader, with the head especially a good deal broader; in other respects it appears extremely similar to *G. pumila*.

A single individual, found on the Rio Purus with *G. pumila* by Dr. Trail.

3. *Gyrophæna parca*, n. sp. Nigro-fusca, nitida, antennis fuscis, basi pedibusque testaceis; parcissime punctata, prothorace valde transverso, disco subtiliter quadripunctato. Long. corp. $\frac{7}{8}$ lin.

Antennæ short and stout; the four basal joints yellow, the others infuscate; 3rd joint much smaller than 2nd, its basal portion constricted; 4th joint extremely small; 5—10 similar to one another, each rather strongly transverse; 11th short and obtuse. Head broad, but a good deal narrower than thorax, shining, black, with a few very fine punctures; the eyes moderately large. Thorax strongly transverse, much rounded at the base, and distinctly narrowed towards the front; pitchy, very shining, with four fine punctures on the middle, placed so as to form the corners of a square. Elytra short, but a good deal longer than the thorax, pitchy, shining, with a very few indistinct punctures. Hind body pitchy, slightly paler at the base and extremity; almost impunctate, but not quite so shining as the anterior parts. Legs yellow.

In the male the hind margin of the dorsal plate of the 7th segment of the hind body is produced in the middle so as to form two short, stout, almost confluent teeth, and it has also at each outer angle a rather longer and more slender pointed spine.

Rio Purus, 13th October, 1874, and Garrao, 11th November, 1874; a single male from each locality, captured by Dr. Trail.

4. *Gyrophæna lævis*, n. sp. Sat convexa, nitidula, nigro-picea, antennarum basi pedibusque testaceis; prothorace disco quadripunctato; clytris parce punctatis. Long. corp. 1 lin.

In the male the hind margin of dorsal plate of the 7th segment of the hind body is armed in the middle with two fine, pointed, approximate spines, and on each side with a stouter, distinctly incurved spine; the hind margin is most prominent in the middle, so that though the outer spines are considerably longer than the two middle ones they do not project farther backwards.

Garrao, 11th November, 1874; two male individuals found in fungus by Dr. Trail.

Obs.—This species is excessively closely allied to *G. parca*, and agrees almost exactly with it in most respects; but it is a little larger and more convex, and is readily distinguished by the different character of the teeth on the 7th segment of the hind body in the male.

5. *Gyrophæna juncta*, n. sp. Picea, nitidula, antennarum basi pedibusque testaceis; clytris abdomineque castaneis, hoc ante apicem piceo, illis versus angulos externos infuscatis; prothorace valde transverso, disco quadripunctato, clytris apicem versus fortiter punctatis. Long. corp. 1 lin.

Antennæ only moderately stout, the first four joints yellow, the others infuscate; 3rd joint small, very much smaller than 2nd; 4th minute; 5—10 differing little from one another, transverse, but not strongly so; 11th obtuse. Head pitchy, with some fine but distinct punctures on each side; it is much narrower than the thorax. Thorax very transverse, the base rounded, the sides only a little narrowed towards the front; it is of a shining pitchy colour, and on the disc has four rather fine punctures. Elytra distinctly longer than the thorax, of a chestnut-yellow colour, but infuscate at the sides towards the hinder angle; they are shining, and have some unevenly distributed coarse punctures, they being most numerous and distinct towards the outer angle. Hind body dark yellowish, but infuscate towards the extremity, extremely obsoletely punctured. Legs yellow.

In the male the hind margin of the dorsal plate of the 7th segment of the hind body bears in the middle two fine approximate spines or teeth, and at each outer angle a stouter acuminate spine.

Garrao; a single male found in fungus by Dr. Trail, 11th November, 1874.

Obs.—The male characters here scarcely differ from *G. lævis*, but in other respects the two species are easily distinguished; the coarse punctures on the elytra of the *G. juncta* will afford an easy means of distinguishing the species from *G. lævis*.

6. *Gyrophæna convexa*, n. sp. ♂. Nitidula, picea, antennarum basi, abdomine pedibusque testaceis; prothorace antrorsum angustato, disco quadripunctato; elytris parcius granulatis; abdomine obsoletissime punctulato. Long. corp. 1⅛ lin.

Mas: abdomine segmento 7° dorsali apice, medio laminâ magnâ triangulari apice fissâ, utrinque spinâ elongatâ.

Antennæ with the four basal joints yellowish, the others blackish; 3rd joint rather shorter than 2nd; 4th much smaller than any of the others; 5—10 scarcely differing from one another, each about as long as broad; 11th joint rather short, quite as stout as, and a good deal longer than, the 10th. Palpi pale yellow. Head much narrower than the thorax, the eyes moderately large and finely granulated, pitchy in colour, and with a series of about four fine punctures on each side the middle. Thorax strongly transverse, rounded at the sides and narrowed towards the front; the base distinctly margined and slightly truncate in front of the scutellum, and a little sinuate on each side near the outer angle; the colour is pitchy, and on the disc are four large equidistant punctures; between these and each side is a finer puncture, and there are also two or three fine punctures very close to the front margin. Elytra short and broad, about as long as the thorax, pitchy, with the shoulders paler, along the suture with a series of fine tubercles and with other tubercles elsewhere. Hind body broad, yellow, almost impunctate. Legs pale yellow.

In the male the 6th dorsal segment is opaque and bears a few obsolete tubercles; the dorsal plate of the 7th segment is transversely depressed along the middle, and beyond the depression is produced as a large triangular plate, the apex of which is divided by a narrow slit; on each side is a long pointed spine, directed inwards, and attaining a similar length to that of the central plate.

A single male individual of this species was found in fungus by Dr. Trail, at Garrao, on the river Jurua, on the 11th of November, 1874.

7. *Gyrophæna sparsa*, n. sp. Convexa, nitidula, picea, antennarum basi, pedibusque testaceis; abdomine basi dilutiore; prothorace valde transverso, disco quadripunctato; elytris parce tuberculatis, versus angulos externos lævibus. Long. corp. vix 1 lin.

Antennæ short and stout, the four basal joints yellow, the others darker; 3rd joint very small, much smaller than 2nd; 4th very minute; 5—10 very similar to one another, rather strongly transverse; 11th short and obtuse. Head a good deal smaller than the thorax, pitchy, shining, scarcely visibly punctured. Thorax strongly transverse, much rounded at the base, which is a good deal emarginate in the middle in the front of the scutellum; the sides also a good deal rounded; it is of a shining, pitchy colour, with four fine punctures on the disc. Elytra a good deal longer than the thorax, rather paler in colour, very shining, with a few rather coarse elevated tubercles, which however do not extend to the outer angles. Hind body impunctate, pitchy yellow, with the penultimate segments pitchy. Legs yellow.

In the male the hind margin of the 7th segment of the hind body forms in the middle a rather large triangular projection, and at each outer angle has a curved, pointed spine, which reaches a little further backwards than the middle projection.

Garrao; a single male found in fungus on the 11th of November, 1874, by Dr. Trail.

8. *Gyrophæna quassa*, n. sp. Picea, nitidula, antennis pedibusque testaceis, illis basi quam apice dilutioribus, abdomine rufo-obscuro, ante apicem piceo; prothorace transverso, fere impunctato; elytris parce tuberculatis. Long. corp. 1¾ lin.

Antennæ moderately long and not stout, yellowish, with the four basal joints paler than the others; 3rd joint very small; 4th minute; 5—10 very similar to one another, each a little transverse. Head a good deal smaller than the thorax, pitchy yellow, quite shining, almost impunctate, the eyes rather large. Thorax a good deal narrower than the elytra, but strongly transverse, the sides and base distinctly rounded, the surface very shining, and with the discoidal punctures scarcely visible, the colour slightly paler than that of the head. Elytra a little longer than the thorax, very shining, almost similar in colour to the head, the basal portion being a little paler than the apical; they bear a few elevated punctures, which are most distinct on the sutural portion of their area. Hind body distinctly narrowed towards the extremity, obscure yellowish, with the penultimate segments darker, almost impunctate. Legs yellow.

In the male the hind margin of the dorsal plate of the 7th segment of the hind body is produced in the middle, so as to form an obtuse angle, and each outer angle possesses a moderately long incurved spine, which projects a little further backwards than does the central prominence.

Garrao; a single male found in fungus on the 11th of November, 1874, by Dr. Traill.

9. *Gyrophæna tridens*, n. sp. Nitidula, castaneo-testacea, elytrorum lateribus abdomineque ante apicem infuscatis; antennarum basi, pedibusque testaceis; prothorace transverso, disco quadripunctato; elytris parce punctatis. Long. corp. ⅞ lin.

Antennæ only moderately stout, yellowish, with the four basal joints paler; 3rd joint small, much smaller than 2nd; 4th minute; 5—10 differing little from one another, each a little shorter than long. Head dark yellowish, rather small, almost impunctate. Thorax slightly narrower than the elytra, strongly transverse, much rounded at the base, and a little narrowed towards the front, shining, similar in colour to the head, with four punctures on the disc. Elytra a little longer than the thorax, infuscate-yellow, with the base paler, shining, sparingly and not very distinctly punctured. Hind body almost impunctate, yellowish, with the penultimate segments infuscate. Legs yellow.

In the male the hind margin of the dorsal plate of the

7th segment of the hind body forms a very acute tooth in the middle, and at each outer angle has a stout, elongate, pointed spine, which is distinctly curved inwards and projects much farther back than the central tooth.

Garrao; a single male found in fungus 11th November, 1874, by Dr. Trail.

10. *Gyrophæna boops*, n. sp. Rufo-testacea, nitida, capite, elytris (humeris exceptis) abdomineque ante apicem nigricantibus; oculis maximis; capite thoraceque fere impunctatis; elytris abdomineque parce punctatis, pubescentiâ sparsâ distinctâ. Long. corp. 1¼ lin.

Antennæ rather slender and moderately long, yellow; 3rd joint more slender and distinctly shorter than the 2nd; 4th joint only a little smaller than the 5th; 5—10 only differing slightly from one another in width and scarcely at all in length, the 5th about as long as broad, the 10th a little transverse; 11th moderately long, pointed. Head blackish, the eyes exceedingly large and convex, occupying the whole side of the head, coarsely facetted; the space between them is quite parallel-sided, shining, and almost impunctate. Thorax very strongly transverse, more than twice as broad as long, slightly narrowed behind, the base rounded; it is of a shining-yellowish colour, and is almost impunctate. Elytra a good deal longer than the thorax, shining, blackish, with the humeral angles yellow, with a very few, fine, setigerous punctures. Hind body yellow, with the penultimate segments blackish, the segments very finely and sparingly punctured, and also sparingly pubescent. Legs clear yellow.

In the male the elytra bear each near their hind margin two tubercles, one near each angle; the dorsal plate of the 6th segment of the hind body bears in the middle, at the base, an oblong, large, shallow impression; the hind margin of the following segment exhibits an obtuse, tubercle-like projection in the middle, and on each side a very short spine.

Rio Purus, 13th October, 1874; a fine series brought by Dr. Trail; also an individual from Ega found by Mr. Bates; and yet another, found at Garrao, on the 11th November, 1874, in fungus, by Dr. Trail.

Obs.—This species varies considerably in size, somewhat in colour and sculpture, and a good deal in the secondary sexual characters of the male: the individual above described is one of the largest and most developed males; the thorax is sometimes infuscate, and even quite pitchy, and the dark colour in other cases reduces the pale colours to smaller areas. The tubercles on the elytra of the male seem in some cases nearly to disappear, and possibly in some individuals of that sex may be quite absent.

11. *Gyrophæna debilis*, n. sp. Rufo-testacea, nitida, parcissime punctulata, capite, elytris apicem versus, abdomineque ante apicem infuscatis; oculis maximis. Long. corp. ⅞ lin.

The male of this species has a minute tubercle at the hinder and inner angle of each elytron; the dorsal plate of the 6th segment of the hind body has an indistinct impression on the middle of the basal portion; the hind margin of the following plate is narrow, and very nearly truncate; the middle being scarcely visibly prominent, and the outer angle on each side bearing only a very short, indistinct, obtuse projection.

Rio Purus, 13th October, 1874; a single male found by Dr. Trail, and a second individual found at Garrao, in fungus, on the 11th November, 1874.

Obs.—I am by no means sure that this individual may not ultimately be shown to be an extremely undeveloped male of the variable *G. boops*. It is, however, rather smaller than the smallest individual of that species, and the colours are paler than the palest thereof; the 4th joint of the antennæ appears to be smaller in proportion to the 5th, and the male peculiarities of the hind margin of the 7th segment are so reduced as to be almost absent. Should this insect ultimately be shown by a series of intermediate individuals to be a mere form of the *G. boops*, we shall then have demonstrated to us that the male secondary sexual characteristic in this species (as is known to be the case with many larger *Coleoptera* of different families) may be in some individuals almost entirely absent; and this will be all the more interesting, as in this case we shall see that this diminution of the male peculiarities extends even to those external parts that are in most immediate proximity to the primary sexual organs; for I have convinced myself, by numerous dissections of *Gyrophænæ*, that the ædeagus and its coverings have their chief attachment to the dorsal plate of the 7th segment of the hind body.

DEINOPSIS.

Only five species of this extremely distinct and peculiar genus have yet been described, but they come from very different parts of the world, so that this genus is probably nearly universally distributed. No species has hitherto been made known from South America; but I have, besides the two species here described, one or two others from that quarter, so that it would seem probable the species will prove to be rather more numerous there than elsewhere.

1. *Deinopsis Matthewsi*, n. sp. Opacus, nigro-fuscus, dense subtilissime punctatus. Antennis, palpis pedibusque rufescentibus, antennis medio infuscatis. Long. corp. 2⅛ lin.

Very similar in form to *D. fuscatus*, but larger, and with the antennæ more slender and elongate; they are rather less than ⅞ lin. in length; 3rd joint a little shorter than 2nd, but a little longer than 4th; 5—9 very similar to one another in length; 10th a little shorter than 9th; 11th joint shorter than 10th, and terminated by a seta-like spine. The palpi and front of the head are obscure red. The margins of the thorax are also very obscurely red. The elytra are very deeply sinuate at their outer angle.

A single individual of this species was found by Mr. Bates, but I have no exact indication of its locality; the specimen is, I believe, a female.

I have named this species in honour of the Rev. A. Matthews, who has displayed a most extraordinary amount of entomological skill in his treatment of the *Trichopterygidæ*, and to whom this genus is of special interest, as he considers it to make a remarkable approach in many points, both of internal and external structure, to the *Trichopterygidæ*.

2. *Deinopsis longicornis*, n. sp. Ferrugineo-nigra, antennis, palpis pedibusque rufis; antennis tenuissimis, valde elongatis. Long. corp. (abdomine extenso) 3 lin.

Very similar in form to *D. fuscatus*, but much larger, and of a more rusty black colour. The antennæ are yellowish-red, very slender and elongate, being just over one line in length. The front of the head, the margins of the thorax, and the extremity of the hind body, are of an obscure rusty colour. The sinuation at the extremity of the elytra is extremely deep.

Tapajos; a single female individual.

Obs.—Though very closely allied to the *D. Matthewsi*, I believe the elongate antennæ indicate this to be a distinct species; it is also a little larger, its head being notably broader.

COPROPORUS.

Under this generic name I have described sixteen new species; all these would, in Erichson's classification, find their position in *Tachinus*, Family I. Kraatz and others have proposed more than one other generic name for insects that would by Erichson have been located as above mentioned; but these names I have not adopted, because I feel extremely doubtful as to what amount of generic differentiation will be found to exist among these insects; for it is evident that a vast number of closely-allied species exist in the warmer parts of the world, and that only an insignificant fraction of their number are as yet known to us; and I consider it is therefore premature to attempt to predict where the limits of aggregation of the species will ultimately be found. It is sufficient for my present purposes to state that all the *Tachyporini* here described as *Coproporus* are comparatively little elongate in form, have the anterior half of the body very shining and glabrous, the mesosternum carinate, and the front tarsi in the male scarcely dilated (this latter character being of course only inferred in the case of those species of which the female alone is known to me). *Tachyporini* possessing these characters appear to be very numerous in species in the warm parts of the world, and South America appears to be specially rich in them, so that their study will not be without difficulty, leaving out of consideration generic questions. In all the cases in which I have observed the ædeagus, it is but small, without appendages, and differs but slightly from species to species; on the other hand, the external sexual characters of the apical segments of the hind body offer remarkable and striking distinctions, so that their examination much facilitates the recognition of the species.

1. *Coproporus rotundatus*, n. sp. Convexus, piceus,

nitidus, antennis tenuioribus fuscis, basi pedibusque testaceis; capite prothoraceque lævissimis; elytris subtilissime punctulatis, lateribus leniter rotundatis, angulo externo rotundato; abdomine piceo-rufo. Long. corp. 2¼ lin. (abdomine extenso).

Very convex; of a pitchy colour. Antennæ with the three basal joints yellow, the rest infuscated; they are rather long and slender, a little incrassated towards the extremity; 3rd joint scarcely so long as the 2nd; 4th considerably shorter than the 3rd; 4—10 differing but little in length, but the apical ones evidently stouter than the others; 11th joint rather stout, longer than the 10th, its extremity paler. Head pitchy-black, the parts of the mouth and palpi yellowish. Thorax pitchy in colour, a little paler at the sides, very much narrowed to the front; the base distinctly sinuate on each side, and the posterior angles projecting backwards, their extremity rounded. Elytra longer than the thorax, but about as broad, their sides gently curved, evidently narrowed to their outer hind angle, which is broadly rounded, very finely and sparingly punctured. Hind body, when extended, very much narrowed to the extremity, pitchy in colour, its punctuation more distinct than that of the elytra. Legs yellowish.

In the male the dorsal plate of the 7th segment of the hind body ends in four spines, of which the two middle ones are closer to one another than to those at the sides, and are more slender and project farther backwards; the lateral notch on each side extends a good deal farther forwards than does the middle one; the ventral plate of the same segment bears at the extremity a deep and extremely broad triangular excision.

Ega; a single male.

2. *Coproporus similis*, n. sp. Convexus, piceus, nitidus, antennis pedibusque rufis, illis articulo ultimo, dilutiore; capite thoraceque impunctatis; elytris subtilissime punctulatis, lateribus apicem versus leviter angustatis, angulo externo obtuso. Long. corp. 2¼ lin.

In the female the dorsal plate of the 7th segment of the hind body terminates in four elongate, almost equidistant spines, the two in the middle being distinctly more slender, but scarcely longer than the lateral; the ventral plate also ends in four spines, of which the two in the middle are rather stouter than the lateral, and the interval between them is distinctly less than that between the middle and lateral ones.

Ega; a single female.

Obs. I.—This species so closely resembles *C. rotundatus*, that I at first supposed them to be the sexes of a single species; careful observation reveals, however, such important differences, that I have no doubt these are specific. *C. similis* is a little smaller; it has the antennæ considerably shorter, the eyes smaller and less prominent, the hinder angle of the elytra much less rounded, and the tarsi a good deal shorter.

Obs. II.—Besides the female individual above described and named, I have a male specimen from the same locality, which is slightly larger, and has the sides of the elytra a little more rounded, so that I am not quite sure whether it belongs to this species or not. I give, however, a description of its sexual characters as follows.

In the male the dorsal plate of the 7th segment of the hind body ends in four almost equidistant spines, the middle ones being much more elongate than the lateral ones; the three spaces between the spines extend about equally far towards the front; the ventral plate has at the extremity a broad notch, the sides of which are sinuate, and its hinder angles somewhat produced, in the form of short teeth.

3. *Coproporus obesus*, n. sp. Latior, convexus, piceoniger, nitidus, antennis tenuioribus pedibusque testaceis; capite prothoraceque lævissimis, elytris parce subtilissime punctatis, angulo externo rotundato, intra marginem lateralem obsolete impresso. Long. corp. $2\frac{1}{2}$ lin. (abdomine extenso).

Very closely allied to *C. rotundatus;* a little broader and darker, the antennæ more slender at the extremity, and the sides of the elytra just a little turned outwards, so as to give them the appearance of being indistinctly impressed along the margin. Antennæ long and slender, slightly stouter at the extremity, yellowish, 2nd and 3rd joints differing little in length; after this the joints to the 10th differ but little in length, the 10th being stouter and a little shorter than the 4th; it is evidently longer than broad; 11th joint rather long. The head is blackish, smooth and shining, the parts of the mouth and the palpi yellowish. The thorax is very much narrowed to the front; it is pitchy black, paler at the sides; it is broadly but slightly sinuate at the base on each side, the extreme hind angle just a little rounded. The elytra are the width of the thorax at the base, almost straight at the side, a little narrowed behind, the hind angle much rounded; they are sparingly and extremely finely punctured. The hind body is pitchy, with the last segment and the hind margins of those before it paler; its punctuation fine, closer and more distinct than that of the elytra. The legs dark yellowish.

In the female the dorsal plate of the 7th segment of the hind body terminates in four very long, equidistant, slender spines, of which the two middle ones are slightly more slender than the lateral ones; the ventral plate of the same segment also ends in four slender spines, of which the two in the middle are rather longer, but not stouter than the lateral ones.

Ega; two female specimens.

4. *Coproporus retrusus*, n. sp. Piceus, convexus, nitidus, antennis apicem versus incrassatis, pedibusque rufis; capite prothoraceque lævissimis, elytris parce subtilissime punctulatis, lateribus subparallelis, angulo externo minus fortiter rotundato. Long. corp. 2¼ lin.

This species is very closely allied to the three preceding ones, and is about the same size, but has the antennæ stouter and more thickened towards the extremity, and its elytra are straighter at the sides. Antennæ moderately long, rather stout, distinctly thickened towards the extremity, of an obscure reddish colour in the middle, with the three basal and the 11th joints paler; 3rd joint longer than the 2nd; 4th to 10th each a little shorter and broader than its predecessor, the 10th scarcely so long as it is broad; 11th joint stout, one and a half times as long as the 10th. Thorax as broad as the elytra, much narrowed to the front; a little sinuate at the base on each side, so that the hinder angles project a little backwards, the extremity of these rounded. Elytra rather straight at the sides, with the external angle moderately rounded, without any lateral impression; very finely and sparingly punctured. Legs dark yellowish.

In the female the ventral plate of the 7th segment of the hind body ends in four spines, of which the two middle ones are very elongate; the lateral ones are a good deal shorter than the middle ones, and but little stouter;

besides these the outer angle on each side is also produced, and forms a distinct short spine, so that the ventral plate is really six-spined, the outside tooth being only half the length of the one next it. The dorsal plate is four-spined, but, as the spines are broken in the individual described, I cannot speak of their relative lengths.

Ega; a single female.

5. *Coproporus curtus*, n. sp. Convexus, nigro-piceus, nitidus, prothoracis lateribus antennarumque medio piceo-rufis, his basi apiceque testaceis, pedibus rufis; antennis crassiusculis, articulis 6—10 leviter transversis. Long. corp. $1\frac{7}{8}$ lin.

Antennæ rather short and stout, the three basal joints yellow; 4—10 pitchy, 11th yellowish; 2nd and 3rd joints subequal in length; joints 4—10 differing but slightly from one another in length, each a little broader than its predecessor, the 4th rather longer than broad, the 10th not quite so long as broad; 11th obtusely pointed, much longer than 10th. Head blackish, impunctate, the palpi yellow. Thorax pitchy, with the sides paler, shining and impunctate, the base but little sinuate, so that the rounded hind angles are very little produced backwardly. Scutellum impunctate. Elytra rather broad and short, a good deal longer than the thorax, only a little curved at the sides, and with the hind angle only moderately rounded; they are very finely but yet distinctly punctured, and have a broad obsolete impression near the lateral margin. Hind body broad and short, pitchy, paler towards the extremity, rather closely and distinctly punctured, but little shining. Legs red.

In the female the dorsal plate of the 7th segment of the hind body ends in four long equidistant spines, of which the middle ones are a little the longer; the three notches between the spines extend about equally far forwards. The ventral plate ends in four long spines and two short lateral teeth; the two middle teeth project slightly farther back than those next them, and the notch separating them is narrower and does not extend quite so far forwards as the notch next them.

A single female, from Parentins to Jurua; 1st to 5th April, 1874; Dr. Trail.

6. *Coproporus politus*, n. sp. Convexus, pernitidus, niger, abdomine piceo, antennis medio piceis basi apiceque testaceis, mediocribus; prothorace basi fere truncato; elytris crebre punctulatis, lateribus minus rotundatis. Long. corp. 1¾ lin.

Antennæ not elongate, moderately stout, slightly thickened towards the extremity; the three basal joints and the apical one yellow, the middle ones pitchy; 3rd joint slender, quite equal in length to the 2nd; 4th—10th each very slightly shorter and very slightly broader than its predecessor, the 4th longer than broad, the 10th about as long as broad. Head impunctate, shining black; palpi yellow. Thorax with the base almost straight, and the hind angles not greatly rounded; it is very shining, quite without sculpture, blackish, with the sides paler. Elytra a good deal longer than the thorax, very finely but yet distinctly and rather closely punctured, the hind angle only a little rounded. Hind body paler and less shining than the front parts, rather finely and not densely punctured. Legs red. Mesothoracic carina extremely elevated, with its front angle rounded.

In the female the dorsal plate of the 7th segment ends in four spines, of which the middle ones are slightly longer than the lateral ones, and the middle notch is slightly broader than the outside one; the ventral plate ends in four spines, and a short lateral tooth on each side; the middle space between the teeth is a good deal narrower than the lateral one; the middle teeth are a little longer than the lateral ones.

Anana, 6th September, 1874; a single female, brought back by Dr. Trail.

7. *Coproporus brevis*, n. sp. Convexus, pernitidus, niger, abdomine piceo, antennis rufo-testaceis, pedibus rufis; prothorace basi fere truncato; elytris crebre punctulatis, lateribus sat rotundatis. Long. corp. 1¾ lin.

Antennæ yellowish, only moderately long, rather stout, distinctly thickened towards the apex; joints 2 and 3 differing but little in length; 4—10 differing little from one another in length, the 10th not quite so long as broad; 11th joint stout, a little paler than the others. Head black, shining, impunctate. Thorax as broad as the elytra, narrowed towards the front; the hinder angles rounded; the base scarcely sinuate, smooth, shining and impunctate; the lateral margins pitchy. The elytra are nearly black, longer than the thorax, very finely punctured, without impression near the side; the hind angle a good deal rounded.

In the male the dorsal plate of the 7th segment ends in four almost equidistant spines, of which those in the middle are the more slender, and a good deal the longer; the ventral plate bears a deep angular notch, the sides of which are a good deal produced, so as to form projecting teeth.

Ega; two male specimens.

Obs.—This species appears to be extremely closely allied to *C. politus*, and has the mesothoracic carina similar, but the antennæ are entirely pale; I am sorry I am unable to compare the sexes of the two species.

8. *Coproporus ignavus*, n. sp. Sat convexus, nitidus, piceus, antennis basi rufo, articulo ultimo apice testaceo, pedibus rufis; prothorace basi utrinque leviter sinuato; elytrorum angulo externo sat rotundato. Long. corp. 2 lin.

Antennæ moderately long, rather slender, but a good deal thickened towards the extremity; the three basal joints yellow, the following ones pitchy, the extremity of the 11th yellow; 3rd joint fully as long as 2nd; 4th joint longer than broad; 10th very nearly as long as broad. Thorax very transverse, the hind angles a little rounded but not produced, the base a little sinuate on each side; it is of a pitchy colour, with the sides paler, almost impunctate. Elytra very finely and not closely punctured, not impressed at the sides, the hind angle moderately rounded. Hind body pitchy, paler towards the extremity, rather closely and distinctly punctured. Legs red.

In the female the dorsal plate of the 7th segment ends in four elongate spines; the middle ones are the more slender, they are a little longer than the lateral ones, and the space between them is not quite so broad as that between them and the side spine; the three spaces extend about equally far forward; the ventral plate ends in four spines about equally stout; those in the middle are distinctly longer than the side ones, and the space between them is scarcely so broad as that between them and the lateral spine; the middle space does not extend so far forward as the lateral space; the outside of this plate is also a little produced, so as to form a short tooth.

Anana; a single female found on the 6th September, 1874, by Dr. Trail.

Obs.—This species appears to be almost intermediate between *C. similis* and *C. politus*; the less-produced hind angles of the thorax readily distinguish it from the former, while from the latter it is separated by the more rounded angles of the elytra and the sexual differences of the apical segment in the female. The mesothoracic carina has been smashed by an accident in the individual described.

9. *Coproporus inclusus*, n. sp. Convexus, nitidus, piceus, antennis pedibusque testaceis; capite thoraceque lævigatis; elytris obsolete punctatis, angulo externo obtuso; thorace basi subtruncato. Long. corp. $1\frac{1}{2}$ lin.

Antennæ yellowish, slender, rather short, a little thickened towards the apex; joints 1, 2, 3, differing but little from one another in length, 3rd the most slender; 4th much shorter than 3rd, a little longer than the 5th; 5—10 differing but little in length, each just a little shorter and a little broader than its predecessor, 10th scarcely so long as broad; 11th a little stouter than the 10th, and twice as long. Thorax ample, of a pitchy colour, very shining, transversely very convex, much narrowed towards the front; hinder angles rounded; the base slightly sinuate on each side near the outer angle. Elytra pitchy red, about one-third longer than the greatest length of the thorax; they are very finely and indistinctly punctured, and the outer angle is not greatly rounded. Hind body reddish. Legs yellow. Mesothoracic carina not much developed.

In the female the dorsal plate of the 7th segment ends in four spines, of which the middle ones are more slender and a little longer than the outer ones; the middle space is but little narrower than the lateral one, but does not extend so far towards the front; the ventral plate also ends in four spines, of which the middle ones are a good deal the longer; the lateral space is rather broader than the middle one, and extends more to the front: the outer angle of this plate is scarcely produced.

In the male the dorsal plate of the 7th segment ends in four short and approximate spines, of which the middle ones are longer than the lateral ones; the ventral plate of the same segment bears a notch in the middle at the extremity, the hinder angles of which are acuminate, but not produced.

Ega; two individuals, ♂ and ♀.

Obs.—The above description is made from the female individual, the male specimen being greatly mutilated; I have some doubts indeed whether it is of the same species as the female, for it has the antennæ a little shorter, and the intermediate joints darker, than in the female described.

10. *Coproporus cognatus*, n. sp. Convexus, piceus, nitidus, antennarum basi, pedibusque rufis; capite thoraceque lævigatis, hoc basi fere truncato; elytris obsolete punctatis, angulo externo obtuso. Long. corp. $1\frac{1}{2}$ lin.

Antennæ slender, moderately long; blackish, with the three basal joints yellowish, the apex of the 11th joint also paler; 3rd joint slightly shorter than 2nd; 4th longer than broad; 10th about as long as broad. Thorax with the base only slightly sinuate on each side, shining, impunctate. Elytra ample, obsoletely and sparingly punctured, with very fine strigosities connecting the punctures; their hind angle moderately distinct. Hind body greatly narrowed towards the extremity. Legs red. Mesothoracic carina only a little elevated.

In the male the dorsal plate of the 7th segment ends in four teeth, of which the middle ones project much farther back than the side ones; the teeth are but short and the lateral interval extends much more to the front than does the middle interval; the ventral plate bears a broad but not deep angular notch, the sides of which are only indistinctly sinuate, and its lateral extremities are not produced.

Anana; a single male found by Dr. Trail on the 6th September, 1874.

Obs.—This species is very similar to *C. inclusus*; it is about the same size, but appears a little broader; the antennæ are darker in colour and more elongate. When both sexes of each species are known, I have no doubt good sexual distinctions will be found.

11. *Coproporus conformis*, n. sp. Sat convexus, nitidus, rufus, capite, prothoracis disco, antennisque piceis, his basi testaceo; elytris obsolete punctatis, lateribus haud impressis, angulo externo obtuso. Long. corp. $1\frac{1}{2}$ lin.

Antennæ moderately long and slender, blackish in colour, with the three basal joints yellowish; 3rd joint nearly equal to 2nd in length, 4th slender; 4—10 each slightly shorter and broader than its predecessor, so that the 10th is a little transverse; 11th rather long. Head pitchy, shining and impunctate; palpi yellow. Thorax with the hind angles only moderately rounded; the base slightly sinuate on each side, shining and impunctate, reddish with the disc broadly pitchy. Elytra dark reddish, ample, very finely punctured and reticulated, the hind angle not much rounded. Legs slender, red. Mesosternal carina only a little elevated.

In the male the dorsal plate of the 7th segment ends in four rather broad sharp teeth, of which the middle ones project much farther backwards than the side ones; the space separating the middle teeth is not broad, and does not extend far forwards; the base of the lateral notch is much nearer to the front. The ventral plate bears a large deep notch, the sides of which are a little sinuate, and the lateral angles acuminate and a little produced.

A single male was taken by Dr. Trail on the 5th November, 1874, but I have no record of the exact locality.

Obs.—Though extremely similar to *C. cognatus*, I believe this will prove to be a distinct species; it is slightly narrower, and not quite so convex; the hind angle of the elytron is a little less rounded, and the spines on the 7th segment in the male are a little longer, and the notch on the ventral plate is a little deeper.

12. *Coproporus rufescens*, n. sp. Rufo-testaceus, nitidus, glaber, transversim sat convexus, elytris marginem lateralem versus late profundeque impressis, angulo externo minus rotundato. Long. corp. 1¼ lin.

Entirely of a reddish colour and impunctate, except that on the elytra are traces of a sparing and very obsolete punctuation. Antennæ a little thickened towards the apex; 3rd joint small; 4—10 differ but little in length, each just a little broader than its predecessor, 10th scarcely so long as broad; 11th nearly twice as long as the 10th. Thorax ample, as broad as the elytra, narrowed to the front, the hind angles rather obtuse and a little rounded; the base nearly truncate, being very little produced near the external angles. Elytra one-third longer than the thorax; close to the outside margin they have a broad, deep impression, commencing a little behind the humeral angle but continued close to the outer angle, which is but little rounded. Hind body short, furnished with exserted black setæ. Legs concolorous. Mesothoracic carina but little elevated.

In the female the dorsal plate of the 8th segment of the hind body terminates in four acute teeth of nearly equal length, the three notches between which are very similar to one another in breadth and depth. The ventral plate of the same segment terminates in four teeth, shorter than those of the dorsal plate; the middle notch between these is much shorter than the lateral ones, and is, in fact, nearly filled up, but is continued forwards as a groove or longitudinal impression.

Ega; two specimens, ♀.

Obs.—I have not been able to restore the hind body to its natural elongation in this species, so that the length mentioned for the species is only an estimate.

13. *Coproporus tinctus*, n. sp. Convexus, nitidus, testaceus, antennis ante apicem clytrisque basi fuscis; elytris evidenter punctulatis, intra marginem lateralem latius impressis, angulo externo obtuso. Long. corp. 1⅓ lin.

Antennæ rather slender, only moderately long, a little thickened towards the extremity; the three basal joints yellow, the following ones infuscate; the apical one again a little paler; 3rd joint about as long as 2nd; 4th slender, a good deal longer than broad; 10th a little transverse. Head yellow, impunctate. Thorax very slightly sinuate at the base on each side; the hind angles not much rounded, yellow, very shining, quite impunctate. Elytra yellow, but with the basal portion largely infuscate; they are distinctly punctured on the basal portion, but towards the apex the punctuation becomes quite obsolete; they have a large, distinct impression near to the lateral margin, and the hind angle is but little rounded. Hind body indistinctly and not closely punctured. Mesosternal carina low.

In the female the dorsal plate of the 7th segment of the hind body terminates in four moderately long teeth; the middle ones project a little farther back than the lateral ones, and are rather the more slender; the middle notch does not extend quite so far forward as the lateral ones, the ventral plate of the same segment ends in four shorter teeth; the middle ones are stout, the lateral ones not quite so stout and not projecting quite so far back as the middle ones; the middle notch is not deep, and does not extend nearly so far forward as the lateral one. In the male the ventral plate of the 7th segment ends in four very short, stout teeth; the middle ones project much farther back than the lateral ones, which are very short; the middle notch is angular and small; the ventral plate bears a broad, not very deep, angular notch at the extremity.

Rio Purus, 13th October, 1874; two individuals (♂ and ♀). Dr. Trail.

Obs.—The male specimen is quite immature, and has no dark colour on the elytra, but I have no doubt whatever but that it is of the same species as the female described.

14. *Coproporus distans*, n. sp. Sat convexus, niger, nitidus, antennarum basi, pedibus, elytrorumque apice testaceis, prothoracis marginibus dilutioribus; antice impunctatus, abdomine omnium subtilissime, dense punctulato, opaco. Long. corp. 1 lin.

Antennæ moderately long, blackish, with the four or five basal joints yellow; of joints 4—10 each is distinctly shorter and very slightly broader than its predecessor, the 4th a good deal longer than broad, the penultimate joints transverse; the 11th short and stout. Head black, shining and impunctate. Thorax blackish, with the margins paler and translucent, the base not at all sinuate, the hind angles not much rounded. Elytra blackish, with the extremity broadly yellowish, shining and impunctate; they are not very convex, but the lateral margin is somewhat explanate and very distinct, the hind angle much rounded. The hind body is blackish, and is very densely and finely punctured, and with an extremely short, fine and delicate pubescence. The legs are yellow. The mesosternal carina very fine and slightly elevated.

In the male the dorsal plate of the 7th segment has in the middle two extremely short teeth, separated by a broad, very shallow notch; the sides of the plate are scarcely prominent; the ventral plate bears a broad, shallow notch at the extremity.

Rio Purus, 25th October, 1874; two individuals found by Dr. Trail, one of which, however, has lost the hind body.

15. *Coproporus duplex*, n. sp. Subdepressus, nitidus, piceus, antennarum basi, pedibus, prothoracis maginibus, elytrisque apicem versus testaceis; antice impunctatus, abdomine crebre subtiliter punctulato. Long. corp. 1½ lin.

Antennæ elongate, slender at the base, distinctly thicker at the extremity; the three basal joints yellow, the rest blackish; 3rd joint a little shorter than 2nd; 4th joint slender; 10th about as long as broad. Head rather small, shining and impunctate, blackish. Thorax very slightly sinuate at the base on each side; the hind angles moderately rounded, the sides and base yellowish, the front portion darker in colour; the surface is shining and impunctate. Scutellum large, impunctate. Elytra but little convex transversely, shining and impunctate, pitchy yellow, the base darker than the apex; the lateral margins explanate; the hind angle a good deal rounded. Hind body slender in proportion to the front parts, blackish, finely, evenly and rather closely punctured. Legs yellow; tibiæ very slender; the tarsi elongate and extremely slender. The mesosternal carina not much elevated, but with the anterior part more elevated than the hinder part, so that an acute projection is formed on its middle.

In the male the dorsal plate of the 7th segment of the hind body ends in the middle in two short, stout, acuminate teeth, and in a scarcely prominent lateral one on each side; the ventral plate bears a deep, angular excision, the sides of which are a little curved.

Conceição, Rio Maulies; a single male, captured by Dr. Trail, May, 1874.

16. *Coproporus scutellatus*, n. sp. Subdepressus, nigropiceus, nitidus, antennis pedibusque testaceis, capite prothoraceque vix visibiliter, elytris subtiliter, abdomine evidentius punctatis; elytris secundum marginem lateralem canaliculatis. Long. corp. 1⅓ lin.

This species differs from the others of the genus here described by its much flatter and more Tachinoid form, and its hind body is evidently less retractile. It appears to be closely allied to *T. brevicollis*, Er. The antennæ are yellowish, they are of a moderate length and stoutness, a little thickened to the apex; 2nd and 3rd joints nearly equal in length; 4–10 differing but little from one another, the 4th considerably stouter than the third;

the 11th joint large, stouter than the 10th, and about twice as long. Head broad and short, black and shining, extremely finely, almost imperceptibly, punctured. Thorax pitchy, paler at the sides, narrowed towards the front, the base slightly sinuate on each side, the hind angles rounded and projecting a little backwards; it is very finely and indistinctly punctured. Scutellum large, smooth and impunctate. Elytra fully a third longer than the thorax, straight at the sides, the outer angle not much rounded; they have a deep, narrow channel close to the external margin, extending from just below the humeral angle to the extremity; they are finely punctured, but more distinctly so than the thorax. Hind body rather closely and distinctly punctured; segments 2—5 distinctly margined, 6th and 7th immarginate. Legs yellowish, short; mesothoracic carina but little prominent.

In the female the dorsal plate of the 7th segment ends in four stout but long acuminate spines; these project about equally far back; the notches between them are only narrow, but are elongate; the lateral ones reach a little farther forwards than the middle one; the ventral plate ends in four shorter and more widely-separated spines; of the notches between these the middle one is not deep, but the lateral notches extend considerably farther forwards than the middle one.

Ega; a single individual.

CONURUS.

The *Tachyporini* bearing this name are very easily recognized from the fine, delicate pubescence with which all the parts of the body are clothed. Species of the genus are probably to be found in nearly all countries, and, though only five species have yet been described from South America, there is but little doubt that these insects will be found to be numerous there, for I have nine species from the neighbourhood of Rio de Janeiro alone in my collection.

1. *Conurus latus*, n. sp. Ferrugineus, convexus, pubescens, antennis articulis 5—10 nigris, ultimo pallido basi nigro. Long. corp. 2½ lin.

A broad species, very convex about the thorax and elytra, with the exception of the antennæ of an uniform tawny colour. The antennæ moderately long, thickened towards the extremity; the four basal joints yellowish, the six following joints and the base of the 11th blackish, the rest of the last joint quite pale. Head small, narrower than in *C. pubescens*. Thorax ample and very convex, the sides much narrowed towards the front, the front margin very distinctly bisinuate; the hind angles are obtuse and rounded, the punctuation and pubescence very fine. Elytra about as long as the thorax, their punctuation and pubescence fine but not dense. The hind body is densely and distinctly punctured; its well-marked pubescence is of a golden colour.

The male has a large angular notch at the hind part of the ventral plate of the 7th segment of the hind body; the hind margin of the dorsal plate is simple.

In the female the dorsal plate of this segment is divided at the extremity by three narrow elongate incisions into four approximate processes, while the hind margin of the ventral plate is furnished with long cilia.

In each sex the front tarsi are rather strongly dilated.

Ega; two individuals.

2. *Conurus setosus*, n. sp. Angustulus, cinnamomeus, subtilissime punctulatus et pubescens, abdomine longius nigro-setosus; antennis gracilibus, basi apiceque pallidioribus; prothorace elongato. Long. corp. 1⅔ lin.

Antennæ slender and elongate, pale yellow, a little darker in the middle; 3rd joint about equal in length to 2nd; 4th slender and elongate; 5—10 each a little broader and shorter than its predecessor, 10th slightly longer than broad; 11th rather elongate, nearly twice as long as 10th. Thorax longer than broad, nearly straight at the base, the hind angles almost rectangular, but a little rounded, the sides curved and a little narrowed towards the front, the surface very finely and indistinctly punctured. Elytra hardly so long as the thorax, similarly but a little more distinctly punctured. Hind body slender, very finely punctured, furnished with remarkably evident long black setæ. Legs pale yellow; tarsi elongate and very slender.

Garrao; a single individual found in fungus by Dr. Trail, 11th November, 1874.

Obs.—This individual is, I have no doubt, a male, as it has the front tarsi slightly dilated. The elongate black setæ with which the very slender extremity of the hind body is armed do not allow me to see with certainty the structure of the apical segments; but any peculiarities, if existent, must be very slight.

TANYGNATHUS.

Of this genus eight species are all that have as yet been described, viz., four from the Old World tropics, two from the New World tropics, one from temperate Europe, and one from the Atlantic Islands. I here add three new species to this number, and can state, moreover, that the genus will be found ultimately rather rich in species, as I have a number of other undescribed ones in my collection, one of which is from Southern Australia, and several from Brazil.

The genus is one of very considerable interest; for it was assigned by Erichson to the *Tachyporini*, by Kraatz to the *Quediini*, and yet possesses certain points foreign to both these groups, which appear to me to indicate a third relationship with the *Aleocharini*. A careful examination of the structural characters of the species seems to me indeed to be urgently needed before its nearest relationship can be satisfactorily decided.

1. *Tanygnathus longicornis*, n. sp. Rufescens, antennis elongatis, apicem versus pallidis, elytris piceis; abdomine fortiter fere irregulariter punctato, longius pubescente. Long. corp. 2½ lin.

Antennæ very slender and elongate, not in the least thickened towards the extremity; the basal joint yellowish: the next five or six darker, and the rest again paler. Head obscure reddish, very narrow and elongate, smooth and shining. Thorax obscure reddish, narrowed to the front, not quite so long as broad, with four very fine punctures placed as usual in this genus. Elytra darker in colour than the head and thorax, scarcely so long as the latter, closely and very finely punctured. Hind body reddish, with faint iridescent reflections; the base of each segment finely punctured; the other part of each segment with sparing, rather large elevated points; its pubescence rather long, and much mixed with black erect setæ. Legs reddish.

One specimen; the only locality indicated being "Amazons."

2. *Tanygnathus nasutus*, n. sp. Fusco-rufus, antennis

sat tenuibus, medio fuscis, basi apiceque rufo-testaceis; thorace læte rufo; elytris fuscis, margine apicali rufo; abdomine dense subtiliter punctato, opaco; pedibus rufis. Long. corp. 2½ lin.

Antennæ moderately long, and not very slender; the basal joint yellow; joints 2—7 infuscate, 8—11 yellow. Head obscure reddish; the clypeus in front acuminate. Thorax very shining bright red. Scutellum reddish. Elytra of a smoky colour, with the hind margin reddish, closely and finely punctured. Hind body very obscure reddish; the base rather darker than the extremity, very closely and densely punctured, and very pubescent, so as to be quite dull. Legs reddish-yellow, underside dull obscure red.

A single individual, without special locality.

Obs.—This species is about the size of the European *T. terminalis*; the antennæ are of about the same length but distinctly stouter; the front part of the clypeus is more prolonged, and the punctuation of the hind body is much denser. It greatly resembles *T. flavicollis*, but has the antennæ stouter, the clypeus more prolonged in front, and the hind body more densely punctured.

3. *Tanygnathus flavicollis*, n. sp. Rufescens, capite piceo, antennis apice thoraceque flavis, elytris piceo-rufis. Long. corp. 1⅓ lin.

Closely allied to *T. ruficollis*, Kr., and about the size of that species. Antennæ slender, rather long; the 1st joint yellowish, the next five darker, the rest paler again. Head pitchy. Thorax bright reddish-yellow, rather broad, of the usual form in this genus, and with the four ordinary punctures. Elytra pitchy; the suture yellowish at the extremity, about the length of the thorax, closely and finely punctured. Hind body dark reddish, rather closely and finely punctured, and distinctly pubescent. Legs reddish-yellow.

Tapajos; one specimen, in bad condition.

ACYLOPHORUS.

Up to the present time sixteen species appear to have been described of this genus; they are from widely different parts of the globe; and the species in my collection enable me to state that the genus is probably to be found in all the warm and temperate parts of the earth's surface; while in Australia there are to be found some very interesting forms, apparently intermediate between this genus and *Quedius*.

The four species here described belong, I think, clearly to the same genus as our European species; at any rate, their facies is so similar to that of our European species that any one acquainted with these would at first sight declare the Amazonian species to be congeneric therewith. I have several allied species from Brazil in my collection, so that the genus will probably prove to be quite as rich in species in South America as in any other part of the world.

1. *Acylophorus puncticentris*, n. sp. Niger, antennarum basi, pedibusque obscure testaceis; elytris fortius punctatis; abdomine subtiliter iridescente, segmento sexto apice extremo, 7° basi apiceque testaceis, segmentis singulis basi crebre subtiliter, apicem versus fortiter parciusque punctatis pubescentibusque. Long. corp. 5 lin.

Considerably larger than *A. glabricollis*. The antennæ are elongate and slender, with only the two or three apical joints a little stouter. The 1st joint dusky yellowish, very long, about as long as the five following; the 2nd and 3rd joints about equal in length; 4th joint considerably shorter than the 3rd and a little shorter than the 5th; from the 5th to the 9th each a little shorter than its predecessor, all of these joints elongate; 10th joint much shorter, but as long as broad; last joint rounded, nearly as long as the 10th. Palpi yellowish; mandibles elongate, slender, crossed, dull yellowish. Head broadly ovate, with three punctures between the eyes, the middle one the most forward. Thorax black and shining, rather broad, about as long as broad, with the usual four punctures. Elytra black, scarcely so long as the thorax, with the scutellum coarsely, moderately closely punctured. Hind body elongate, a little iridescent, with a very rigid pubescence; each segment at the extreme base closely and finely punctured, the rest of each sparingly and rather coarsely punctured; segments 2—5 with the hind margin furnished with a row of very coarse setæ projecting backwards; the extreme margin of the 6th segment, the base and apex of the 7th, reddish. Legs reddish, a little infuscate.

Ega, one; Tapajos, two individuals.

2. *Acylophorus angusticeps*, n. sp. Niger, nitidus, tarsis fulvis, elytris sat crebre fortiterque punctatis; capite angusto. Long. corp. 4½ lin.

Closely allied to *A. glabricollis*, rather larger, the head narrower; the antennæ (the basal joints at any rate) longer; and the 6th and 7th segments of the hind body entirely black. The antennæ are destroyed with the exception of the first four joints; these are longer than in *glabricollis*, blackish, with the extreme base of the 1st joint a little yellowish; 3rd joint not quite so long as the 2nd, a little longer than the 4th. Palpi pitchy. Head elongate and very narrow, with the usual six larger punctures, the two forming the obliquely placed pair (near the neck on each side) very close together. Thorax as in *glabricollis*. Scutellum broader than in *glabricollis*, strongly punctured. Elytra scarcely so long as the thorax, their punctuation as in *glabricollis*. Hind body much narrowed to the extremity, black, the apex of the 6th segment concolorous; each segment is, at the extreme base, closely and strongly punctured, the hinder part of each much more sparingly. Legs pitchy black; tarsi dark reddish.

Tapajos; one specimen.

3. *Acylophorus acuminatus*, n. sp. Niger, nitidus, antennarum basi, pedibusque testaceis, elytris sat crebre fortiterque punctatis. Long. corp. 3 lin.

Much smaller and narrower than *A. glabricollis*. Antennæ with the basal half of the 1st joint yellowish, the rest pitchy; 2nd joint much longer than the 3rd, 4th shorter than the 3rd, 5th longer than 4th; from this to the 10th each a little shorter and stouter than its predecessor, 10th distinctly transverse; last joint stout and rounded at the extremity, a little longer than the 10th. Head suborbiculate, being comparatively both shorter and broader than in *glabricollis*, with the usual punctures. Thorax pitchy, rather broad but of the usual form. Elytra scarcely so long as the thorax, black, together with the scutellum rather closely, moderately coarsely punctured. Hind body very pointed at the extremity; the segments at the base and sides of each closely and not coarsely punctured. Legs yellow.

Ega; three specimens.

4. *Acylophorus iridescens*, n. sp. Piceo-rufus, antennis capiteque piceis, illarum summo basi, pedibusque testaceis; abdomine fortiter punctato, iridescente; elytris crebre, subtiliter punctatis. Long. corp. 2½ lin.

Much smaller than *A. glabricollis*, and very different in colour. Antennæ pitchy; the basal portion of the 1st joint yellowish; they are moderately long, scarcely thickened towards the extremity; 3rd joint much shorter than the 2nd, longer than the 4th; 5th again longer than the 4th; up to the 9th each joint longer than broad; 10th and 11th joints rather stout, each about as long as broad. Palpi yellowish. Head pitchy black, suborbiculate. Thorax pitchy, a little narrowed to the front. Elytra reddish or pitchy red, scarcely so long as the thorax; together with the scutellum finely and rather closely punctured. Hind body reddish, with iridescent tints, roughly and strongly punctured, the points appearing elevated, its pubescence coarse. Legs yellow.

Tapajos; one specimen.

QUEDIUS.

Of this extensive and widely distributed genus it is remarkable that I have received but a single species from the Amazons, and I have only a few others from South America in my collection: as only about five species have been described from tropical America, it seems probable that the species are not numerous there.

1. *Quedius clypealis*, n. sp. Nitidus, rufo-testaceus, capite pectoreque piceis, illo antice rufo, abdomine iridescenti-nigro, parce punctato basi lævi, apice testaceo; elytris fere lævigatis, punctis paucis subseriatis impressis. Long corp. 4 lin.

Antennæ short and rather stout, clear yellow; 3rd joint longer than 2nd; 4—10 each a little shorter and broader than its predecessor, the penultimate joints rather strongly transverse, and with their upper and inner angle a little produced, so as to be subserrate. Palpi yellow. Head a good deal smaller than the thorax, rather short and broad; the eyes very large, and occupying very nearly the whole side of the head, shining, blackish, with the front part broadly yellow, impunctate, except for two or three punctures at the margin of the eye. Thorax curved at the sides and a little narrowed in front, about as long as broad, shining reddish-yellow, impunctate, except for some punctures along the margins. Scutellum large, shining red, impunctate. Elytra as long as the thorax, shining red, with two or three not very distinct punctures along the suture, with two other punctures near these, with a discoidal series of four or five punctures, and with a few lateral punctures. Hind body blackish, with iridescent metallic reflection, the hind part of the 6th and all the following segment yellow, the lateral styles of the terminal segment black. Legs reddish-yellow, stout.

In the male the ventral plate of the 7th segment of the hind body has a shallow emargination in the middle of the hind margin.

Ega.

Obs. I.—I have before me eight specimens, which I believe to be conspecific, and one of which I have described as above. Three of these individuals are males, and agree closely with one another, except that in one of them the breast is red. The five females differ from the males, inasmuch as they have the elytra and thorax black, and the legs more or less infuscate; whether these differences in colour will prove to be sexual, I am unable to say.

Obs. II.—This species is, to judge from Erichson's description of *Q. labiatus*, very closely allied thereto, and I had at first considered it a variety thereof, but on careful examination I think it will more probably prove to be a distinct species.

CORDYLASPIS.

This genus was proposed by Nordman for a most remarkable insect, and it has hitherto remained without any known near allies; the extremely rare *Scariphœus luridipennis* connects it unmistakably with *Hæmatodes*, and I have one or two other undescribed allies in my collection. The only species yet distinguished is,—

1. *Staphylinus pilosus*, Fab.
Found by Mr. Bates at Pará, Tapajos and St. Paulo.

PLATYPROSOPUS.

This genus up to the present time consists of nine or ten described species found in the warm portions of the Old World. I here add another ten species from the Amazons, and consider that they form a most unexpected addition to the South American fauna; except these

Amazonian specimens, I have never seen another individual of the genus from the New World.

These new species appear to exhibit the peculiar characteristics of the genus very highly developed. The structure of the front of the head and the insertion of the antennæ approaches in these species even more to what exists in the *Xantholini* than it does in the Old World *Platyprosopi*; the antennæ are even more approximate in their insertion than in the Old World species, and moreover the part of the head to which they are attached is more prominent, and is a little emarginate on each side of the middle, so that the front of the head and the attachment of the labrum have very much the appearance presented by the same parts in the *Xantholini*.

The genus is one of the most interesting of the *Staphylinidæ*; it is located by Erichson and Kraatz as a peculiar member of the *Xantholini*, but I cannot consider that this is a correct mode of treating it. The points of structure I have already alluded to, viz., the antennal insertion and the attachment of the labrum, are almost the only points the genus has in common with the *Xantholini*, while it wants some of the most important points of structure of that group, and in certain respects approaches to the *Quediini* and even to the *Pinophilini*. As the group *Xantholini* appears to me one of the most specialized portions of the *Staphylinidæ*, and as *Platyprosopus* is pretty clearly of a synthetic or little specialized character, it seems to me that it will be very much more suggestive of the truth if the genus be considered to form of itself a group, to be located in the neighbourhood of the *Quediini*; for I cannot but think that the purposes of inquiry are very much better served by the establishment of a considerable number of provisional groups, than by slumping together (if I may use such a term) under one name a number of heterogeneous forms, having probably very different genetic histories.

1. *Platyprosopus major*, n. sp. Parallelus, nigropiceus, capite subopaco, dense punctato, medio spatio angusto lævi, nitido; thorace parce punctato, nitido, marginibus lateralibus dense fortiter punctatis; elytris abdomineque dense subtiliter punctatis, opacis, fusco-pubescentibus; pedibus fuscis. Long. corp. extens. 10—12 lin.

Antennæ pitchy, stout, about as long as the head and half the thorax; 3rd joint longer than 2nd; 4—10 differing little from one another, each a little shorter than broad; last joint longer than the 10th, obtusely pointed on one side. Head large, quite as broad as the thorax, above densely and coarsely punctured, a space along the middle free from the coarse punctures, but with a few fine and indistinct ones; besides this there are three or four still larger punctures on each side mixed with the others, and in front of the middle there is a transverse impunctate space; on the underside it is extremely dull, densely and finely rugulose-punctate. Thorax as broad as the elytra, as long as broad, black, very shining, with fine punctures scattered over it, with a dorsal series of six punctures on each side the middle, with seven or eight other punctures on each side near the front part, and just inside the lateral margins, with a narrow strip of coarse dense punctuation extending also some way along the front and hind margins. Elytra scarcely longer than the thorax, densely and finely punctured, nearly opaque, and with a very fine fuscous pubescence. Hind body opaque, densely and finely punctured, the apical segments more coarsely than the others. Legs dusky reddish, very pubescent.

Ega; two specimens, ♂ and ♀.

Obs.—Besides these two individuals, there is another ♂ specimen from Pebas, which differs in several slight particulars and may possibly be a distinct species, but more probably is only a local form of *P. major*.

2. *Platyprosopus laticeps*, n. sp. Nigro-fuscus, capite prothoraceque disperse punctatis, nitidis, elytris abdomineque dense subtiliter punctulatis, fusco-pubescentibus, opacis; pedibus obscure rufis. Long. corp. 7—8 lin.

Much smaller than *P. major*, and without the marginal punctuation of the thorax. The antennæ are moderately stout, and reach about half-way to the back of the thorax; they are of an obscure dull-reddish colour; the 3rd joint much longer than the 2nd; 4—10 differing but little from one another, the 10th about as long as broad. The head is quite as broad as the thorax; above it is coarsely, irregularly and rather sparingly punctured, the punctures less numerous about the middle than at the sides; scattered with the coarser punctures are numerous very fine ones; on the underside it is quite dull, densely and finely rugulose-punctate, and with a fine fuscous pubescence.

Thorax quite as broad as the elytra, slightly longer than broad, on each side the middle with an irregular and not very distinct dorsal row of six or seven punctures, and between this and the sides with numerous other coarse punctures, from which, however, the hinder part is free; besides these it is covered with numerous other very fine and distant punctures. The elytra are about as long as the thorax, dull, densely and finely punctured, and with a very fine pubescence. The hind body is pitchy, with the extremity as well as the hind margin of each segment ferruginous; it is densely punctured, and with a fine fuscous pubescence. The legs are dull yellowish, very pubescent.

Ega; three specimens.

3. *Platyprosopus parallelus*, n. sp. Angustus, piceo-ferrugineus, antennis pedibusque obscure rufis, abdomine segmentorum marginibus, anoque ferrugineis; capite prothoraceque disperse punctatis, nitidis; elytris abdomineque opacis, dense punctatis, fusco-pubescentibus. Long. corp. 6 lin.

Closely allied to *P. laticeps*, smaller and narrower. The antennæ are moderately stout, and reach about half-way to the back of the thorax; joint 3 much longer than 2; 4—10 differing but little from one another, each about as long as broad. Head slightly narrower than the thorax, of a pitchy colour, with two kinds of punctuation on the upper surface, viz., a fine punctuation visible on the middle as well as elsewhere, and some other larger and scattered punctures wanting on the middle part. Thorax as broad as the elytra, of a pitchy colour, about as long as broad, on each side of the middle with an irregular and indistinct row of six or seven punctures, and between this and the sides with some other punctures, wanting towards the base; besides this numerous extremely fine and small punctures are scattered on the upper surface. Elytra about as long as the thorax, of a pitchy colour, quite dull, densely and finely punctured, and with a very fine fuscous pubescence. Hind body quite dull, pitchy, the extremity and the hind margin of each segment dull reddish, extremely finely and densely punctured, and with a very fine pubescence. Legs yellowish, very pubescent.

Ega; one specimen.

4. *Platyprosopus puncticeps*, n. sp. Angustus, piceus,

thorace magis rufo, elytris, abdomine segmentorum marginibus anoque ferrugineis, antennis pedibusque testaceis; capite supra crebre fortiter punctato, nitido, vertice medio impunctato. Long. corp. 4¼ lin.

Antennæ rather slender, a little shorter than head and thorax, dull yellow; 3rd joint one and a half times the length of the 2nd; 4—10 each just a little shorter and stouter than the preceding one, 4th longer than broad, 10th about as long as broad. Head pitchy, above coarsely and rather closely punctured, the punctures closest about the hind angles and front part, the middle of the vertex free, but with a few very fine and obsolete punctures; beneath it is very opaque, from its very fine and dense rugulose punctuation. Thorax pitchy red, very shining, about as long as broad, sparingly punctured, the punctures consisting of a row of five or six on each side the middle, and fifteen or twenty others on each side of these towards the front part. Elytra quite as long as the thorax, dull red, closely and finely punctured. Hind body of a pitchy colour, the extremity and margins of the segments reddish; it is very finely punctured, and with a very dense fine pubescence. Legs yellow.

Tapajos; two specimens.

5. *Platyprosopus rectus*, n. sp. Angustus, parallelus, rufescens, capite piceo-rufo, crebre fortiter punctato, nitido; pedibus testaceis; thorace nitido, medio utrinque parce seriatim punctato, et versus latera anterius punctis nonnullis; elytris abdomineque dense subtiliter punctatis, opacis. Long. corp. extens. 4¼ lin.; lat. ¾ lin.

Very narrow and parallel. Antennæ rather slender, 1 lin. in length; red. Head rather darker than the other parts, narrow, coarsely punctured, the punctures wanting on a space down the middle except at the anterior parts. Thorax longer than broad, straight at the sides, just as broad as the elytra; reddish, shining, bearing only a few punctures, viz., a series of about four large punctures on each side of the middle, and a few other large punctures between these and the outside towards the front. Elytra scarcely longer than the thorax, of an obscure reddish colour, densely and finely punctured and densely pubescent. Hind body rufo-fuscous, densely and finely punctured and densely pubescent. Legs yellow; underside of head very densely sculptured and opaque.

A single female of this species was brought back by Dr. Trail; it was attracted by light at Manaos, in August, 1874.

Obs.—This species is extremely closely allied to *P. puncticeps*, but is a little smaller, is narrower and more parallel; the elytra and hind body are a little more closely and finely punctured, while the punctures on the head and thorax are slightly coarser than in the larger species.

6. *Platyprosopus minor*, n. sp. Angustus, obscure rufus, antennis pedibusque testaceis; capite dense subtiliter punctato pubescenteque, subopaco; thorace subnitido, crebre sat fortiter punctato, elytris abdomineque opacis, dense subtiliter punctulatis, fusco-pubescentibus. Long. corp. 4 lin.

Antennæ yellow, not stout, not reaching quite to the back of the thorax; 3rd joint longer than 2nd; 4—10 each just a little shorter and stouter than the preceding one, the 10th about as long as broad. Head dull reddish, finely and densely punctured; the middle of the vertex almost free from punctures; the sides behind the eyes especially densely and finely punctured and pubescent. Thorax rather longer than broad, red, moderately closely but not coarsely punctured, a middle longitudinal line impunctate. Elytra about as long as the thorax, opaque red, densely and obsoletely punctured, and very finely pubescent. Hind body pitchy red; the hind part and the extremities of the segments paler, very dull, very densely and finely punctured, and with a dense fine pubescence. Legs yellow.

Ega; one specimen.

7. *Platyprosopus rufescens*, n. sp. Obscure rufo, abdomine piceo, segmentorum marginibus anoque rufescentibus, antennis pedibusque testaceis; capite dense subtiliter (medio parcius) punctato, pubescenteque, subopaco; prothorace parcius disperse punctato, nitido. Long. corp. 5 lin.

Allied to *P. minor*, but larger and with the thorax more sparingly punctured.

Antennæ rather slender, not so long as head and thorax, dull yellow; 3rd joint much longer than 2nd; 4—10 each a little shorter than the preceding one, 4th much longer than broad, 10th about as long as broad. Head very nearly as broad as the thorax, the sides very densely and finely punctured and pubescent, the middle parts much more sparingly so; the underside quite dull and finely pubescent; it is of a pitchy or pitchy-red colour. Thorax as broad as the elytra, rather longer than broad, of a dark-reddish colour; a space along the middle, and another of about equal width at the base, free from punctures; the other parts rather sparingly and irregularly punctured. Elytra about as long as the thorax, reddish, very densely and finely punctured, with a dense very fine fuscous pubescence. Hind body pitchy; the extremity and the margins of the segments reddish, very densely and finely punctured and densely pubescent. Legs yellow.

Ega, Tapajos, St. Paulo; six individuals.

8. *Platyprosopus opacifrons*, n. sp. Piceus, antennis pedibusque obscure testaceis; elytris, abdomine segmentorum marginibus, anoque ferrugineis; capite omnium dense subtilissimeque punctato, pubescenteque, peropaco. Long. corp. 5½ lin.

The sculpture of the upper surface of the head at once distinguishes this species from the others here previously described. The antennæ are rather long and slender, about as long as head and thorax, dull yellow; 3rd joint much longer than 2nd; 4—10 each a little shorter than the preceding one, the 10th rather longer than broad. Head about as broad as the thorax, nearly black, quite dull, very densely, evenly and finely punctured and pubescent. Thorax pitchy, slightly longer than broad, rather closely punctured, a narrow line along the middle free from punctures. Elytra quite as long as the thorax, dull reddish, very finely and densely punctured and pubescent. Hind body dusky, with the margins of the segments and its extremity reddish, very densely and finely punctured and pubescent. Legs dull yellow.

Ega; one specimen.

9. *Platyprosopus frontalis*, n. sp. Fusco-rufus, opacus, thorace nitido, pedibus testaceis; capite, elytris, abdomineque dense subtilissimeque punctatis et flavescenti-pubescentibus; thorace ad angulos anteriores dense subtiliter, disco parcius fortiter punctato, medio longitudinaliter impunctato. Long. corp. 5 lin.

Antennæ obscure red, rather long; 3rd joint much longer than 2nd. Head extremely densely covered with a very fine punctuation and pubescence, which render it quite opaque. Thorax dark red, only slightly longer than broad, straight at the sides, shining, but at the front angles densely and finely punctured, near the middle sparingly and rather coarsely punctured, along the middle itself a rather broad, but not sharply-limited, space, free from punctures; this space bounded on each side by an irregular longitudinal patch of coarse punctures; the basal portion of the surface free from punctures. Elytra dark reddish, as broad as and only a little longer than the thorax, densely and finely punctured, and clothed with a very fine, short, dense, yellowish pubescence. Hind body fuscous, becoming redder towards the extremity, very densely, finely and evenly punctured, and clothed with an extremely fine and dense-yellow pubescence. Legs reddish-yellow.

A single individual, captured by Mr. Bates, and bearing no special locality, but probably from Tapajos.

Obs.—This species is extremely similar to *P. opacifrons*, but has the thorax much less densely and regularly punctured, and the pubescence with which the upper surface is clothed is rather denser, finer, and brighter in colour.

10. *Platyprosopus similis*, n. sp. Rufo-fuscus, thorace nitido, pedibus testaceis; capite, elytris, abdomineque dense subtilissime punctatis, et griseo-flavescenti pubescentibus; thorace ad angulos anteriores dense, subtiliter, disco parcius fortiter punctato; medio, spatio longitudinali minus discreto, impunctato. Long. corp. 6 lin.

Thorax just as broad as long; elytra a little longer than the thorax.

A pair, ♂ and ♀, of this species were brought from Manaos by Dr. Traill; they were attracted by light in August, 1874.

Obs.—This species is so extremely close to *P. frontalis* that a special description is unnecessary; it is rather larger and distinctly broader, and its colour is not so bright; the impunctate area on the middle of the thorax is also not quite so distinct. The different punctuation of its thorax will distinguish it from *P. opacifrons*, to which species it is also extremely similar.

BRACHYDIRUS.

This genus consists at present of five described species, to which I now add nine others. It is quite peculiar to South America, and was established by Nordmann for a

Brazilian species, but was not considered valid by Erichson, who relegated Nordmann's species to the genus *Staphylinus*, and described two or three allied species. Kraatz, however, has re-affirmed the validity of the genus, and pointed out some of its important structural points. In point of fact, the genus seems at present to me a really distinct and isolated one. The structure and form of the head, and insertion of the antennæ, as well as some points in the formation of the prothorax, bring the genus into proximity with the *Aleocharini*, but it seems probable that the points alluded to indicate a functional, and not a genetic, relationship.

The species appear to be very rare in collections, so that it is quite possible they may have some peculiar mode of life.

1. *Brachydirus maculiceps*, n. sp. Niger, nitidus, antennis, ore, ano, femoribus anterioribus et intermediis apice, tibiis tarsisque anterioribus testaceis; fronte maculis duabus obscuris rufis. Long. corp. 5 lin.

Mas: abdomine segmento sexto ventrali, medio leviter, lateque emarginato, 7º medio profunde inciso.

Antennæ yellow, distinctly incrassated from the 5th to the 10th joint; 2nd and 3rd joints about equal; 4th much shorter than 3rd; 5th a little shorter than the 4th; 6—10 differing but little in length, but each a little broader than its predecessor, 6 and 7 about as long as broad; 8—10 transverse; 11th joint rather large, about twice as long as the 10th. Mandibles and palpi yellowish. Head black, near the vertex with two indistinct reddish spots; it is as broad as the thorax; all the front is coarsely and very densely punctured, the vertex more sparingly so; it is clothed with very fine and rather scanty, but longish, yellow-grey pubescence. Thorax narrower than the elytra, about as long as broad; the base and hinder angles rounded, the sides a little sinuate; so that it is a little broader in the front part; it is black and shining, with large, irregularly placed punctures, so arranged as to leave a narrow space at the base, a longitudinal space along the middle, and an obscure elevation near the front angles, free from punctures. Scutellum closely and rather finely punctured, and with a grey pubescence. Elytra a little longer than the thorax, black, with a faint bluish tinge; moderately coarsely and not closely punctured, with a very fine pubescence. Hind body narrowed towards the apex, black, with the extremity of the 6th segment and the whole of the 7th bright yellow; the 2nd segment is nearly impunctate; 3—6 moderately closely and distinctly punctured, 7th very finely and sparingly; 3rd, 4th and 7th segments with a yellowish, the others with a blackish pubescence. The front legs are yellowish, with their coxæ and the base of the femora pitchy, the middle legs blackish, with the lower half of the femora yellowish. Hind legs black.

In the male the 6th segment is, on the underside, a little emarginate in the middle; the 7th with a rather deep and narrow notch, on each side of which the hind margin is just a little prominent. Lateral lobes of the 8th segment broader towards the extremity, instead of being pointed as usual.

Ega; eight individuals.

Obs.—This species is probably rather closely allied to *B. xanthoceros*, Nord. I have a female specimen from Peru in my collection, which I believe is conspecific with the Amazonian individuals.

2. *Brachydirus antennatus*, n. sp. Niger, nitidus, ano testaceo, pedibus anterioribus, intermediisque ex parte testaceis, antennis fuscis, testaceo-variegatis. Long. corp. 4 lin.

Mas: abdomine segmento sexto ventrali apice medio emarginato, 7° triangulariter producto, apice ipso emarginato.

Antennæ short, distinctly thickened towards the end; 1st joint yellowish at the base, fuscous towards the extremity; 2—5 fuscous, 6—9 pale yellow, 10 and 11 fuscous, 2nd and 3rd of about the same length; 6th joint about as long as broad, 7—10 transverse, the 10th rather strongly so; 11th long and pointed. Head with the eyes a little broader than the thorax, the front half very densely and coarsely, the hinder half much more sparingly punctured, finely and sparingly pubescent; mandibles and palpi pitchy. Thorax much narrower than the elytra, about as long as broad, its width greater in front than behind; it is black and shining, irregularly and very coarsely punctured, the punctures so disposed as to leave a narrow portion at the base; a line along the middle, and an obscurely elevated space near the front angles, free. Scutellum rather closely punctured. Elytra longer than the thorax, rather coarsely punctured. Hind body distinctly narrower towards the extremity, black, with the hinder portion of the 6th and the whole of the 7th segment bright, reddish-yellow; segments 3—6 sparingly punctured, 7th nearly impunctate; the pubescence is scanty, and is yellowish on segments 3—5. Front legs yellowish, with the coxæ and base of the femora pitchy; middle legs dusky yellow, their tibiæ rather darker; hind legs nearly black, with the tarsi ferruginous.

In the male the 6th segment beneath has the hind margin emarginate in the middle; the 7th is triangularly produced in the middle, but instead of being pointed has a small notch at the extremity.

Ega; four specimens, 3 ♂, 1 ♀.

3. *Brachydirus styloceros*, n. sp. Fulvo-testaceus, abdomine nigro, ano pedibusque testaceis. Long. corp. 4½ lin.

Mas: abdomine segmento sexto ventrali apice medio sat profunde semicirculariter emarginato, 7° carinato-compresso, in stylo tenui apice bifido producto.

Antennæ yellow, rather short, thickened towards the extremity; 2nd and 3rd joints about similar in length; 6th joint about as long as broad, 7—10 transverse; last joint moderately long and pointed. Head a little broader than the thorax, very short; the front part extremely densely, the vertex more sparingly punctured. It is of a tawny-yellow colour, with an extremely fine pubescence. Thorax rather shorter than long, a little broader in front than behind, the sides a little sinuate; like the head and elytra, it is of a tawny-yellow colour, coarsely and irregularly punctured, with the usual smooth spaces. Elytra considerably longer than the thorax, rather coarsely punctured, with a distinct concolorous pubescence. Hind body narrower at the extremity, black, with the hind margin of the 6th segment and the whole of the 7th yellow; lateral lobes of the 8th black; 2nd segment impunctate, 3 – 6 moderately punctured, 3 and 4 with a yellowish pubescence. Legs yellow.

In the male the 6th segment of the hind body beneath has a broad and rather deep semicircular incision; the ventral plate of the 7th segment is most remarkably formed, being compressed into a sort of keel, and produced behind as a slender tongue, the extremity of which is divided into two still more slender, short styles.

In the female the ventral plate of the 7th segment is a little produced, and is pointed at the extremity; in this sex also the head is rather less closely but more coarsely punctured than in the male.

Ega; eleven individuals.

Obs.—The peculiar structure of the ventral plate of the 7th segment shows no variation in a series of six ♂ individuals.

4. *Brachydirus cribricollis*, n. sp. Fulvus, abdomine nigro, ano testaceo; prothorace omnium fortissime punctato. Long. corp. $4\frac{1}{2}$ lin.

Mas: abdomine segmento sexto ventrali apice medio obsolete emarginato, segmento 7° subtriangulariter inciso.

Antennæ yellowish, rather short and stout, much thickened towards the extremity; joints 2 and 3 about equal, 4th longer than the 5th and 6th, these two short; the 6th transverse, 7—10 strongly transverse; 11th long, stout and pointed. Head tawny, coarsely and densely punctured, the punctuation more sparing on the vertex. Thorax not quite so long as broad, distinctly broader in front, and with the sides a little sinuate, its upper surface extremely closely and deeply punctured; the punctures are numerous and close, and leave a very narrow line at the base, a central line, and a spot near the front angles, smooth. Elytra considerably longer than the thorax, of a similar colour to it, and the head coarsely and rather closely punctured, with their pubescence rather coarse and distinct. Hind body narrowed towards the apex, black, with the extremity of the 6th segment, and the whole of the 7th, yellow; the 2nd segment is impunctate, segments 3—6 closely and finely punctured and densely pubescent; the pubescence on segments 3—5 yellow, 7th segment nearly impunctate; 8th segment with its lateral lobes pale at the base, tawny brown at the extremity. Legs yellow.

In the male the hind margin of the 6th segment of the hind body is slightly emarginate, and the 7th has a rather narrow and moderately deep notch in the middle of the hind margin; in the female the hind margins of the 6th and 7th segments are simple, that of the 7th being gently rounded; the last joint of the antennæ seems to be shorter in this sex than in the male.

Ega and St. Paulo; seven specimens.

5. *Brachydirus simplex*, n. sp. Fulvus, abdomine nigro, apice antennisque testaceis, capite superne viridi-æneo, vertice fulvo. Long. corp. 5 lin.

Mas: abdomine segmento 6° ventrali margine posteriore emarginato, 7° apice medio anguste sat profundeque exciso.

Antennæ yellow, short, a good deal thickened towards the extremity; 4th and 5th joints similar to one another, not at all transverse; 6th joint a little transverse, 8—10 strongly transverse, 11th moderately long. Mandibles pitchy. Head with the upper surface largely of a metallic-green colour; the vertex however tawny; densely and coarsely punctured, the punctuation distinctly less dense on the vertex than on the anterior portion. Thorax a good deal narrower than the elytra, with rather coarse and numerous punctures, distributed in the usual manner. Elytra a good deal longer than the thorax, and, like it, of a tawny colour. Hind body black, with the hind part of the 6th and with the 7th segment yellow; the lateral lobes of the 8th segment black; the 2nd segment is impunctate, and has a few yellow hairs on each side; 3rd segment sparingly punctured, and with yellow hairs on each side; 4th and 5th rather sparingly punctured, much clothed with pale hairs; 6th segment more closely, 7th very sparingly, punctured.

In the male the hind margin of the ventral plate of the 6th segment is emarginate, while the next segment bears a narrow, rather deep notch in the middle. In the female the hind margin of the ventral plate of the 7th segment forms a very obtuse angle in the middle.

Three individuals, 2 ♂, 1 ♀, found by Mr. Bates; the male specimen described is labelled Pará. I have also a male individual of the species labelled Peru in my collection; it differs only very slightly from the Pará individual.

6. *Brachydirus amazonicus*, n. sp. Fulvus, abdomine nigro, ano testaceo, antennis articulis 8—10 fuscis. Long. corp. 4½ lin.

Mas latet.

Allied to the *B. cribricollis*; antennæ longer and less stout, and with the three joints before the last one dark, and the thorax much more sparingly punctured. Antennæ with 3rd joint slightly longer than the 2nd, slender to the 6th joint, 7th slightly transverse, 8—10 distinctly so; last joint much pointed. Head broad, rather broader than the thorax, closely and very coarsely punctured in front, more sparingly on the vertex. Thorax rather broader than long, yellowish, coarsely and irregularly, but rather sparingly punctured, with the usual smooth spaces; it is a good deal broader in front than behind, and considerably sinuate at the sides. Scutellum large, rather finely and closely punctured and pubescent. Elytra longer than the thorax, rather coarsely and closely punctured, and with a coarse pale pubescence. Hind body narrowed towards the extremity, black, with the hind part of the 6th segment, and the whole of the 7th, yellow; 2nd segment impunctate; 3rd to 6th rather closely and finely punctured and densely pubescent, the pubescence on segments 3—5 yellowish. Legs pale yellow.

Ega; one specimen (♀).

7. *Brachydirus Batesi*, n. sp. Fulvus, antennis (basi excepto), capite areâ pone mandibulas, abdomineque nigris, hoc ano testaceo; antennarum basi, pedibusque pallidis. Long. corp. 4—4¼ lin.

Mas: abdomine segmento sexto ventrali medio emarginato, segmento 7° triangulariter producto, apice carinato-compresso.

Antennæ rather short, thickened towards the apex; first four or five joints pale yellow, the rest blackish, the extremity of the 11th joint being again paler; 6th joint considerably stouter than the 5th, about as long as broad, 7—10 strongly transverse; 11th joint pointed, stout and rather long. Head broader than the thorax, with the palpi pale yellow: the mandibles pitchy; the labrum, and a space behind it (not reaching to the eyes on each side), black, the rest tawny; it is coarsely punctured, the punctures on the front part not so dense as in the other species, and rather irregular, the hinder part (broadly) more sparingly punctured. Thorax about as long as broad, much sinuate at the sides, the front markedly broader than the hind part, very coarsely and rather sparingly punctured, with the usual smooth spaces. Scutellum closely and moderately finely punctured. Elytra considerably longer than the thorax, coarsely, deeply and rather closely punctured. Hind body black, with the 7th segment and hind part of the 6th pale yellow, finely but not densely punctured, the yellow portion almost impunctate; the 3rd and 4th segments and the sides of the 5th with the pubescence yellow. Legs pale yellow.

In the male the 6th segment of the hind body beneath is a little emarginate behind. The 7th segment has the hind part much produced and pointed, and is compressed in a keel-like manner as it approaches the extremity.

Ega; four male individuals.

Obs.—Besides these four males, I have a single individual from the same locality, which I believe to be the female of *B. Batesi*; the ventral plate of the 7th segment is distinctly produced, and its hind margin is simply rounded; the antennæ are slightly shorter and more clavate than in the other sex.

8. *Brachydirus longipes*, n. sp. Fulvus, abdomine capiteque superne nigris, illo apice testaceo, hoc vertice fulvo; antennis fuscis. Long. corp. 4½ lin.

Mas latet.

Antennæ short, thickened towards the extremity, blackish, the base of the 1st joint yellowish; 5th and 6th joints small and short, but scarcely transverse, 7—10 rather strongly transverse. Palpi and mandibles dark. Head above black, with the vertex fulvous, coarsely and very densely punctured, the vertex more sparingly than the front part. Thorax tawny, with the very coarse and close punctures distributed in the usual manner. Elytra rather longer than the thorax, rather coarsely punctured, similar in colour to the thorax. Hind body black, with the 7th segment and hind part of the 6th yellow, the styles of the 8th segment dark tawny; segments 4, 5 and 6 are rather closely punctured, and 4 and 5 bear pale hairs, as also do 2 and 3 near the lateral margins. Legs yellowish, with the femora, except at the knees, slightly infuscate; front legs rather long and slender.

In the female the hind margin of the ventral plate of the 7th segment is simply rounded.

Pará; a single female.

Obs.—Though closely allied to *B. Batesi*, I have no doubt this is a distinct species; the antennæ are shorter and darker in colour at the base, the black colour covers a larger portion of the head, which also is more densely punctured, the coarse punctures of the thorax are more crowded together, and the anal styles are paler in colour.

9. *Brachydirus æneiceps*, n. sp. Fulvus, capite supra (vertice excepto) ænescente, abdomine nigro, apice testaceo. Long. corp. 3½ lin.

Mas: abdomine segmento sexto ventrali apice leviter emarginato, segmento 7° medio triangulariter exciso.

Antennæ rather short, a little thickened towards the extremity; 3rd joint scarcely so long as the 2nd, 6th about as long as broad, 7—10 transverse, 11th pointed. Palpi and parts of the mouth pitchy yellow. Head broader than the thorax, its upper surface greenish-brassy, with the hinder part tawny, the punctuation of the metallic part very dense, and finer than in the other species here described; the tawny part much more sparingly punctured. Thorax about as long as broad, slightly sinuate at the sides, and with the front part a little broader; its upper surface very coarsely, and, with the exception of the usual smooth spaces, rather closely punctured. Elytra coarsely and moderately closely punctured, like the thorax of a tawny colour. Hind body narrowed towards the extremity, black, with the 7th segment and hind part of the 6th yellow; 2nd segment impunctate, 3—6 rather sparingly punctured, 3—5 with a yellow pubescence at the side parts. Legs yellow, with the femora infuscate; the sternum pitchy.

In the male the hind body beneath has the 6th segment a little emarginate at the extremity, and the 7th segment has a rather deep and narrow triangular notch.

Ega; one specimen (♂).

PLOCIOPTERUS.

This genus consists of six described species, and I here characterize ten new ones. The five species known to Erichson were described by him under the generic name of *Staphylinus*, and it is to Kraatz that we are indebted for the name and some of the characters of the genus; these insects are undoubtedly most allied to *Brachydirus*, but the structure of the antennæ and front feet seem to afford satisfactory points of distinction.

The species are confined to tropical America, and those I possess are easily referable to two sections, in one of which the front tarsi and tibiæ are simple in each sex, while in the other section they are more or less dilated, at any rate in the males, and the hinder face of the front tibiæ is cut away in a peculiar manner near the extremity. Three of the species here described were indicated as being captured in fungus, which, however, they probably frequent for predaceous purposes.

In this genus the sexual characters become remarkable, and are well worthy of study, for they appear to me to suggest the functional result of some very remarkable modifications. The specimen of *P. Traili* here described was received by me in spirit in very fresh condition; and by dissecting out the apical segment, and mounting it immediately in Canada balsam, the structure of the hard and soft parts of the intromittent organ are finely displayed, as well as their position in relation to the lateral valves.

1. *Plociopterus tricolor*, n. sp. Niger, nitidus, antennis articulis ultimis quatuor pallidis, abdominis apice rufotestaceo; elytris cyaneis fasciis duabus, abdomineque fasciâ singulâ cinereo-tomentosis. Long. corp. 6—7 lin.

Mas: abdominis segmento sexto ventrali lineâ mediâ transversali dense longeque testaceo-pilosâ instructo, margineque apicali leviter emarginato, segmento 7° apice medio late triangulariter emarginato; tarsis anticis simplicibus.

Obs.—Mares majores, elytris lineâ laterali elevatâ longitudine variabile instructis, insignes.

Antennae slender, elongate, black, with joints 8—11 pale yellow; 3rd joint very long, about twice as long as the 4th; 4th about equal to 2nd in length; 11th joint oblong, longer and narrower than the preceding one. Head broader than the thorax, black, shining, coarsely and irregularly punctured; the punctures so disposed as to leave a small triangular space behind the labrum, and a large irregular space on the disc, free; the punctures armed with fine greyish setæ. Mandibles and palpi elongate, black or pitchy-black. Thorax about as long as broad, much narrower than the elytra, a little narrowed behind, the hind angles quite rounded, the sides scarcely sinuate; it is black and shining, coarsely and irregularly punctured, a broad irregular smooth space in the middle free; the punctures are finer at the sides. Scutellum large, dull, closely and finely punctured, clothed with a dense grey pubescence. Elytra longer than the thorax, blue, rather closely and finely punctured, clothed at the base and apex with a dense grey pubescence. Hind body narrower towards the extremity, black, with the 6th and 7th segments bright yellow; 2nd segment impunctate, 3rd finely and rather closely punctured, and clothed with a grey pubescence; 4 and 5 finely and not closely punctured, with a fine black pubescence; 6th finely punctured, 7th more sparingly and finely punctured, these two with a concolorous yellow pubescence.

Legs black, with the tarsi obscure reddish.

In the male the 6th segment on the underside is furnished in the middle with a line of very long projecting hairs, and the hind margin is slightly emarginate; the 7th segment has a large triangular notch; the ædeagus is large, and is furnished with a stout ligula projecting far beyond the body of the organ and bifid at the extremity. In the larger individuals of this sex the mandibles and palpi are more elongate, and the elytra are furnished near the outside with a longitudinal fold or plica of variable length; this is quite absent in the smaller males. The anterior tarsi are quite simple.

Ega; nine specimens.

2. *Plociopterus fungi*, n. sp. Niger, nitidus, abdominis apice rufo-testaceo; elytris cyaneis, fasciatim cinereo-pubescentibus; antennis articulis duobus ultimis albidis. Long. corp. 7 lin.

This species is almost the same as *P. tricolor* in most respects, but it has only two joints at the apex of the antennæ white, and the male characters are a little different.

In the male the 6th segment is on the underside, furnished across the middle with a curved line of long projecting hairs, and its hind margin is rather deeply emarginate; the 7th segment bears a large triangular notch. The ædeagus is similarly formed to that of *P. tricolor*, but its elongate ligula is less produced, and the bifid processes at its extremity are shorter and more rounded.

Pará: a single specimen found by Mr. Bates; it is labelled "stump fungus."

Obs.—As I have dissected out the ædeagus in three males of *P. tricolor*, and find its form to be completely similar in the three, I cannot consider this individual to be a mere variety of the Ega species.

3. *Plociopterus nigripes*, n. sp. Niger, nitidus, capite, prothorace, elytrisque cyaneis; antennis articulis ultimis tribus testaceis, abdomine apice rufo-testaceo; elytris fasciis duabus, abdomine fasciâ singulâ cinereo-tomentosis. Long. corp. 5½ lin.

Mas: abdomine segmento sexto ventrali medio basin versus foveâ transversâ setigerâ ornato, segmento 7° apice medio, late minus profunde emarginato; tarsis anticis simplicibus.

Antennæ moderately long, blackish, the last three joints yellow; 3rd joint twice as long as 2nd, 4th a little longer than the 2nd. Mandibles and palpi pitchy. Head above blue, slightly broader than the thorax, very coarsely punctured, with a triangular space behind the labrum, and a small central one, impunctate; pubescence and setæ rather long, underside black and impunctate. Thorax bluish above, a little longer than broad, a little sinuate at the sides, a little narrowed behind, very coarsely punctured, with a medial line impunctate; pubescence long, scanty and fine, grey. Scutellum dull, finely punctured and pubescent. Elytra longer than the thorax, moderately closely and not finely punctured, at the base and apex with grey pubescence. Hind body narrowed towards the extremity, black, with the 6th and 7th segments bright yellow; 2nd segment impunctate, 3—5 rather sparingly punctured, 3rd with a grey pubescence. Legs black, tarsi pitchy.

In the male, on the underside, the 6th segment of the hind body has in the middle, near the base, a short transverse impression or fovea, bearing some long, fine hairs; the 7th segment is furnished in the middle of the hind margin with a broad shallow notch or emargination. The ædeagus is small, and furnished with a ligula shorter than the body of the organ, and so slender as to be easily overlooked. The front tarsi are simple.

St. Paulo; one ♂ specimen.

Obs.—Notwithstanding the extreme resemblance of this species to *P. tricolor*, the ædeagus is so different as to suggest that the two insects may possibly have to be referred to distinct genera.

4. *Plociopterus affinis*, n. sp. Niger, nitidus, antennis articulo ultimo ferrugineo, elytris cyaneis, fasciatim cinereo-pubescentibus; abdominis segmentis apicalibus rufo-testaceis, stylis analibus nigris. Long. corp. 6½ lin.

Mas latet.

Antennæ moderately long, black, with the apical joint reddish. Mandibles and palpi pitchy. Head shining black, coarsely punctured, with a large space on the middle impunctate. Thorax shining black, distinctly shorter than broad, coarsely punctured, with a broad, irregular, impunctate space along the middle. Elytra rather longer than the thorax, blue, with ashy pubescence at the base and extremity, rather coarsely punctured. Hind body black, with the 6th and 7th segments bright reddish-yellow; the anal styles are yellow at the base, but their apical half is quite black; the segments are rather closely punctured, and the 2nd and 3rd bear an ashy pubescence. The legs, including the coxæ, are black, with the tarsi pitchy, but having the apical joint reddish.

Pará; a single female taken two or three years ago.

Obs.—This unique specimen, though in bad condition, represents, I have no doubt, a distinct species, which will be easily distinguished from *P. tricolor* and *P. fungi*, by the colour of the antennæ and the anal styles. Compared with a female of *P. tricolor* it is seen that the antennæ are much shorter, that the thorax is shorter and more sinuate at the sides, and more narrowed behind, and that the elytra are shorter and more coarsely punctured.

5. *Plociopterus dimidiatus*, n. sp. Niger, nitidus, abdomine antennarumque articulo ultimo rufo-testaceis; elytris cyaneis, fasciis duabus cinereo-pubescentibus. Long. corp. 5½ lin.

Mas: abdominis segmento 6° ventrali apice medio obsolete emarginato, 7° triangulariter inciso.

Antennæ black, with the last joint yellow; 3rd joint not quite twice as long as the rather long 2nd joint, 4th about as long as 2nd, 5th long, but shorter than 4th; last joint long, rounded at the extremity. Head scarcely broader than the thorax, shining black, with a very faint bluish tinge; the upper side with coarse, irregular punctures, those in the front very large, the disc free from punctures. Thorax much narrower than the elytra, about as long as broad, slightly narrowed behind, and a little sinuate at the sides, with coarse scattered punctures, which are neither so numerous nor so coarse as in the other species here described; a broad middle space free from punctures. Scutellum dull, the lower half finely punctured, and with a fine grey pubescence. Elytra blue, moderately closely and moderately finely punctured, with two bands of grey pubescence. Hind body yellow, narrowed towards the extremity; 2nd segment almost impunctate, 3—6 sparingly punctured, especially in their centres, 7th still more finely and sparingly punctured. Legs black, with the tarsi reddish. The basal joint of the front tarsi quite as long as the three following together.

The male has the hind margin of the 6th ventral segment slightly emarginate in the middle, and the 7th with a triangular notch in the middle. The ædeagus is rather broad and has the ligula flat and broad, similar in length and breadth to the body of the organ, and closely applied thereto. The front feet simple.

Tunantins; one ♂.

Obs.—This species must be closely allied to *Staphylinus scenicus*, Er., but I have no doubt a comparison will prove it distinct therefrom.

6. *Plociopterus lætus*, n. sp. Niger, nitidus, antennis (articulis 7—9 fuscis), pedibus, abdomineque rufo-testaceis; elytris cyaneis, fasciis duabus cinereo-pubescentibus. Long. corp. 5½ lin.

Mas: tarsis anticis leviter dilatatis, abdominisque apicis structurâ complicatâ insignis.

Antennæ rather long and slender, yellow, with joints 7—9 darker; 3rd joint elongate, 4th about as long as 2nd; from 4—10 each is a little shorter than its predecessor; last joint rather long, rounded at the extremity. Parts of the mouth yellow. Head a little broader than the thorax, black and shining, coarsely and irregularly punctured, with the middle part smooth, the punctures behind the labrum extremely large and confluent. Thorax much narrower than the elytra, about as long as broad, distinctly narrowed behind; the sides a little sinuate, the upper surface black and shining, very coarsely and irregularly punctured, a middle longitudinal space impunctate. Scutellum closely and finely punctured, with a grey pubescence. Elytra longer than the thorax, blue, with two bands of grey pubescence, one at the base the other a little before the extremity, their punctuation moderately fine, not close. Hind body narrowed towards the extremity, yellow, with the 2nd segment pitchy; segments 3—6 finely and rather sparingly punctured. Legs yellow, with the coxæ pitchy. The anterior tibiæ are dilated behind (especially in the male), and furnished just below the middle with three or four coarse setæ placed close together; thence they are narrowed to the extremity. First joint of front tarsus but little longer than the 2nd.

In the male—joints 1–3 of the front feet are a little dilated. The dorsal plate of the 7th segment of the hind body is produced in the middle, the apex of the produced part a little emarginate; on each side of this large middle projection is a much smaller projection. On the underside the hind margin of the 6th segment is slightly trisinuate; the 7th segment is a little produced in the middle, and has a deep, rather narrow notch in the middle. The 8th segment has the lateral lobes modified and irregular, each is corneous, bluntly pointed at the extremity, and there furnished with two stout black setæ; before the extremity each is distorted and has an irregular broad projection, which is black at its extremity, the rest of the lobe being pale yellow. The ligula of the ædeagus is compressed and keel-like, its hinder half furnished with fine, black, file-like asperities, it reaches much beyond the body of the organ; from the latter projects a small fine appendage, extending about to the apex of the ligula.

Ega; 2 ♂s, 1 ♀; also a male, taken at Garrao by Dr. Trail on the 11th November, 1874, in fungus.

7. *Plociopterus ventralis*, n. sp. Niger, nitidus, antennis (articulis 7—9 fuscis), pedibus, abdomineque testaceis; elytris cyaneis, fasciis duabus cinereo-pubescentibus. Long. corp. 4½ lin.

Mas: tarsis anticis leviter dilatatis; abdomine segmento 7° ventrali producto, medio exciso.

This species is extremely similar to *P. lætus*, but it is smaller and has the head a great deal smaller, and the male characters very different. The front tibiæ and tarsi are similarly formed to those of *P. lætus*, but the abdominal characters are very different. The dorsal plate of the 7th segment has the hind margin simply rounded. The hind margin of the 6th ventral segment is slightly emarginate in the middle, and the ventral plate of the 7th segment is distinctly produced, and has a well-marked notch in the middle at the extremity. The styles of the 8th segment are yellow, and are broad at their extremity, which is densely fringed with short, black, file-like setæ, and on the under face of the style there is an additional short series of such setæ very close to the extremity.

Ega; two males.

8. *Plociopterus Traili*, n. sp. Niger, nitidus, antennis

(articulis 6—9 fuscis), pedibus, abdomineque testaceis; elytris viridi-cyaneis, minus discrete bifasciatim cinereotomentosibus. Long. corp. 5 lin.

Mas: abdomine segmento 7° dorsali margine posteriore medio obtuse angulato, angulo ipso exciso; ventrali sat producto, apice lato, leviter emarginato.

Antennae quite as long as head and thorax, pale yellow, with joints 6—9 infuscate. Palpi pale yellow; mandibles pitchy. Head black, with a very slight greenish reflection, very coarsely punctured, but with a large impunctate space on the middle; it is just about as broad as the thorax. Thorax narrower than the elytra, just about as long as broad, shining black, with an irregular series of coarse punctures along each side of the middle, and with some other coarse punctures between these and the sides, and with a short longitudinal impression in front of the base in the middle. Elytra slightly longer than the thorax, rather coarsely punctured, clothed at the base and extremity with pale hairs, which form two not very distinct transverse fasciae. Hind body yellow, with the basal segment blackish; the basal segments impunctate, and the 6th only sparingly and finely punctured. Legs yellow, with the coxae black.

In the male the three basal joints of the front tarsi are distinctly dilated; the hind margin of the dorsal plate of the 7th segment is a little produced, so that it would form in the middle a very obtuse angle, but where the angle would be there is a very small excision; the ventral plate is distinctly produced in the middle, but the middle part is not acuminate, but forms a rather broad lobe, which has its hind margin a little emarginate. The styles of the 8th segment are yellow, and are broad at their extremity, which is very densely set with short, black, file-like setae, and both on the upper and under side there is an additional short row of such setae very near the extremity.

Garrao: a single male, found in fungus by Dr. Trail, on the 11th November, 1874.

Obs.—This species is very closely allied to *P. ventralis*, but the male characters are a little different, and the thoracic punctuation is rather coarser.

9. *Plociopterus virgineus*, n. sp. Niger, nitidus, antennis ex parte, pedibus, abdomineque rufo-testaceis; elytris

cyaneis, fasciis duabus cinereo-pubescentibus, abdomine supra segmentis 4 et 5 subtiliter punctatis. Long. corp. 5 lin.

Mas latet.

Fem.: abdomine segmento 7° et supra et infra medio leviter producto, acuminatoque.

This species is very closely allied to the preceding, so that it is only necessary to point out the characters distinguishing it therefrom. The antennæ have the first five joints pale yellow, 7—10 nearly black, the 11th again pale. The punctures on the thorax are coarser and rather more numerous in *P. virgineus*; the upper side of the hind body is distinctly though finely punctured, this being especially evident on the 3rd and 4th segments. The hind body is also narrowed and pointed at the extremity, but this is probably a character peculiar to the female. I have no doubt the discovery of the male will prove this to be a good and distinct species.

Fonteboa; one ♀ specimen.

Obs.—Though this species greatly resembles *P. lætus*, *P. ventralis* and *P. Traili*, I feel no doubt it will prove distinct from all of them: the female is readily distinguished by its sexual characters from the same sex of *P. lætus*; the females of the other two species being unknown to me, I can of course make no comparison with them.

10. *Plociopterus mirandus*, n. sp. Niger, nitidus, antennis, pedibus, abdomineque rufo-testaceis; elytris cyaneis, fasciis duabus cinereo-pubescentibus; abdomine apicem versus vix angustato, fere impunctato. Long. corp. 4¼ lin.

Mas: tarsis anticis leviter dilatatis, abdominisque apicis structurâ valde complicatâ insignis.

Fem. latet.

Antennæ slender and rather long, yellow, with the 7th and 8th joints infuscate; 3rd joint not quite twice the length of the 2nd; 4th not quite so long as 2nd; 11th joint rounded at the extremity. Palpi pale yellow, mandibles pitchy. Head about as broad as the thorax, above of an obscure-greenish colour, and not very shining; all the disc free from punctures, and a little convex; the sides coarsely punctured; a quadrate space behind the labrum depressed, and with some very coarse but not well-defined punctures. Thorax black and shining, narrower than the elytra, transversely convex, its length hardly greater than its width; just a little broader in front, and the sides a little sinuate; on each side of the middle is an irregular line of about nine punctures, and other punctures are scattered along the sides, especially near the front part. Elytra about the length of the thorax, blue, with a grey pubescence at the base and near the extremity, moderately closely punctured. Hind body yellow, with the 2nd segment darker; it is almost impunctate, and very sparingly pubescent above, beneath distinctly but rather sparingly punctured, and with a fine long pubescence. Legs yellow, with the coxæ pitchy.

The characters of the male are very complicated and most remarkable. The front tibiæ are dilated towards the apex, and somewhat concave on the inner side; the three basal joints of the front tarsi are a little dilated. The dorsal plate of the 7th segment of the hind body has three sinuses at the hind margin; the middle one is the broader, but is not formed by the margin being cut away, but by its being turned downwards; at the base of this turned-down portion are two sharp teeth, placed near to one another; from the extremity of this turned-down part project two vertical processes. The ventral plate of the same segment is a little produced, and has a deep incision or notch in the hind margin; along each side of this notch it is broadly impressed. The lateral lobes of the 8th segment are modified in a most extraordinary manner; each terminates in three processes,—a broad, truncate, central one, armed on the inner side with two rows of file-like asperities, and a long, slender, somewhat curved process on each side. The body of the ædeagus terminates in a produced point or beak, and is furnished beneath with a ligula longer and broader than the beak, and densely set with black asperities on each side, towards the end.

Ega; two males.

Obs.—This species is undoubtedly closely allied to *St. venustus*, Er., from Cayenne, but I cannot make the ♂ characters agree with Erichson's description. These male characters are the most extraordinary I have met with in any Coleopterous insect.

XANTHOPYGUS.

This genus, like the two preceding ones, is due to Dr. Kraatz; but unlike them, it seems to be composed of heterogeneous species, and will not improbably undergo other changes. The species known to Erichson were described by him in part as belonging to the genus *Staphylinus*, and in part to the genus *Philonthus*. About sixteen species are described, all from South America. I here refer eleven Amazonian species to the genus, of which I consider seven new. The species, however, present great difficulties, and I have no doubt some time must elapse, and considerable discussion and comparison take place, before their limits and characters are fully ascertained.

1. *Staphylinus sapphirinus*, Er.

This appears to be a common species in the Amazon Valley; a fine series before me indicate it as being found at Obydos, Tapajos, Ega and Pebas. I think I am correct in the name I have assigned to these specimens, for they agree well with Erichson's description (Gen. et Spec. p. 364), except that the male has the hind margin of the 6th segment beneath rather deeply emarginate, while no allusion is made to this in the description above mentioned. Erichson records the species from Columbia and from the Pará in the north of Brazil, but it does not occur, I believe, so far south as Rio de Janeiro.

2. *Xanthopygus Solskyi*, n. sp. Niger, nitidus, abdominis segmentis duobus ultimis rufis; elytris cyaneis, antennis testaceis; abdomine apicem versus crebre punctato. Long. corp. 7 lin.

Antennæ yellow, 1½ lin. in length; 4th joint much longer than broad, 10th about as long as broad; palpi yellow; labrum pitchy yellow. Head very nearly as broad as the thorax, black, with a rather large impunctate space on the middle, elsewhere punctured; the punctures not coarse nor close. Thorax shining black, just about as long as broad, the sides rather sparingly and not coarsely punctured, with a rather broad impunctate space along the middle, and also in front of the base at the sides. Scutellum punctured. Elytra rather longer than the thorax, of a dark blue colour, moderately closely and coarsely punctured. Hind body rather slender, black, with the two basal segments entirely reddish-yellow; the segments rather coarsely but not altogether densely punctured; the punctuation much denser on the basal than on the apical portion of each segment; the 6th rather more sparingly punctured than the preceding one. Legs black; front tarsi with the apical joints reddish, and clothed beneath with tawny hairs.

In the male the hind margin of the ventral plate of the 7th segment is emarginate in the middle, and in front of the emargination the surface is shaved away so as to form an angular depression.

Pará, one specimen, ♂; Ega, one ♀.

Obs. I.—This species closely resembles the preceding one, but is decidedly narrower, and has the head and the thorax and the 6th abdominal segment less densely punctured, and the male characters different; besides the external differences, I may add that the apical portions of the ædeagus are very much less elongate than in *X. sapphirinus*.

Obs. II.—Besides the specimens above mentioned, I have several other individuals from Tapajos, Ega and St. Paulo, which are, perhaps, varieties of this species, but as they are all females I cannot speak certainly; they are generally a little larger than the individual described, in two of them the elytra are more purple, and in the larger specimens the head and thorax are more coarsely punctured, and the antennæ a little stouter. I have also a male individual, found by Mr. Buckley in Ecuador, which I have no doubt is conspecific with the Pará male from which I have drawn up my description; this specimen has the hind margin of the 6th segment underneath a little emarginate in the middle; this point I cannot ascertain for the Pará individual, as just that part of the specimen is slightly broken.

Obs. III.—I have named this species in honour of Mr. Solsky, of St. Petersburg, who has of late years published the descriptions of many interesting species of South American *Staphylinidæ*.

3. *Xanthopygus cyanipennis*, n. sp. Niger, nitidus, antennis testaceis; elytris cyaneis, abdomine segmento 6° dimidio apicali, segmentoque 7° toto rufo-testaceo. Long. corp. 7 lin.

Mas: segmento 6° ventrali margine apicali medio obsolete emarginato, segmento 7° sat profunde triangulariter emarginato.

Allied to *X. sapphirinus*, but considerably narrower, the punctuation of head and thorax much more sparing, the hind body not nearly so densely punctured, and with the base of the 6th segment black.

The male differs from that of *sapphirinus* by wanting the transverse pilose line on the 6th segment beneath, by having the hind margin of the same segment only obsoletely emarginate, and by the different shape of the notch of the 7th segment.

Ega; four male specimens.

Obs.—This species may, perhaps, ultimately prove to be only a variety of *X. Solskyi*, from which it differs almost solely by the dark basal portion of the 7th segment of the hind body. In the male the notch of the 7th segment is deeper, and the ædeagus itself is larger than in *X. Solskyi;* but the former of these characters must be, I think, liable to variation, for in one of the individuals above mentioned the notch scarcely differs from that of *X. Solskyi*, and yet I can scarcely anticipate that this will prove a different species, for the resemblance in other respects is very great.

4. *Xanthopygus apicalis*, n. sp. Niger, nitidus, antennis testaceis; elytris cyaneis, abdomine segmento 6º dimidio apicali, segmento 7º toto rufo-testaceo. Long. corp. 5 lin.

Mas: abdominis segmento 6º ventrali margine apicali medio obsoletissime emarginato, segmento 7º late minus profunde triangulariter emarginato.

Much smaller than *sapphirinus*, with head, thorax and hind body more sparingly punctured. Very close to *X. cyanipennis*, and differing therefrom only by being considerably smaller and more slender, and by the broader and less deep notch of the 7th segment in the male. I have not examined the ædeagus.

Ega; two specimens, ♂, ♀.

5. *Xanthopygus violaceus*, n. sp. Niger, nitidus, antennis testaceis, capite thoracoque violaceis; elytris cyaneis, abdomine segmento 6º dimidio apicali, segmentoque 7º toto rufo-testaceo. Long. corp. 6 lin.

Mas: abdominis segmento 6º ventrali medio lineâ transversâ longe pilosâ, marginique apicali minus evidenter emarginato, seg. 7º apice sat profunde inciso.

Much smaller than *sapphirinus*, and readily distinguished by the beautiful violet colour of the head and thorax. Antennæ and palpi entirely yellow. Head rather smaller than the thorax, with the disc broadly impunctate, the punctures rather coarse and moderately numerous. Thorax about as long as broad, nearly straight at the sides, the punctures numerous and rather coarse, the middle smooth space rather narrow. Scutellum rather coarsely punctured. Elytra blue, broader and rather longer than the thorax. Hind body densely and rather coarsely punctured, the hinder half of the 6th segment as well as all the 7th reddish-yellow. Legs black, front tarsi ferruginous.

The male has on the underside a transverse line of long erect hairs in the middle of the 6th segment, the hind margin of the same segment slightly emarginate; in the middle of the hind margin of the 7th segment is a rather deep, abruptly cut-out notch.

Conceição, Rio Mauhes, May, 1874, one male; Tunantins, 24th November, 1874, one female, found by Dr. Trail; also one specimen of each sex brought from Ega by Mr. Bates.

Obs. I.—The female of this species has the antennæ a little shorter and their penultimate joints more transverse than the ♂.

Obs. II.—Though this species is closely allied to *X. sapphirinus*, there can be no doubt it is quite distinct therefrom. It is worthy of remark that not only do the external abdominal characters of the ♂ greatly resemble those of *X. sapphirinus*, but that also the structure of the ædeagus in the two species is very similar, the ligula being in both more detached from the body of the organ than in the other species here described. The front tarsi are, on the other hand, sufficiently dissimilar in this sex of the two species to afford of themselves satisfactory characters by which the two may be distinguished; they are not so broad and patellated in *X. violaceus*, and are less densely pubescent beneath.

6. *Xanthopygus depressus*, n. sp. Subdepressus, niger, nitidus, elytris vel viridibus vel cyaneis, abdominis segmentis ultimis duobus flavis, antennis pedibusque ex parte rufo-testaceis. Long. corp. 5 lin.

Mas: abdominis segmento 7° ventrali margine apicali medio, late haud profunde triangulariter emarginato.

Antennæ dull yellow at the base, infuscated in the

middle, the last joint again paler; 3rd joint rather long and slender, considerably longer than the 2nd; from the 4th to the 10th each joint is a little shorter than its predecessor, the 4th considerably longer than broad, 7th about as long as broad, 8—10 rather transverse; last joint pointed, nearly twice as long as the 10th. Palpi yellow, mandibles pitchy. Head broad, quite as broad as the thorax, coarsely and irregularly punctured, with a broad impunctate space in the middle. Thorax about as long as broad, a little narrowed behind, with two irregular lines of ten or twelve coarse punctures along the middle, separated by a rather broad, impunctate space, and with other coarse, irregular punctures, especially numerous near the anterior angles. Scutellum large, rather strongly punctured, with a narrow impunctate margin. Elytra broader than the thorax, and about as long, greenish or bluish, rather sparingly punctured. Hind body narrowed towards the extremity, black, with the last two segments yellow; segments 2—5 moderately closely and distinctly punctured, 6th more finely, 7th very finely punctured. The four front legs yellow, the hinder ones pitchy.

The male has a shallow, broad notch in the middle of the hind margin of the 7th segment beneath.

Pará, Ega, St. Paulo, Rio Purus; sixteen individuals.

Obs.—I judge from the specimens before me that this is a variable species; the individual from which the above description is taken is a large male from St. Paulo, having the head and thorax more coarsely punctured, and the antennæ more elongate than in the other individuals. The individual from Rio Purus is a small female, having the head and thorax sparingly punctured, and is a little smaller, narrower, and less depressed than the other specimens. The two individuals from Pará have the front legs black, or nearly so, and the antennæ rather shorter, while one of them has the elytra of a pitchy colour, with blue reflections. In the absence of any definite characters to separate these forms, I have considered them all as one species.

7. *Xanthopygus nigripes*, n. sp. Niger, nitidus, antennis fusco-testaceis, elytris viridi-cyaneis, abdomine segmentis duobus ultimis flavis. Long. corp. 5 lin.

Closely allied to *X. depressus*, and distinguished only by the following characters. The head is smaller, being a little narrower than the thorax; the antennæ are dusky yellow, and are rather shorter, joint 4 being about as long as broad, joints 5—10 transverse; all the legs are black. The 6th segment of the hind body slightly darker at the extreme base; the head and thorax rather more finely punctured.

St. Paulo; one specimen, ♀.

Obs.—A second female individual, labelled only Amazons, departs still more from *X. depressus*, its head and thorax being still more finely and sparingly punctured; but I believe it to be only a variety of *X. nigripes*.

8. *Staphylinus xanthopygus*, Nord.

I refer to this name a series of individuals, from Ega and Pebas; they appear to me to be quite conspecific with other specimens from Mexico and central America, and I have two or three other closely allied species from other parts of South America. The characters of the male are not described either by Nordmann or Erichson; in that sex the hind margin of the ventral plate of the 7th segment of the hind body has a broad but shallow notch in the middle, and the front tarsi are slightly more dilated than in the female. Erichson names the species *Philonthus xanthopygus*, but his description does not accord very satisfactorily with Nordmann's, and I judge it to have been drawn up from more than one species. In the Munich Catalogue, Nordmann's species is recorded under the name *Xanthopygus abdominalis*, and it is probable that the appellation of the species will be again changed.

9. *Xanthopygus cognatus*, n. sp. Subdepressus, niger, nitidus, abdomine segmentis duobus ultimis rufo-testaceis. Long. corp. 6 lin.

Closely allied to *Staphylinus xanthopygus*, Nord. (*Philonthus xanthopygus*, Er.), but not half the size of that species. Antennæ rather short and stout, not thickened towards the extremity; the first 3 joints black, the rest fuscous; 3rd joint almost shorter than 2nd, 4th and 5th slightly transverse, 6—10 evidently so; last joint sinuate at the extremity and pointed. Head short, about as broad as the thorax, coarsely and irregularly punctured, the middle parts without punctures. Thorax about as long as broad, the sides scarcely sinuate behind and very little rounded towards the front. Along the middle are two irregular rows of large punctures, leaving a broad space between them free from punctures; scattered about the sides are also numerous large irregular punctures, especially numerous towards the front. Scutellum closely punctured. Elytra longer than the thorax, moderately closely and finely punctured. Hind body black, with the extreme hind margin of the 5th segment, and the whole of the 6th and 7th segments, reddish-yellow; the punctuation moderately close and fine. The legs are stout, pitchy black, with the tarsi pitchy red; the four hinder tibiæ strongly spinulose.

Ega; one specimen, ♀.

10. *Philonthus analis*, Er.

Pará, Obydos, Tapajos, Ega, St. Paulo.

This appears to be one of the most widely distributed and abundant of the South American *Staphylinidæ*; one of Mr. Bates's specimens is labelled as found in dung.

The male characters are omitted by Erichson; in that sex the 6th segment of the hind body has, on the underside on its middle, a small fovea, from which projects a slender tuft of elongate hairs, and the following plate has a deep but rather narrow notch at the extremity; the front tarsi are moderately dilated in each sex, in the male only slightly more than in the female.

11. *Staphylinus bicolor*, Lap. (*Philonthus bicolor*, Er.).

Ega and St. Paulo.

The male characters in this species also have not been recorded; in that sex the ventral plate of the 6th segment of the hind body has, near the base in the middle, a transverse impunctate space, in front of which is another transverse space which is very slightly depressed and finely punctured; the hind margin of the 7th segment is very slightly emarginate in the middle; the front tarsi are rather broadly dilated and are apparently similar in the two sexes.

PHILOTHALPUS.

The species referred to this genus are at present nine in number, and are confined to South America; they were most of them known to Erichson, and divided by him among his genera *Staphylinus* and *Philonthus*. Three others, considered by me as new, are here added.

I have had great difficulty in dealing with this genus and its allies, *Gastrisus* and *Eugastus*, and feel far from satisfied with the course I have adopted. I would have preferred considering them all as one genus, containing a number of heterogeneous forms, but the characters on which the now accepted genera of the *Staphylinini* are based would not allow me to do this; to have dealt with them in a satisfactory manner would have necessitated a fresh re-grouping of the South American *Staphylinini*, a step which is at present out of the question. On the other hand, to have gone backwards and applied to the whole of these insects the name *Staphylinus*, would, I think, have been too retrograde a step. Had there been in use a collective name to designate all those *Staphylinini* in which the lateral pieces of the thorax are not abbreviated, I would gladly have used it for all these insects; but such a name has never existed, for Kraatz, to whom we owe the indication of this very important character, when he pointed it out, at the same time distributed the species possessing it among a number of new genera, while the species I am here describing are, many of them, intermediate between the genera he then characterized.

The three species here described as appertaining to the genus *Philothalpus* differ considerably from one another in facies, and no doubt many entomologists would be inclined to consider them as belonging to three distinct genera.

1. *Philothalpus luteipes*, n. sp. Capite thoraceque obscure æneis; scutello, elytris, pectore abdomineque testaceo-ferrugineis, hoc segmentis 4—7 nigro-signatis; pedibus testaceis. Long. corp. 4½ lin.

Mas: tarsis anticis dilatatis; abdomine segmento 7° ventrali medio triangulariter inciso.

Fem. tarsis anticis leviter dilatatis.

Antennæ about as long as the head and thorax, not thicker towards the extremity, blackish, the basal joints indistinctly paler, the last joint also obscurely paler; each joint longer than broad, 3rd considerably longer than the 2nd. Head as broad as the thorax, orbiculate, very closely and coarsely punctured, with an impunctate space in the middle; a broad depression between the antennæ, the punctured parts with stiff, outstanding setæ. Thorax rather longer than broad, a little narrowed behind, and the sides a little sinuate behind the middle, dull brassy above, numerously and moderately coarsely punctured, with an impunctate line in the middle, the punctures with outstanding grey hairs. Scutellum tawny, closely and finely punctured. Elytra tawny, a little longer and much broader than the thorax, rather closely and finely punctured and with a concolorous pubescence. Hind body tawny, 4th segment slightly marked with black at the base, 5th and 6th broadly black at the base, basal half of the 7th yellow, extremity blackish, lateral lobes of the 8th segment blackish; the base of each segment is finely and moderately closely punctured, the extremity of each sparingly punctured; the upper surface with coarse black hairs, the basal part of the 7th segment without these hairs and scarcely punctured. Legs yellowish: basal joint of hind tarsi nearly as long as the three following together.

Pará; Ega, seven specimens.

Obs.—This species I anticipate will prove closely allied to *Staphylinus segmentarius*, Er.; indeed, I should have referred these individuals to that species had it not been that Erichson describes the apical segment of the hind body by the words "toto nigro," whereas in *P. luteipes* it is yellow, with the hind margin black. Erichson's locality for *S. segmentarius* is Columbia, and I have an individual of *P. luteipes* from Venezuela; should it prove that the words I have quoted from Erichson's description are erroneous, it may be probable that *P. luteipes* is conspecific with *S. segmentarius*.

2. *Philothalpus latus*, n. sp. Fulvus, capite antennisque nigris, his articulo ultimo ferrugineo; abdomine segmentis 2—5 late piceis. Long. corp. 5½ lin.

Mas: tarsis anticis fortiter dilatatis; abdomine segmento sexto ventrali apice medio late emarginato, 7º triangulariter inciso.

Fem. latet.

Broader than usual in this genus. Antennæ nearly as long as the head and thorax, blackish, with the 1st one or two joints pitchy, and the last joint obscure reddish; 3rd joint considerably longer than 2nd, 4—10 each a little shorter than its predecessor, 4—6 longer than broad, 8—10 a little transverse, 7—10 each slightly produced on the inside; 11th joint sinuate and pointed at the extremity. Mandibles and palpi pitchy red. Head as broad as the thorax, black, scarcely brassy, coarsely and irregularly punctured, an ill-defined space in the middle smooth, with a kind of triangular depression in front. Thorax narrower than the elytra, tawny, shining, a line in the middle smooth, the rest of the upper surface covered with rather fine and not close punctures. It is a little narrowed behind, but the sides are scarcely sinuate behind the middle. Scutellum moderately closely and finely punctured. Elytra tawny, a little paler than the thorax, finely and not closely punctured. Hind body above with segments 2—5 pitchy black, 6th and 7th yellow, hind part of the styles on the 8th black; segments 2—6 finely and not densely punctured, 7th very finely and sparingly punctured at the base, more coarsely on the hind part. Legs yellowish.

In the male the front tarsi are broadly dilated; the hind margin of the 6th segment of the hind body is broadly and shallowly emarginate in the middle, and the 7th segment has a triangular notch.

St. Paulo; one ♂ individual.

3. *Philothalpus incongruus*, n. sp. Fulvus, nitidus, capite brevi, nigro sub-æneo, oculis magnis; antennis, basi excepto, fuscis: abdomine segmentis 2—4 sine lineis curvatis impressis, segmentis 5 et 6 leviter infuscatis. Long. corp. 4 lin.

Mas latet.

Fem. tarsis anticis haud dilatatis.

Antennæ about as long as head and thorax, moderately stout, a little thickened towards the extremity, four or five basal joints dusky fulvous, the rest infuscated; 3rd joint rather longer than 2nd, 4th a little shorter than 2nd; from this to the 10th each joint shorter and stouter than its predecessor, 5th longer than broad, 9th and 10th a little transverse, 11th about twice as long as the 10th, pointed. Mandibles short, palpi pitchy. Head about as broad as the thorax, broad and short, shining and blackish except behind the labrum, where it is reddish; it is covered with large punctures, except in the middle, where it is smooth; the eyes are large and prominent and extend very nearly to the back of the head. The thorax is about as long as broad, rather narrowed behind, the anterior angles deflexed and a little rounded, the sides a little sinuate behind the middle; it is of a shining tawny colour, the upper surface covered with numerous but not coarse punctures, leaving a central line smooth. The scutellum is large, densely and distinctly punctured. The elytra are about as long as, and rather broader than the thorax, of a tawny colour, moderately closely and rather finely punctured. The hind body is rather narrowed to the extremity, of a tawny colour, the 5th and 6th segments darker and the 7th yellowish; segments 2—4 finely and rather closely, 5th and 6th densely and finely, 7th very sparingly and finely punctured. Legs yellowish.

Ega; one specimen.

Obs.—This species differs from *segmentarius* and its allies, in that the curved impressions on the basal segments of the hind body are so obsolete that they might almost be correctly described as absent. The insect differs, however, from *Eugastus* completely in its facies, and can therefore scarcely be considered intermediate between it and *Philothalpus*.

GASTRISUS, n. gen.

Ligula integra.

Palpi labiales articulo ultimo suboblongo, apice truncato; maxillares articulo ultimo præcedente longiore, apice acuminato.

Thorax lineis marginalibus lateralibus utrinque haud conjunctis, lateribus membranâ stigmaticâ instructis.

Abdomen segmentis 2 et 3 sine lineâ incurvatâ.

Genus *Philonthi* staturâ similis, sed prothoracis lineis lateralibus haud conjunctis, lateribus pone coxas membranâ instructis, differt. Generis *Philothalpi* quoque affinis, sed ab illo, prothoracis membranâ stigmaticâ, abdomineque sine lineis incurvatis, discedit. Generis typus *G. lævigatus*.

The three new species I refer to this genus are very discrepant by their sculpture; *G. obsoletus* and *G. lævigatus* are peculiarly smooth, while *G. punctatus* is remarkable for its coarse sculpture. *G. obsoletus* and *lævigatus* may possibly prove to be only one species; they suggest, at first sight, a comparison with *Quedius*, but differ therefrom by the less abruptly-inflexed lateral pieces of the thorax. *G. punctatus* has quite the facies of a diminutive *Philonthus analis*, Er.

1. *Gastrisus obsoletus*, n. sp. Rufo-testaceus, antennis nigris, basi piceo, capite æneo; thorace disco fusco-æneo;

elytris opacis, obsolete parceque punctatis; abdomine segmentis basalibus medio infuscatis, parce punctatis. Long. corp. 4½ lin.

Mas: tarsis anticis sat dilatatis, abdomine segmento 7° ventrali apice medio minus profunde exciso.

Antennae short and moderately stout, blackish; 1st joint pitchy, the extreme base of each of the following joints reddish; 3rd joint longer than 2nd, 4th rather longer than broad, 8—10 rather strongly transverse; palpi reddish. Head brassy, moderately shining, with two punctures behind the labrum, each placed in a depression, with two or three along the margin of the eye, and some others at the extreme hind angles; all the middle part impunctate; it is distinctly narrower than the thorax. Thorax distinctly longer than broad, a good deal narrower than the elytra, along the middle of an obscure æneous colour, the sides and base yellowish; it is distinctly sinuate at the sides and a little narrowed behind; on each side the middle, at some distance from the front, is a single puncture, and two or three others at each front angle, elsewhere impunctate. Elytra not longer than the thorax, of a yellow colour, opaque, sparingly and very obsoletely punctured, and with scanty fine hairs. Hind body yellowish, with the middle of the basal segments infuscate, sparingly and finely punctured, and scantily pubescent, at each hind angle of each segment with a long black seta; anal styles tawny yellow. Legs yellow, with concolorous spines; basal joint of hind tarsus as long as the three following together.

In the male the front tarsi are moderately dilated and furnished beneath with pale pubescence; the hind margin of the ventral plate of the 7th segment bears a small notch in the middle.

A single specimen was brought back by Mr. Bates without any special indication of its locality.

2. *Gastrisus lævigatus*, n. sp. Subdepressus, rufotestaceus, capite supra æneo-micante, antennis fuscis. Long. corp. 5 lin.

Mas latet.

Fem. tarsis anticis sat dilatatis.

Antennæ shorter than the head and thorax, moderately stout, the basal joint pitchy red, the rest fuscous; 3rd joint a little longer than 2nd; 4th and 5th each about as long as broad, 6th transverse ; from this to the extremity no broader; joints 7—10 rather strongly transverse; 11th joint rather short, sinuate and pointed at the extremity. Mandibles and palpi reddish. Head nearly as broad as the thorax, above brassy and smooth, with two punctures in the front behind the labrum, and several large and smaller punctures behind the eyes. Thorax reddish, about as long as broad, but little narrowed behind ; the sides slightly rounded in front, and a little sinuate behind, smooth and impunctate, with the exception of a puncture on each side behind the neck and two or three others near the front angles. Scutellum obsoletely punctured. Elytra yellowish, about as long as the thorax, dull, obsoletely and very sparingly punctured. Hind body yellowish, the basal segments a little infuscated in the middle, rather finely and rather sparingly punctured. Legs yellow; tarsi rather slender.

Ega ; one specimen, ♀.

Obs.—This is very closely allied to *G. obsoletus*, but is a little larger, and has the thorax unicolorous; the head is rather larger and a little more elongate, so that the punctures at the hind angles are rather more numerous and more conspicuous.

3. *Gastrisus punctatus*, n. sp. Capite, prothorace, elytrisque obscure cyaneis; abdomine nigro, apice flavo; antennis pedibusque nigro-fuscis. Long. corp. 4½ lin.

Mas: tarsis anticis minus dilatatis ; abdomine segmento 7° ventrali apice medio late triangulariter emarginato.

This insect is remarkable by the deep and close punctuation of the fore parts, in opposition to the very fine sculpture of the hind body. The antennae are shorter than the head and thorax, moderately stout, blackish ; 3rd joint considerably longer than 2nd; from this to the extremity very slightly thickened ; 4—6 each about as long as broad, 7—10 a little transverse ; 11th joint moderately long and pointed. Mandibles and palpi pitchy. Head rather small, a little narrower than the thorax, above bluish-green, closely, deeply and coarsely punctured, with a well-limited quadrate space on the middle impunctate. Thorax rather longer than broad, a little narrowed behind and slightly sinuate at the sides; the whole upper surface deeply, closely and evenly punctured, with a narrow impunctate line along the middle. Scutellum large, densely

punctured. Elytra about as long as, and a good deal broader than the thorax, of a dull greenish colour, closely and deeply punctured, the punctures being much finer than on the head and thorax. Hind body not much narrowed at the extremity, black, with the hind margin of the 6th and the whole of the 7th segment bright yellow. It is very finely and rather closely punctured. Legs pitchy black.
St. Paulo; one specimen, ♂.

EUGASTUS, n. gen.

Antennæ sat longæ, filiformes, articulo ultimo apice obliquo.
Palpi filiformes, articulo ultimo præcedente longiore.
Thorax lineis lateralibus haud conjunctis, sine membranâ stigmaticâ.
Abdomen segmentis 2—4 sine lineis transversis incurvatis; segmentis 2 et 3 basi utrinque lineâ brevi obliquâ impressis.
Tarsi intermedii et postici graciles, articulo primo lineari, elongato.
Labrum medio incisum. Mandibulæ breves. Palpi maxillares articulo ultimo apice acuminato, labiales articulo ultimo lineari. Pedes graciles. Habitu *Staphylino* et *Philontho* quasi intermedium.
Locus systematicus prope genus *Philothalpum*.

This genus is undoubtedly very close to *Philothalpus*, but as Kraatz specially bases that genus on the curved lines of the hind body, and as these insects do not exhibit that character, and as they present a facies strikingly peculiar, I have decided on giving a new generic name, though with much hesitation.

1. *Eugastus bicolor*, n. sp. Rufo-ferrugineus, elytris cyaneo-nigris; abdomine minus nitido, segmentis 4—6 late infuscatis, 7° apice fusco. Long. corp. 6¼ lin.; lat. (elytrorum) 1¼ lin.
Mas: tarsis anticis leviter dilatatis: abdomine segmento 7° ventrali margine apicali medio leviter emarginato.

Narrow and elongate, the front parts dull, and only the hind body somewhat shining. The antennæ are rather slender, not quite so long as head and thorax, scarcely thicker towards the extremity. The three basal joints reddish, the rest pitchy; 3rd joint one and a half times the length of the 2nd; from 4—10 each is a little shorter than its predecessor; 4th joint much longer than broad, 10th about as long as broad; last joint about as long as the 4th, pointed on one side. Parts of the mouth red. Head as broad as the thorax, orbiculate; the eyes large, above red and opaque, with numerous very obsolete punctures, each bearing a fine upright hair; beneath smooth and shining, with a few very fine punctures. Thorax a little narrower than the elytra, the length one and a half times the width, narrowed behind, transversely very convex at the front angles, which are very rounded, the sides a little sinuate behind, the hinder angles quite rounded; above it is of a very dull-reddish colour, extremely obsoletely punctured, but with a rather coarse and evenly distributed black pubescence. Scutellum very densely pubescent. Elytra scarcely the length of the thorax, of an obscure dull-bluish colour, with an obsolete but rather rugulose sculpture. Hind body slender and elongate, a little narrowed towards the extremity, of a reddish colour; a large part of the 5th and 6th segments pitchy, the extremity of the 7th also dark, and the basal segments are also a little infuscate at the extreme base; the segments are moderately coarsely but not closely punctured; the hind half of the 6th and the basal part of the 7th very finely and sparingly punctured, the styles of the 8th segment blackish. The legs are reddish-yellow and rather long, the hind tarsi long and slender, the basal joint about twice the length of the second.

St. Paulo; one specimen, ♂.

2. *Eugastus mundus*, n. sp. Opacus, fulvus, antennis basi excepto nigris, elytris viridi-opacis, dense punctatis; abdomine nigro-fulvoque variegato, dense punctato. Long. corp. 7 lin.

Antennæ with the three basal joints red, the rest blackish; joints 4—10 each distinctly shorter but scarcely broader than its predecessor; 4th a good deal longer than broad, 10th scarcely so long as broad. Head rather narrow, with the eyes very large, of a tawny-red colour, dull, covered with an obsolete punctuation, and with a fine scanty golden pubescence. Thorax elongate, similar in colour to the head, very dull, very obsoletely punctured, and with a fine, very depressed, reddish pubescence. Scutellum velvety-black. Elytra as long as the thorax, of a dull-greenish colour, the basal part of their inflexed portion reddish; they are densely and rather coarsely punctured, the punctuation being confluent and rough. Hind body with the basal segments blackish, but their hind portions tawny; 5th segment entirely black; 6th black, but with the hind margin broadly and abruptly yellowish; 7th yellowish, with its hind margin black; anal styles black. Legs tawny yellow.

A single female of this remarkable species was found by Dr. Trail at Lages, near Manaos, on the 5th January, 1875. A very mutilated individual of the same sex was also sent by Mr. Bates, but without any indication of locality.

Obs.—This species may be readily distinguished from *E. bicolor*, by its larger size, broader form, and shorter-jointed antennæ, as well as by the more densely punctured elytra and hind body; the two females have the front tarsi distinctly dilated, but scarcely so broad as in *E. bicolor*, ♂.

ISANOPUS, n. gen.

Antennæ tenues, elongatæ.

Palpi filiformes, elongati, maxillares articulo ultimo præcedente duplo longiore.

Thorax lineis lateralibus haud conjunctis, sine membranâ stigmaticâ.

Abdomen segmentis 2—4 sine lineis incurvatis.

Tarsi antici dilatati, intermedii et posteriores articulis 2—4 sublobatis.

Labrum medio incisum. Mandibulæ breves, acutæ; tibiæ posteriores tarsis fere duplo longiores.

Genus præcedenti affinis, differt palpis longioribus tarsorumque structurâ aliâ.

The insect to which I apply this new name is remarkable for the structure of the four posterior tarsi; these have joints 2—4 somewhat lobed and dilated, the dilatation being chiefly on their inner sides, so that each of these joints is unsymmetrical in shape.

1. *Isanopus tenuicornis*, n. sp. Niger, nitidus, elytris obscure cyaneis, antennis articulis 4—11 testaceis, abdomine apice rufo-testaceo. Long. corp. 6 lin.; lat. (elytrorum) 1⅓ lin.

Mas latet.

Femina tarsis anticis dilatatis.

About similar in size to *Philonthus cribratus*, but less

depressed, and with the hind body narrower. The antennæ are longer than the head and thorax, slender, not in the least thickened towards the extremity; the three basal joints are pitchy, the others pale yellow; the 3rd joint is longer than the 2nd, 4th about as long as the 2nd; from this to the extremity the joints differ but little from one another; 11th joint rather shorter than the 10th, its length two or three times its breadth. Mandibles and palpi pitchy red. Head about the width of the thorax, suborbicular, covered with numerous closely placed, large punctures, except the disc and a transverse space behind the labrum, which are free from punctures. The eyes rather large. Thorax nearly one and a-half times as long as broad, narrowed behind; the front angles much deflexed and rounded, strongly sinuate at the sides; the hinder angles obtuse and rounded, the upper surface with two irregular lines of large punctures along the middle, leaving a broad space between them impunctate, and with numerous other large punctures at the sides, scarcely leaving the two middle lines of punctures distinct from the others, the punctures more numerous about the front than at the hinder part. Scutellum black, large, densely and distinctly punctured. Elytra about as long as the thorax, and considerably wider, dark bluish, rather roughly and moderately closely punctured. Hind body elongate, a little narrowed to the extremity, black, with very faint bluish reflections; the 6th segment reddish, the 7th yellow; the base of each segment is closely and finely punctured, the apex more sparingly; the hind part of the 6th and the whole of the 7th segment finely and sparingly punctured. Legs black, rather long, the hind tibiæ especially long; the four hinder tarsi with the joints a little dilated and uneven; the 1st joint rather stout and nearly linear, about as long as the two following together; the second joint triangular, the internal angle more produced than the outer one; 3rd joint only half as long as the second, distinctly lobed, especially on the inner side; 4th joint rather narrower than, but nearly as long as the 3rd, its outer angle distinctly produced; the tarsi are pitchy, their terminal joints reddish.

Ega; one specimen, ♀.

TRIGONOPSELAPHUS (*Trigonophorus*, Nord.).

This genus is also peculiar to South America; it consists at present of about a dozen species; some of its species are amongst the most brilliant of natural objects. Kraatz has remarked that the species fall into three distinct groups, adding, that it will be well to leave them together in one genus till more species are known. The species I here describe as *T. mutator* adds yet another form to those previously included under this generic name. It is well to add, to prevent misconception, that though the dilated terminal joint of the labial palpi is given as one of the most important characters of the genus, *S. venustus* and *violaceus* have that joint quite simple, and are allied to the *Philonthus pretiosus*, Er., in this respect.

1. *Trigonopselaphus opacipennis*, n. sp. Capite thoraceque viridibus, nitidis; elytris obscure ænescentibus, opacis; abdomine opaco, nigro, apice rufo-testaceo; antennis pedibusque nigris, illis articulo ultimo ferrugineo. Long. corp. 9 lin.

Antennæ the length of the head and half the thorax, not thickened towards the extremity, black, with the 10th joint pitchy and the 11th dull yellowish; 3rd joint considerably longer than the 2nd, quite twice as long as the 4th; from the 4th to the 10th each a little shorter than its predecessor; 10th joint about as long as broad, the others longer than broad; 11th joint about as long as the 9th, its extremity rounded, but pointed on one side. Palpi pitchy; mandibles black. Head scarcely as broad as the thorax, rather quadrate, above shining green, with large, coarse punctures irregularly scattered, but leaving a rather broad irregular space, extending from the labrum to the neck, free; below black and dull. Thorax a little narrower than the elytra, rather longer than broad, very slightly narrowed behind, with the sides but little sinuate on the upper side, with two lines of about eight punctures, with a rather broad space between them, and besides these with about twenty other punctures on each side near the front part and the outsides; it is of a shining, bluish-green colour; the margins below black and provided with a stigmatic membrane. Scutellum large, black, dull, very obsoletely and sparingly punctured. Elytra about as long as the thorax, of a dull black colour with a slight metallic tint, with a very peculiar sculpture consisting of large shallow punctures, placed at a good distance from one another, and with faint, irregular, wandering lines between them. The hind body is narrowed towards the extremity;

it is of a dull black colour, with the hind margin of the 6th segment, as well as the whole of the 7th and 8th, dull-orange colour, it is sparingly punctured, the punctures more numerous on the 6th segment than elsewhere, and each bearing a rather stout hair. Legs dull blackish.

St. Paulo; one specimen, ♀.

2. *Trigonopselaphus mutator*, n. sp. Niger, antennis piceis, capite thoraceque viridi-cyaneis, elytris æneis fortiter punctatis; abdomine nigro-æneo, apice testaceo. Long. corp. 5¼ lin.

Antennæ short and rather stout, 1¼ lin. in length, of an obscure-reddish colour; 3rd joint about as long as 2nd, 4th—10th each a little shorter than its predecessor, each much narrowed towards the base, especially on its inner side; the 4th about as long as broad, 8—10 rather strongly transverse, 11th rather short, acuminate, its apical portion yellowish, being paler than the basal portion. Palpi blackish, the maxillary ones slender; the last joint of the labial much dilated, the middle joint not so long as broad. Mandibles short, very thick, but at the extremity slender and very acuminate; labrum large, quite bifid, pitchy. Head small, a good deal narrower than the thorax; the eyes large and occupying most of the side; the vertex very truncate, so that the neck is abrupt and rather slender; it is of a greenish-blue colour, not much shining; it is coarsely punctured, the punctures absent about the middle; on the underside it is black, shining and impunctate, and the genæ are distinctly margined. Thorax distinctly narrower than the elytra, about as long as broad, truncate in front, rounded at the base, the sides a little curved; it is similar in colour to the head, and is coarsely and sparingly punctured, the punctures being absent from a space along the middle. Scutellum large, dull blackish, rather coarsely punctured. Elytra slightly longer than the thorax, of a shining-brassy colour, coarsely punctured, and only sparingly pubescent. Hind body blackish, with the 7th segment and the hind margin of the 6th yellow; the basal segments rather brassy, the 5th and 6th very densely punctured and clothed with a remarkably dense, coarse, black pubescence, the basal segments more sparingly punctured; the anal styles nearly black except at the base. Legs black, with the tibiæ and tarsi pitchy; the front tibiæ broad, their tarsi strongly dilated; hind tarsi moderately long, their basal joint rather longer than the three following together.

Pebas; a single female, collected by Mr. Hauxwell.

Obs.—This is a very remarkable species, and one which has at first sight the facies of the *Xanthopygi* with metallic elytra; but the labial palpi, and the absence of a stigmatic membrane to the thorax, forbid its being associated with them. I have thus been compelled either to establish a new genus for it, or to call it a *Trigonopselaphus*, and I have preferred the latter course, as that name has already scarcely any definite meaning, owing to the heterogeneous nature of the few species associated under it.

3. *Trigonopselaphus violaceus*, n. sp. Violaceus, opacus, antennis pedibusque nigris, elytris sparsim fortiter punctatis. Long. corp. 9 lin.

Mas: tarsis anticis sat dilatatis, abdomine segmento 6° ventrali apice medio leviter emarginato, 7° late minus profunde inciso.

N.B.—Hac specie palpi labiales articulum ultimum haud dilatatum, apice truncatum præbent; tarsi postici articulum primum elongatum, ceteros breves.

Antennæ nearly as long as head and thorax, rather slender, not in the least thickened towards the extremity; they are blackish, the three basal joints indistinctly violet; 3rd joint long, one and a half times the length of the 2nd, 4th not quite so long as 2nd; from 5—10 each is a little shorter than its predecessor, the first of them much longer than broad, and even the last longer than broad; 11th joint rather longer than the 10th. Mandibles black; palpi pitchy, last joint of the maxillary twice as long as the preceding one. Head rather narrower than the thorax, a little narrowed to the front; the eyes moderately large, extending quite half-way to the back of the head; above it is of a beautiful dull-violet colour, and has a few large punctures scattered irregularly over it. The thorax is considerably longer than broad, the sides slightly sinuate behind the middle, and a little narrowed towards the front angles, so that it is scarcely broader at the front than at the hind angles; it is similar in colour to the head, and has two lines formed of three or four indistinct punctures along the middle, and a few other punctures near the front part. The scutellum is large, blackish, sparingly and ob-

solctely punctured. The elytra are about as long as the
thorax, very dull, of a blackish colour with a slightly
violet tinge, covered with rather large and deeply impressed but distant punctures. The hind body is distinctly narrowed towards the extremity; it is of a dull-
violet colour on the upper side, and is sparingly but distinctly punctured. The legs are blackish; the basal joint
of the posterior tarsi as long as the three following together. The under surface of the insect is of a dull-
blackish colour faintly tinged with violet; the margins of
the thorax without stigmatic membrane.

Ega; one specimen, ♂.

4. *Trigonopselaphus venustus*, n. sp. Violaceus, sat
nitidus, antennis nigris articulis nullis transversis. Long.
corp. 12 lin.

Mas: abdomine segmento 6° ventrali apice medio leviter
emarginato, 7° minus profunde exciso, 8° lobo medio apice
leviter emarginato.

Closely allied in structure to *Philonthus pretiosus*, Er.,
and belonging really to the same genus. Antennæ not quite
so long as the head and thorax, not thickened towards the
apex; 2nd joint long, but shorter than the 3rd; 4th joint
twice as long as broad, 10th longer than broad, 11th
rather longer than 10th, pointed on one side; they are
black, with the basal joints violet. Mandibles black, violet
at the base. Head a little narrower than the thorax, not
rounded at the sides, on the upper side of a beautiful
violet colour, irregularly sprinkled with large punctures,
the central part being free. Thorax longer than broad,
much narrower than the elytra, the sides a little narrowed
at the front and sinuate behind the middle; above similar
in colour to the head, and with an irregular dorsal row of
ten or eleven punctures on each side the middle, and
also with other scattered punctures near the front. Scutellum moderately closely and rather obsoletely punctured.
Elytra scarcely longer than the thorax, of a dark-violet
colour, rather finely and closely punctured. Hind body
distinctly narrowed to the extremity, of a greenish-violet
colour, rather sparingly and finely punctured, the punctures evenly distributed. Legs greenish-violet; the front
tarsi dilated in both sexes, the basal joint of the hind
tarsus not greatly longer than the 2nd.

Ega and Tapajos; three specimens.

Obs.—*Philonthus cyanescens*, Guérin, is an ally of this species, but is very much smaller and has the antennæ much more slender.

GLENUS.

This genus appears to me one of the most distinct of the subfamily *Staphylinini*; the elongate terminal lobe of the maxillæ, and the mandibles, which though elongate are but little curved, taken in conjunction with the subapproximate antennæ, and the peculiar form of the front of the head, give it a peculiar isolation. It consists at present of five species, four of which were known by Erichson, and assigned by him to his genus *Staphylinus*, and it is to Kraatz that we owe the establishment of the genus.

I here describe four new species, two of which are very closely allied to others already known, while the other two, *G. amazonicus* and *G. vestitus*, form a distinct section by reason of the unrounded sides of the thorax.

1. *Glenus Kraatzi*, n. sp. Rufo-testaceus, thorace cupreo, nitido; elytris aureo-tomentosis, fasciâ mediâ ad suturam abbreviatâ fuscâ; abdomine nigro, segmentis singulis apice rufis. Long. corp. 12 lin.

This species is so closely allied to *G. biplagiatus*, that it is only necessary to point out its distinctive characters. The punctuation of the head behind the eyes is less close. The thorax has the hinder angles less completely rounded off, and is a little more sinuate at the sides; the impunctate medial line is broader, and extends quite to the front of the thorax; its punctuation is less dense, and the small punctures mixed with the large ones in *biplagiatus* are in *Kraatzi* nearly absent. The 3rd segment of the hind body is without the transverse curved line which is apparent in *biplagiatus*, and the colouring of the 4th and 5th segments is different; in *biplagiatus* the black marks thereon consist of a central spot and another on each side, these being united at the base, but distinctly recognizable; in *Kraatzi* these spots are replaced by a broad, transverse band, but slightly sinuate behind.

Ega; one specimen, ♀.

Obs.—I have much pleasure in dedicating this fine species to the learned author of the second volume of the "Insecten Deutschlands," and the founder of the genus to which the species belongs.

2. *Glenus Batesi*, n. sp. Rufus, capite rufo-testaceo, thorace elytrisque sanguineo-tomentosis, abdomine nigro-subæneo, segmentis singulis transversim vix sinuatim rufo-marginatis, crebre subtiliter punctatis. Long. corp. 13 lin.

Closely allied to *G. Chrysis*, Grav., but distinguished therefrom by the redder colour of the pubescence on the thorax and elytra, and by the much more closely and finely punctured hind body; the transverse red markings of segments 3—5 are nearly straight.

Tapajos; two specimens.

Obs.—I have also an individual of this fine species in my collection, labelled " Brazil." A fully extended large individual attains sixteen or seventeen lines of length.

3. *Glenus amazonicus*, n. sp. Opacus, pube aureo subtili parcius vestitus, capite rufo-testaceo, thorace sericeo-æneo; elytris obscure violaceo-brunneis, ad latera maculâ nigrâ notatis; abdomine rufo-brunneo, maculis fasciâque nigro-æneis. Long. corp. 9 lin.

Antennæ a little shorter than head and thorax; 3rd joint considerably longer than 2nd; 7th joint a little transverse, 8—10 distinctly so; 11th joint truncate and sinuate at the extremity; they are of an obscure-reddish colour, darker towards the extremity, but with the last joint again a little paler. Mandibles reddish at the base, black towards the extremity; palpi reddish. Head dull reddish, obscurely metallic between the antennæ, above rather convex; sparingly and very finely punctured, with a fine golden pubescence. Thorax much narrower than elytra, considerably longer than broad, distinctly narrowed behind, the front angles rounded and much deflexed, the sides a little sinuate; above it is of a very dull-brassy colour, with a very silky lustre; it is sparingly and very obsoletely punctured, with a fine golden pubescence, apparently very easily removed. Scutellum black, dull and velvety. Elytra not longer than the thorax, of a peculiar dull-reddish colour, with a violet tinge, at the outside with a blackish, ill-defined mark; not visibly punctured, and with a scanty, depressed, golden pubescence, apparently very easily removed. Hind body narrowed to the extremity, of a reddish colour, marked with black, of a faint brassy tinge; 2nd segment nearly entirely black, 3rd with a broad mark in the middle and a small one on each side, 4th and 5th each with a mark in the middle reaching the hind margin, but not the extreme base; 6th brassy black, reddish at the base, 7th entirely yellowish: it is sparingly and rather finely punctured, the red parts with golden hairs, the black parts with blackish hairs. Legs yellowish-brown.

Ega; two specimens, ♀.

4. *Glenus vestitus*, n. sp. Obscure rufus, pubescentiâ rufâ vestitus, capite testaceo, subtiliter aureo-tomentoso, inter antennas aenescente; abdomine segmentis longitudinaliter aeneo-lineatis. Long. corp. 10 lin.

Mas: segmento 6" ventrali medio fasciculâ parvâ pilorum longorum, margine apicali medio leviter emarginato, 7° apice medio late satque profunde inciso.

Antennae long and slender for this genus, nearly as long as head and thorax, of a dull-reddish colour; 3rd joint longer than 2nd, none of the joints transverse, 10th about as long as broad, 11th about as long as 10th; the extremity truncate, with the internal angle much produced. Palpi yellowish; mandibles red at the base, pitchy black towards the extremity. Head yellowish, with a metallic mark in front between the antennae; it is about as broad as the thorax, rather convex above, and clothed with a very fine but rather dense golden pubescence. Thorax considerably narrower than the elytra, rather longer than broad, a little narrowed behind, the anterior angles not much deflexed, and but little rounded; the sides slightly sinuate, the hind angles obtuse; it is covered above with a dense ferruginous-red pubescence, and has at the anterior angles seven or eight long erect setae. Scutellum densely clothed with black velvety pubescence. Elytra about as long as the thorax, densely covered with a reddish pubescence, hiding their colour and sculpture. Hind body narrowed towards the extremity, reddish, with a brassy line along the middle, formed by an elongate spot down the middle of each segment; it is closely and finely punctured, and rendered dull by a dense, depressed, concolorous pubescence. Legs yellowish; tarsi pitchy. Breast clothed with golden pubescence.

In the male the 6th segment of the hind body beneath has a very short line of long, erect hairs, and the hind margin is a little emarginate; the 7th segment has a broad and rather deep notch in the middle of the hind margin.

Pará, Ega, St. Paulo; three specimens ♂, one ♀.

LEISTOTROPHUS.

This genus consists of a few species, but is of very wide distribution. South America possesses but a single species, which, however, is the most developed and remarkable of the genus. The two species described by Motschoulsky under the generic name of *Trichoderma*, which, in the Munich Catalogue, are recorded as South American species of *Leistotrophus*, belong clearly, from Motschoulsky's description, to the genus *Staphylinus*.

1. *Staphylinus versicolor*, Grav.

Pará, Ega, Tapajos.

One of the individuals is labelled as found in cow-dung; the species, like its European congeners, frequents, no doubt, putrescent substances for predaceous purposes.

STAPHYLINUS.

I have used this name with the same extension as that given to it in the Munich Catalogue of Coleoptera, where it includes about 100 species, found in all parts of the world. It is a genus of which the species are extremely closely allied, but yet, studied on the European ones, have proved to be incontestably distinct. The exotic species are probably extremely numerous, and their discrimination will be no easy task. I here enumerate nine species from the Amazon Valley, seven of which I have described as new; of these the first two, viz., *S. subcyaneus* and *S. parviceps*, are quite distinct, by their combinations of colour and sculpture, from any others I am acquainted with. The same remark applies to *S. gratiosus* and *S. gratus*, but *S. priscus* and *S. vetustus* are very closely allied to the *S. antiquus* and some other undescribed South American forms, and thus appertain to what is undoubtedly a most difficult group; while the *S. amazonicus* perhaps finds its nearest ally in the North American *S. tomentosus*. It is worthy of notice that Mr. Bates brought back nothing to represent the very remarkable *S. Buquetii* group, of which species are found in Mexico, Peru and Brazil; it will be remarkable if no allied species is found in the Amazon Valley, and yet so large and striking are they, that if present one would think they would scarcely have been neglected by Mr. Bates during the whole of his long residence there.

1. *Staphylinus subcyaneus*, n. sp. Niger, capite thoraceque nigro-cyaneis, abdomine segmentis duobus ultimis flavis, antennis rufo-testaceis. Long. corp. 8 lin.

Mas: abdomine segmento 7º ventrali apice emarginato.

About the size of *S. chalcocephalus*, but with the head smaller. Antennæ about the length of the head and one-third of the thorax; they are of a yellowish colour; 3rd joint longer than 2nd, 4—10 transverse, differing but little from one another, 11th sinuate and acuminate at the extremity. Mandibles pitchy; palpi yellow. Head smaller than the thorax, narrowed in front, blackish-blue, moderately coarsely and moderately closely punctured, clothed with a fine pubescence. Thorax not quite so long as broad, a little narrowed towards the front, the sides straight, the base and hind angles rounded, above of an obscure-bluish colour like the head, neither very closely nor coarsely punctured, with a short smooth line in front of the scutellum, and clothed with a dark fuscous pubescence. Elytra about as long as the thorax, dull bluish-black, rendered black and opaque by a fine depressed pubescence, under which they are alutaceous but not punctured. Scutellum clothed with black pubescence. Hind body narrowed towards the extremity, quite dull and densely clothed with a very fine concolorous pubescence, black, with the 6th and 7th segments yellow. Legs black, tibiæ strongly spinulose.

Ega and Tunantins; two specimens, ♂ and ♀.

2. *Staphylinus parviceps*, n. sp. Opacus, niger, antennis rufis, capite thoraceque subcyaneis, abdomine segmentis 6º, 7ºque late testaceis, 6º basi nigro; abdomine tomento haud variegato. Long. corp. 7 lin.

Mas: abdomine segmento 7º ventrali apice medio leviter emarginato.

Antennæ short, reddish, with the basal joints yellow; 3rd joint scarcely longer than 2nd, 4—10 transverse, 6—10 scarcely at all differing from one another either in length or breadth. Mandibles pitchy red; palpi yellow. Head a good deal narrower than the thorax, subtriangular, the punctures only moderately coarse, those on the anterior part not dense; it is of a black colour, with a pale blue reflection and very slightly shining. Thorax just as long as broad, nearly as broad as the elytra, slightly narrowed in front, covered with a rather coarse but not deep punctuation, which is close, but the interstices are quite distinct and not at all rugose; its depressed pubescence is of a dark colour, but not altogether black; in colour it is similar to the head. The elytra are just as long as the thorax, and are covered with a blackish tomentum, which makes them appear quite dull and without sculpture, and they bear besides a rather close depressed pubescence. Hind body black, with the two apical segments bright yellow, but the base of the 6th black; the black part is covered with a dense dark tomentum, which is quite unicolorous; the anal styles are yellow. The legs are short and stout, black. The under face of the hind body is distinctly punctured, and bears a rather scanty yellow pubescence; it has a slight metallic reflection.

Ega; seven individuals.

Obs.—I at first considered this species a variety of *S. subcyaneus*, but, after examination of a series of seven individuals, I consider it likely to prove a distinct species; it is considerably smaller, rather darker in colour, has the head smaller, and the basal portion of the 6th segment of the hind body black. The male characters appear very similar.

3. *Staphylinus ochropygus*, Nord. var.

Tapajos, Ega, St. Paulo.

I identify this species from the descriptions of Nordmann and Erichson; the specimens agree therewith except that they have the legs and antennæ paler in colour.

4. *Staphylinus gratiosus*, n. sp. Fulvus, capite thoraceque viridi-cyaneis, nigro-pubescentibus, scutello atro-tomentoso, abdomine aureo-tomentoso, antennis (basi excepto) infuscatis. Long. corp. 7¼ lin.

Mas: abdomine segmento 7° ventrali apice medio late sed minus profunde emarginato.

A very pretty species, and remarkable for the peculiar sculpture of the head and thorax, the interstices of the coarse punctures thereon being finely punctured. Antennæ stout, about as long as head and one-third of thorax; the two basal joints dark yellowish, the others infuscate; 2nd and 3rd joints rather long, the 3rd longer than 2nd, 5—10 transverse; last joint truncate at the extremity and pointed on one side. Palpi yellowish; mandibles pitchy. Head narrowed in front, narrower than the thorax, shining greenish, rather coarsely and sparingly punctured, with numerous fine punctures mixed with the large ones. Thorax nearly as long as broad, a little narrower than the elytra, distinctly narrowed in front, of a shining bluish colour, but dulled by a dark-fulvous pubescence which appears black in most lights; it is moderately coarsely and not densely punctured, fine punctures are distributed over the interstices of the larger ones. Scutellum densely clothed with black tomentum. Elytra about as long as the thorax, of a rich tawny colour, quite dull from a depressed tomentum, and furnished besides this with rather stiff depressed concolorous hairs. Hind body narrowed to the extremity, of a tawny-brownish colour, the segments furnished on the upper side with a rather scanty golden tomentum, and with rather numerous, stiff, concolorous hairs. Legs tawny yellow; front coxæ a little infuscate. The under surface tawny, with a scanty golden pubescence.

Ega; one specimen, ♂.

5. *Staphylinus gratus*, n. sp. Fulvus, capite thoraceque cyaneis, fulvo-pubescentibus, scutello fulvo-tomentoso, abdomine segmento sexto nigricante, apice segmentoque septimo testaceo. Long. corp. 7 lin.

Mas latet.

Allied to the preceding but very distinct. Antennæ longer than the head, the three basal joints tawny yellow, the rest infuscated; joints 2 and 3 rather short, 3rd longer than 2nd, 4—10 transverse, but the 4th narrower than the 5th; 11th joint sinuate at the extremity, one angle being pointed. Palpi yellowish. Head small, narrower than the thorax, much narrower than the elytra, narrowed in front, rather coarsely but not densely punctured, with a considerable impunctate space in the middle, shining blue, with a dark-reddish pubescence of erect hairs. Thorax about as long as broad, the sides a little arched and slightly narrowed in front, coarsely but by no means densely punctured, with a very narrow smooth line along the middle, shining blue, with a long, erect, reddish pubescence. Scutellum with a dense fulvous pubescence. Elytra scarcely so long as the thorax, dark tawny, clothed with a dense concolorous pubescence, and also with depressed fine hairs. Hind body narrowed to the extremity, reddish-brown; the 5th segment darker at the base, the

6th nearly black, but its hind margin and the 7th segment yellowish; it is above clothed with a nearly concolorous pubescence, which is arranged so as to make it appear indistinctly variegated, besides this with stiff reddish hairs, and on the 6th segment with black hairs. Legs reddish.

Tunantins; one specimen, ♀.

6. *Staphylinus amazonicus*, n. sp. Niger, opacus, nigro-tomentosus, scutello densius atro-tomentoso; abdomine supra bifariam nigro-maculato, segmentis singulis summo basi, medio maculâ cinereâ minus conspicuâ. Long. corp. 11 lin.; lat. (elytrorum) 2½ lin.

Mas latet.

Antennæ short and slender, rather longer than the head, black, with the extremity of the last joint rusty; 3rd joint long, one and a half times the length of the 2nd; joints 6—10 much narrowed to the base, but none of them transverse (in the ♀ at any rate). Mandibles black; palpi pitchy. Head smaller than the thorax, greatly narrower than the elytra, narrowed to the front, dull black, densely and moderately coarsely, but not deeply punctured, covered with a dense, fine and short, erect, black pubescence. Thorax narrower than the elytra, fully as long as broad, the sides nearly straight, not narrowed in front, but with the front angles deflexed and rounded; it is of a very dull black, densely and rather coarsely but very shallowly punctured, densely clothed with a pubescence similar to that of the head. Scutellum with a dense black velvety pubescence. Elytra slightly longer than the thorax, quite dull black, scarcely punctured but rugulose, covered with a concolorous pubescence. Hind body black; the sides of each segment with coarse shallow punctures; at the base of each segment is a middle spot of scanty yellowish or ashy hairs, on each side of which there is a velvety-black pubescence. The legs are black; the tibiæ and tarsi with rusty hairs. The wings dull yellowish.

Ega; one specimen, ♀.

Obs.—I regret that while mounting this insect I lost the 7th segment of the hind body (which had become detached). I cannot describe it fully but only say that it was quite black.

7. *Staphylinus antiquus*, Nord., Er.

Pará, Tapajos, Ega.

This appears to be one of the most widely distributed

of the South American species of *Staphylinus*. I have specimens which I consider conspecific with Amazonian ones, from Nicaragua, Columbia and Rio de Janeiro, as well as from intermediate localities.

8. *Staphylinus priscus*, n. sp. Capite thoraceque æneis, elytris obscure æneis, obsolete variegatis, abdomine tessellato, ano rufo-testaceo. Long. corp. 7 lin.

Mas: abdomine segmento 7° ventrali apice late sed haud profunde emarginato.

Closely allied to *S. antiquus*, Nord.; the thorax not longer than broad, not at all narrowed in front, the pubescence of the hind body darker, &c. It resembles *Ocypus cupreus*, but is a little broader and less elongate. Antennæ reaching half-way back the thorax; first three joints reddish, the rest nearly black; 2nd and 3rd moderately long, 3rd longer than 2nd; 4th joint narrower than the 5th, slightly transverse, 5—10 rather strongly so, 11th sinuate at the extremity and pointed on one side. Mandibles pitchy; palpi reddish. Head rather narrower than the thorax (smaller and more triangular in the ♀ than in the ♂), coarsely but not densely punctured, brassy, the interstices shining; a very small, narrow space in the middle free; clothed with a fuscous pubescence. Thorax but little narrower than the elytra, almost straight at the side, about as broad in front as behind, scarcely so long as broad; coarsely and closely punctured, with a line in front of the scutellum smooth, and with slight traces of the continuation of this in front; furnished with a dense and fine fuscous pubescence. Scutellum velvety black. Elytra about as long as the thorax, not so long as their breadth taken together, dull brassy; finely pubescent, and very indistinctly tessellated. Hind body narrowed at the extremity, pitchy in colour, the 7th segment yellowish; each segment has in the middle, at the base, an indistinct ashy mark, on each side of which the pubescence is more closely placed, so as to appear darker; besides this tomentum they are clothed also with numerous coarse hairs of a dark fuscous, nearly black colour. Legs pitchy red.

Ega; five specimens (♂ and ♀).

Obs.—Besides the above-mentioned five individuals, four others from the same locality represent, I believe, a variety; they are rather smaller, and have the head a little smaller, and the setæ of the hind body reddish. In another individual from Pará the head and thorax are rather more sparingly and a little more coarsely punctured.

9. *Staphylinus vetustus*, n. sp. Niger, capite, thorace, elytrisque obscure æneis, his obsolete variegatis, abdomine supra tessellato, ano rufo-testaceo; thorace fere elytrorum latitudine. Long. corp. 7½ lin.

Mas: abdomine segmento 7° ventrali apice medio minus profunde emarginato.

Closely allied to the preceding; the head and thorax more densely punctured; the thorax both longer and broader, and the 3rd joint of the antennæ longer. Antennæ reaching nearly half-way down the thorax, pitchy; 2nd and 3rd joints rather long, 3rd considerably longer than 2nd; joints 4—10 differing but little from one another, transverse, but not strongly so; 4th joint sinuate at the extremity, and pointed on one side. Head small, narrower than the thorax, narrowed in front, dull brassy, closely and rather coarsely punctured, with a fuscous pubescence. Thorax scarcely narrower than the elytra, about as long as broad, very slightly narrowed in front, dull brassy, coarsely and very densely punctured, with a very short and very narrow smooth line in front of the scutellum, and clothed with a fuscous pubescence. Scutellum velvety black. Elytra about as long as the thorax, dull brassy, finely pubescent and indistinctly tessellated. Hind body narrowed to the extremity, pitchy; 7th segment and hind margin of the 6th yellow, obscurely tessellated, with a dark brown and scanty ashy pubescence, and besides this with coarse, nearly black hairs. Legs pitchy; femora marked with yellow towards the extremity.

Tunantins; one specimen; also four other individuals without special locality.

Obs.—This species is closely allied to *S. antiquus*, but is larger and broader; the legs, the antennæ and the pubescence are darker in colour; the antennæ are thicker, and the carina-like space along the head and thorax is absent.

BELONUCHUS.

About thirty species are at present referred to this genus, and all of them are indigenous to its warmer parts, one or two extending their range to the United States of

North America. I here enumerate twelve species from the Amazons, of which nine are new.

The genus has no sufficient characters pointed out to distinguish it from the great genus *Philonthus*; Erichson was in doubt as to whether he should accept it as distinct therefrom, and indicated, as the only character peculiar to it, the arming of the front and hind femora with seta-like spines. This character, however, differs in certain species in the two sexes, as will be seen from my descriptions of *B. decipiens* and *B. setiger*. Species possessing this character are moreover by no means confined to the New World, for I have several undescribed *Philonthus*-like species displaying it from Papua and the Malay Archipelago. I have, therefore, only used this generic name as a matter of convenience, to avoid increasing the enormous number of species already registered under the generic name of *Philonthus*.

1. *Staphylinus hæmorrhoidalis*, Fab.

Pará, Ega, Pebas. Numerous specimens.

2. *Philonthus xanthopterus*, Nord.

Ega, St. Paulo. Also found at Barreiras de Janarape, Rio Solimoes, on the 9th January, by Dr. Trail.

3. *Belonuchus Batesi*, n. sp. Depressus, niger, nitidulus, abdomine dense punctato, apice rufo ; prothorace seric dorsali 4-punctato. Long. corp. 4½ lin.

Mas: abdomine segmento 7º ventrali apice anguste triangulariter inciso.

Very similar in appearance to *B. hæmorrhoidalis*. Antennæ with the basal joint pitchy, the rest black, 3rd joint slightly longer than 2nd, 5—10 transverse, 5th broader than the 4th. Palpi pitchy. Head large, as broad as the elytra, black and shining, with a row of six punctures in front, placed one on each side, close to the eye, and two pairs near to each other in the middle, these separated by a depressed line ; behind this front row of punctures is a second irregular row across the middle of the head, and there are also some other scattered punctures at the sides and back, but there is no raised line at the hinder angle. Thorax longer than broad, the front angles greatly depressed and rounded, the sides strongly sinuate, the base truncate, but the hind angles rounded, on each side of the middle with a row of four punctures, and with a few other punctures scattered near the front angles. Scutellum thickly punctured. Elytra as long as the thorax, rather finely and closely punctured. Hind body narrowed towards the extremity; hind portion of the 6th and the whole of the 7th segment reddish; it is closely punctured, with the exception of the 7th segment, which is finely and sparingly punctured. The legs are black, the front and the hind femora spinous beneath.

Ega; one specimen, ♂.

4. *Belonuchus grandiceps*, n. sp. Rufo-testaceus, nitidus, capite abdomineque nigris, hoc apice flavo, antennis nigris, basi testaceo. Long. corp. 6 lin.

Mas: capite majore, clypeo sub-bidentato, abdomine segmento 7° ventrali, margine apicali, utrinque fortiter inciso.

Allied to *B. xanthopterus*, but with the thorax ree. instead of black. Antennæ with the three or four basal joints yellow, the rest black, the last joint rusty; 3rd joint rather longer than the 2nd, 6—10 transverse. Palpi yellow; mandibles red or pitchy red. Head black and shining, with a deep longitudinal impression in front, with numerous very coarse punctures behind the eyes, and with four or five near the inner margin of the eyes. Thorax reddish-yellow, narrower than the elytra, very slightly narrowed behind, and with the sides nearly straight; it is about as long as broad, and has on each side the middle a dorsal row of five punctures, the three middle ones approximated, and besides this with about six other punctures near the front angles. Scutellum reddish, closely punctured. Elytra about as long as the thorax, reddish-yellow, rather finely and moderately closely punctured. Hind body black, with the hind margin of the 6th and the whole of the 7th segment yellow; it is closely and distinctly punctured, and has a coarse black pubescence; the yellow portion is much more finely and sparingly punctured. The legs are yellow.

In the male the head is larger than the thorax, and the clypeus is obtusely elevated on the inside of the insertion of each antenna; the 7th segment of the hind body has a deep notch on each side of the middle of the hind margin; the anterior femora are obtusely dilated in the middle, and the hind femora are more strongly spinulose than in the female.

Ega; Tapajos, St. Paulo, five specimens, 3 ♂, 2 ♀.

5. *Belonuchus decipiens*, n. sp. Niger, nitidus, antennarum basi pedibusque piceis, ano late testaceo, prothoracis serie dorsali 5-punctato, abdomine crebre punctato. Long. corp. 5½ lin.

Mas: capite majore clypeo sub-bidentato, abdomine segmento 7° ventrali apice leviter emarginato, femoribus posticis subtus fortiter biseriatim spinosis.

Fem.: femoribus posticis uniseriatim setosis.

Antennæ blackish, with the two or three basal joints pitchy, and the last joint rusty; 3rd joint but slightly longer than 2nd, 4th quadrate, 5—10 transverse but not strongly so. Palpi reddish; mandibles pitchy. Head black and shining, with two points on each side between the eyes, and with several other coarse punctures on each side near the hind angles, the front portion with an impressed line extending half-way to the back. Thorax rather narrower than the elytra, quadrate, not or scarcely narrowed behind, the anterior angles a little depressed and rounded, the sides scarcely sinuate; it is black and shining, and has on each side a dorsal row of five punctures; of these the hinder one is placed at a distance from the others. The elytra are about as long as the thorax, shining blackish, rather finely and not closely punctured. The hind body is black, with the hind portion of the 6th and the whole of the 7th segment yellow; segments 2 and 3 are rather sparingly punctured, 4—6 more closely punctured and with a coarse black pubescence, 7th segment very finely and sparingly punctured. Legs pitchy.

The individual described above is a male (from Ega), and has the head rather larger than the thorax; the mandibles and palpi elongate, the clypeus obtusely projecting on each side between the insertion of the antennæ; the back part of the under face of the anterior femora dilated from the base to near the extremity, and furnished with short spines, the hind femora beneath with two rows of stout spines, and the hind margin of the 7th ventral segment of the hind body a little emarginate.

Ega; two males.

Obs.—A series of female specimens from Pará, Tapajos and Ega, are, I have no doubt, the other sex of this species. They have the head not broader than the thorax, the mandibles and maxillary palpi shorter, the clypeus simple, the front femora not dilated beneath, and the hinder femora destitute of the two rows of stout spines, but furnished with a single row of fine setæ. A male individual from Pará agrees with the males from Ega, but has the antennæ a little stouter and the penultimate joints more strongly transverse, and the middle joints each a little produced on the inner side. I consider this individual is only a variety of *B. decipiens*.

6. *Staphylinus formosus*, Grav.

Pará, Tapajos, Ega.

This is one of the commonest and most widely distributed of the New World *Staphylinidæ*; it extends from Pennsylvania to Rio de Janeiro.

7. *Belonuchus clypeatus*, n. sp. Niger, nitidus, ano testaceo, prothorace serie dorsali 5-punctato, lateribus subrectis, abdomine minus crebre punctato. Long. corp. 5 lin.

Mas: clypeo antice vere bidentato, femoribus posticis uniseriatim spinulosis, abdomine segmento 7º ventrali apice leviter emarginato.

Fem. latet.

Allied to *B. decipiens*, but distinguished by its more sparingly punctured hind body and the two teeth of the clypeus in the male. The antennæ are short and stout, black, with the base pitchy; 3rd joint slightly longer than 2nd; 4th joint small; 5th broader than 4th, but also small; 6th joint transverse, 7—10 rather strongly transverse. Mandibles nearly black; palpi pitchy. Head about as broad as the thorax, black and shining, with two short, stout teeth (in the ♂ if not in the ♀) projecting forwards but not upwards from the front part; between the eyes with four punctures, also with four or five other punctures close to to the back part of the eyes, and a few others near the hinder angles. Thorax narrower than the elytra, quadrate, about as long as broad, not narrowed behind, and scarcely sinuate at the sides, the front angles not rounded; it is black and shining, has on each side the middle a dorsal row of five punctures—of these the hind one is a little the more remote,—and with four or five other punctures near the front angles. Scutellum rather coarsely punctured. Elytra a little longer than the thorax, black and shining, rather finely and sparingly punctured. Hind body black, with the hind portion of the 6th and the whole of the 7th segment reddish-yellow; it is rather coarsely and not closely punctured, and sparingly pubescent. The legs are black.

Ega; one specimen, ♂.

8. *Belonuchus holisoides*, n. sp. Angustus, depressus, niger, nitidus, antennis elytrisque fuscis, illarum basi pedibusque testaceis. Long. corp. 2 lin.

Mas: abdomine segmento 7° ventrali margine apicali medio minus profunde triangulariter inciso.

Antennæ short, three basal joints yellow, the rest infuscate; joints 2 and 3 short, about equal in length, 6—10 differing little from one another, slightly transverse. Mandibles and palpi yellowish. Head large, quite as broad as the elytra, black and shining, with an irregular row of six punctures between the eyes, and three or four other punctures behind these on each side, and with a fine impressed line on the front part. Thorax narrower than the elytra, longer than broad, narrowed behind, the front angles quite rounded and deflexed, the sides sinuate; it is black and shining, with a dorsal row of three punctures on each side the middle, and besides these with only two or three other small punctures. Scutellum finely and closely punctured. Elytra about as long as the thorax, blackish, finely and not closely punctured. Hind body slender; segments 2—5 a little depressed at the extreme base, and there coarsely and closely punctured, the other part sparingly punctured, the pubescence fine and scanty. The legs are dirty yellow. The anterior and posterior femora very sparingly furnished with spines beneath.

Ega; two specimens, ♂ and ♀.

9. *Belonuchus æqualis*, n. sp. Elongatus, depressus, capite, thorace antennisque nigris, illarum basi, pedibus, pectore anoque testaceis; elytris rufis, abdomine dense subtiliter punctato. Long. corp. 3½ lin.

Mas: abdomine segmento 7° ventrali apice leviter emarginato.

About the size of *Xantholinus lentus*. Antennæ rather short, yellowish at the base, the rest pitchy; 3rd joint a little longer than the 2nd; joints 5—10 each slightly stouter than its predecessor, the 5th slightly transverse, 10th distinctly so. Mandibles pitchy; palpi yellowish. Head rather large, broader than the thorax, with an irregular row of punctures between the eyes, a longitudinal impression along the front part, and a few punctures scattered about the hind part on each side. Thorax narrower than the elytra, longer than broad, narrowed behind, the front angles depressed and rounded, the sides sinuate; it is black or pitchy; on each side the middle it has a dorsal row of three or four punctures, and has two or three other punctures near the front angles. Scutellum pitchy, closely and finely punctured. Elytra dull reddish, slightly longer than the thorax, closely and finely punctured. Hind body densely and finely punctured and pubescent, black, with the hind portion of the 6th and the whole of the 7th segment yellow. Legs yellow, the front femora with two or three spines near the extremity, the hind ones with a row of spines few in number. Breast and under portions of the prothorax yellowish.

Pará, Ega, St. Paulo; eight specimens, ♂, ♀; also found by Dr. Trail at Conccicão, Rio Mauhes, in May, 1874.

10. *Belonuchus impressifrons*, n. sp. Capite thoraceque nigro-æneis, nitidis; elytris rufis, abdomine nigro, apice testaceo; pectore piceo, antennarum basi, articulo ultimo pedibusque rufo-testaceis; prothorace serie dorsali 5-punctato; femoribus anticis spinulosis, posticis fere muticis. Long. corp. 4¼ lin.

Mas: abdomine segmento septimo ventrali margine apicali leviter emarginato.

Femina latet.

Antennæ longer than the head, the three basal joints reddish-yellow, the rest dark, with the last joint again paler; the 3rd joint is rather longer than the 2nd; the 5th joint broader than the 4th, rather longer than broad; 6—10 differing little from one another, each about as long as broad. Mandibles pitchy; palpi reddish. Head with the front distinctly produced between the insertion of the antennæ, with a deep longitudinal impression on the front part, with four points between the eyes, and several others at the back on each side; it is very shining and brassy black. The thorax is narrower than the elytra, longer than broad, very slightly narrowed behind; the front angles, seen from above, nearly right angles; it is brassy black, and has on each side the middle a dorsal row of five large punctures, and several other punctures on each side. Scutellum black, closely punctured. Elytra about as long as the thorax, deep red, rather sparingly and moderately closely punctured. Hind body rather closely punctured, and with the pubescence rather long; it is black, with the hind portion of the 6th and the whole of the 7th segment yellow; beneath it is closely and rather coarsely punctured. The legs are reddish; the front femora with black spines (in the male, at any rate), the hind femora with only one or two short spines near the base. The breast is pitchy.

In the male the head is large, being much broader than the thorax; the female is unknown to me.

Ega; one male.

11. *Belonuchus armatus*, n. sp. Subdepressus, niger, nitidus, antennarum basi apiceque, et pedibus obscure testaceis, elytris rufis; abdomine crebre punctato, segmento ultimo, præcedentisque apice piceo-testaceis. Long. corp. 4¼ lin.

Mas: abdomine segmento 7° ventrali apice medio exciso, trochanteribus posterioribus elongatis.

Fem. latet.

Antennæ pitchy black, with the basal joints pitchy, and the apex again paler. Palpi reddish, mandibles pitchy red. Head broad and rather short, much broader than the thorax, with two deep punctures on the longitudinal impression behind the labrum, with two others on each side between the eyes, and again others towards the hind angles. Thorax narrower than the elytra, rather longer than broad, distinctly narrowed behind, with a series of five coarse punctures on each side the middle, and outside these with about six punctures on each side; it is, like the head, of a shining-black colour, faintly tinged with brassy. Scutellum large, blackish, closely punctured. Elytra red, about as long as the thorax, moderately closely and not coarsely punctured. Hind body blackish, with the hind part of the 6th segment, and with the base and apex of the 7th, pitchy yellow; it is not very densely punctured, but the black pubescence on the penultimate segments is dense and very coarse. The legs are dark yellow, with the coxæ still darker.

In the male the front femora bear rather long black spines, the hind femora appear to be without spines, but the trochanters project as a long sharp tooth; the ventral plate of the 7th segment of the hind body has a small but distinct notch at the extremity.

A single male, taken at Abacaxis by Dr. Trail on the 11th April, 1874.

Obs.—Though the resemblance between this species and *B. impressifrons* is very great, the darker extremity of the hind body and the elongate trochanters of the male in *B. armatus* readily distinguish it.

12. *Belonuchus setiger*, n. sp. Minus depressus, nigerrimus, nitidus, ano testaceo, prothorace serie dorsali 5-punctato, abdomine minus dense punctato, supra et infra setis erectis crebre vestito; femoribus posticis muticis. Long. corp. 3¼ lin.

Mas: femoribus anticis evidenter spinulosis, abdomine segmento 7° ventrali margine apicali leviter emarginato.

Fem.: femoribus anticis fere muticis.

Very black and shining, with the extremity of the hind body yellow. The antennæ are a little shorter than head and thorax, distinctly thickened towards the extremity; 3rd joint rather shorter than 2nd; 4th joint about as long as broad, 5—10 transverse. Head with the front considerably produced between the antennæ, with a broad and deep longitudinal impression on the front part, with two punctures on each side near the front part of the eyes, and with some other punctures near the sides behind. The thorax is quite as long as broad, scarcely narrowed behind, convex transversely, very black and shining, with the five punctures in a row on each side the middle deep, and with six other punctures on each side. The scutellum is finely and closely punctured. The elytra are longer than the thorax, sparingly and finely punctured. The hind body is narrowed towards the extremity; it is black, with the hind margin of the 6th segment and the whole of the 7th segment pale yellow, the anal styles black; segments 2—4 are rather sparingly punctured, 5 and 6 more distinctly and closely so; the erect setæ, both on its upper and under side, are unusually numerous and long. The legs are quite black.

Ega; two specimens, ♂ and ♀.

Obs.—This species has greatly the form and appearance of a *Philonthus*.

PHILONTHUS.

This generic name at present designates nearly four hundred species found in all parts of the world. I here describe nineteen new Amazonian species. These nineteen species belong to several very different groups. *P. amazonicus* is allied to our European *P. scybalarius*; *P. corallipennis* to the group comprising the *P. fulvipes* and its allies; while *P. deletus* seems quite allied to our *P. prolixus*, and, like it, has the appearance of a small *Lathrobium*. The other species are very different from any we have in Europe. *P. muticus* is very like the depressed *Belonuchi*, and *P. gracillimus* is very remarkable by its elongate, narrow prothorax. The next seven species belong to a brightly-coloured group of species which is peculiar to South America, and of these seven *P. palpalis* is remarkable by the dilated terminal joint of the labial palpi. The next five species belong also to a group confined to South America: the species in colour much resemble those of the preceding group, but whereas in the first of the two groups the anterior angles of the thorax are distinct and rather prominent, in the second they are rounded and very depressed. The *P. longipes* is in form similar to the species I have last named, but it has a very peculiar punctuation along the margins of the thorax, and its elytra are densely punctured. *P. serraticornis* is a remarkably aberrant species, which both in appearance and structure approaches the insects I describe in this paper under the generic name *Gastrisus*.

1. *Philonthus amazonicus*, n. sp. Niger, pedibus fuscotestaceis, abdomine subversicolore, apice indeterminate rufo; capite minore subovato, prothorace serie dorsali 5-punctato. Long. corp. 5 lin.

Mas: tarsis anticis dilatatis, abdomine segmento 7° ventrali margine apicali medio minus profunde triangulariter exciso.

Fem.: tarsis anticis simplicibus.

Allied to *P. scybalarius*, but much larger. Antennæ quite as long as head and thorax, black, the first joint pitchy: 3rd joint considerably longer than 2nd, 4 and 5 much longer than broad, 6—9 slightly produced on the inner side, 10th rather longer than broad; last joint a little longer than the 10th. Head small, sub-ovate, narrower than the thorax, with two punctures near the front part on each side, close to the inner margin of the eye, and very near to one another, and with several other punctures on each side behind the eye. Thorax longer than broad, narrowed to the front, straight at the sides, with a row of five punctures on each side the middle, and outside this row with a row of three other punctures on the front part. Scutellum closely and finely punctured. Elytra about as long as the thorax but broader, closely and finely punctured. Hind body narrowed towards the extremity, blackish and a little versicolorous; the 6th segment black at the base, the rest of it and the 7th segment rusty yellow; the hind margin of the 5th segment also rusty; it is rather closely punctured on the upper side, the punctuation being denser on the basal halves of the 5th and 6th segments than elsewhere; its pubescence is rough and coarse. The legs are of a dirty-yellow colour, with the tibiæ and tarsi darker, the intermediate coxæ approximated.

Pará, Ega; six specimens, ♂ and ♀.

2. *Philonthus corallipennis*, n. sp. Niger, elytris anoque rufis, dense subtilissime punctulatis; antennarum basi pedibusque testaceis, prothorace serie dorsali subtiliter 8-punctato. Long. corp. 3 lin.

Mas: tarsis anticis dilatatis.

Allied to *P. salinus*, but much narrower, and with the prothoracic series of punctures more numerous. The antennæ are long and slender, the basal joint entirely and the 2nd and 3rd partly yellowish; 3rd joint longer than 2nd, all the joints longer than broad. The palpi are yellowish. The head is rather narrower than the thorax; at the back and between the eyes it is finely, rather closely punctured, the middle part impunctate. The thorax is longer than broad, nearly straight at the sides, and not (or scarcely) narrowed in front; it has on each side the middle a dorsal row of eight punctures, not very regularly placed, and has besides four or five others in a line outside these. The scutellum is smoky, densely punctured. The elytra are about as long as the thorax, but a little broader, of a red colour, very densely and finely punctured, quite dull. Hind body rather long and slender, but little narrowed behind, very densely and finely punctured, the hinder portion reddish; the limit of this colour not well marked. The legs are pale yellow.

Pará, Obydos, Ega, Tapajos; numerous specimens.

3. *Philonthus deletus*, n. sp. Rufo-testaceus, nitidus, capite elytrisque obscurioribus, horum apice, antennis pedibusque testaceis; prothorace subtiliter multipunctato, lineâ latâ mediâ lævi. Long. corp. 2 lin.

Quite of the structure of *P. prolixus*. Antennæ yellow, rather stout; 3rd joint about equal to the 2nd; 10th joint scarcely so long as broad. Head dark reddish, about as broad as the thorax, rather finely and not closely punctured, the middle part impunctate. Thorax narrower than the elytra, one and a-half times as long as broad, scarcely narrowed behind, the sides a little sinuate; it is of yellowish colour, shining, a broad middle space impunctate: the sides punctured, the punctuation not deep and not very close. Scutellum yellowish, very finely and indistinctly punctured. Elytra rather longer than the thorax, fuscous; the apex pale yellow, finely and rather sparingly punctured. Hind body yellowish, not narrowed to the extremity, its punctuation and pubescence extremely fine and not dense. Legs yellow, rather stout.

The male has a moderately large notch at the extremity of the ventral plate of the 7th segment of the hind body, and the hind margin of the 6th segment is slightly emarginate.

Tapajos; four individuals.

4. *Philonthus muticus*, n. sp. Depressus, obscure rufo-testaceus, capite abdomineque nigricantibus, hoc crebre subtiliter punctato, apice testaceo, prothorace serie dorsali 5-punctato; coxis intermediis distantibus, tarsis anticis omnino simplicibus. Long. corp. 3½ lin.

Mas: capite majore, abdomine segmento 7° ventrali apice leviter emarginato.

This species resembles greatly *Belonuchus æqualis*, but has the anterior and posterior femora entirely without spines, except the three or four at the extremity of the front femora found in most species of *Philonthus*. The antennæ are inserted quite at the anterior margin of the front, which is not at all produced in the middle; they are a little shorter than the head and thorax, the three basal joints reddish, the rest infuscated; the 3rd joint longer than the 2nd, the 10th about as long as broad. The mandibles and palpi are reddish. The head is blackish, with two points on each side near the front, at the inner side of the eyes, and between these with a punctiform longitudinal impression, and with several other punctures on each side, near the back part of the head. The thorax is rather narrower than the elytra; it is rather longer than broad, the front angles much depressed and rounded, the sides sinuate; it is of a reddish colour, besides the dorsal series of punctures with four or five others on each side. The scutellum is similar in colour and sculpture to the elytra. Elytra reddish, scarcely so long as the thorax, finely and closely punctured. Hind body blackish, rather paler at the base, with the hind margins of the segments narrowly reddish; the hinder portion of the 6th segment and the whole of the 7th yellowish; it is rather closely and finely punctured. Legs and breast yellowish.

Ega and St. Paulo; four specimens, ♂, ♀.

5. *Philonthus gracillimus*, n. sp. Elongatus, perangustus, rufo-testaceus, capite nigro, elytris abdominisque segmentis 5 et 6 infuscatis; antennis elongatis, fuscis, articulo ultimo ferrugineo, prothorace serie dorsali subtiliter 5-punctato. Long. corp. 3 lin.; lat. (abdominis basi) vix ½ lin.

A singular Lathrobioid species. The antennæ are long and slender, and reach nearly to the extremity of the elytra; 3rd joint long and slender, much longer than the 2nd; all the rest of the joints much longer than broad, the last paler than the rest. The mandibles and palpi are yellowish. The head is narrow, but rather broader than the thorax; it is narrowed behind the eyes; it is of a black colour, rather finely and closely punctured, with the disc impunctate. The thorax is very peculiar in form; it is much narrower than the elytra, and very elongate, being more than twice as long as broad; it is not narrowed behind, but appears somewhat narrowed in front, owing to the front angles being greatly inflexed; it is strongly sinuate at the sides before the hind angles; it is of a reddish colour, and has on each side the middle a dorsal row of five fine punctures, and has also a few other fine punctures on each side. The scutellum is finely punctured. The elytra are not so long as the thorax, rather finely and closely punctured. The hind body is elongate and slender, yellowish, with the 5th and 6th segments broadly infuscated; it is closely, finely and evenly punctured. The legs are yellow, long and slender.

Ega; one specimen.

Obs.—This unique individual is in bad condition, being quite deprived of its pubescence.

6. *Philonthus æneiceps*, n. sp. Rufus, capite æneo, nitidissimo, abdomine ante apicem nigro. Long. corp. 3½ lin.

Mas: capite majore, femoribus anticis spinulosis, abdomine segmento 7° ventrali medio producto, ante apicem leviter transversim impresso, tarsis anticis simplicibus.

Antennæ nearly as long as the head and thorax, yellowish; 2nd and 3rd joints about equal, 4—10 each a little shorter than its predecessor, none of them transverse. Mandibles and palpi yellowish. Head on the upper side shining brassy, the front with a medial longitudinal impression, on which are placed two large confluent punctures, and on each side of this with three punctures, forming an irregular row between the eyes, and with eight or ten other large punctures on each side at the back part. Thorax rather longer than broad, slightly narrowed behind, the sides a little sinuate; it is of a shining reddish-yellow colour, with a dorsal row of five coarse punctures on each side the middle, and with six or seven other large punctures on each side, near the front. Scutellum pitchy, the basal part rather coarsely punctured, the apex impunctate. The broad elytra are about as long as the thorax, of a shining-yellowish colour, coarsely and sparingly punctured. The hind body is but little narrowed to the extremity; it is of a brownish-yellow colour, with the 5th and 6th segments black, the extremity of the latter and the 7th segment pale yellow, the hind portion of the anal styles brownish; segments 2—4 are sparingly and finely punctured, 5 and 6 much more closely punctured, these latter with a coarse, depressed, black pubescence. The legs are yellow, the intermediate coxæ distant.

In the male the hind portion of the ventral plate of the 7th segment is produced in the middle; the produced part is truncate at the extremity, and a little transversely impressed before this; in front of this depression are inserted six or eight fine depressed setæ.

In the female the head is only as broad as the thorax, and the front femora are unarmed.

Ega; two specimens, ♂ and ♀.

7. *Philonthus cognatus*, n. sp. Rufus, capite æneo, nitidissimo, abdomine ante apicem nigro. Long. corp. 3½ lin.

Mas: capite majore, femoribus anticis spinulosis, abdomine segmento ventrali apice leviter producto medio anguste sed profunde triangulariter inciso, tarsis anticis simplicibus.

This species is in appearance exactly like the preceding one, but differs by the abdominal characters of the male, and also by these one or two points, which may perhaps be individual rather than specific. The antennæ have the third joint a little longer, so that it is slightly longer than the 2nd, and joints 4—11 are a little more dusky. The elytra are rather more closely and finely punctured.

In the male the middle part of the ventral plate is a little produced backwards and has at its extremity a narrow but rather deep triangular notch. The head is broader than the thorax.

In the female the head is only as broad as the thorax, and the front femora are unarmed.

Ega; two specimens, ♂, ♀.

Obs.—Besides the specimens above described, there are four others from Ega and Tapajos (one ♂ and three ♀), about which I cannot feel sure whether they belong to *P. cognatus* or an extremely closely allied but distinct species.

8. *Philonthus Traili*, n. sp. Rufo-testaceus, nitidus, antennis (basi excepto) abdomineque ante apicem nigricantibus, capite æneo; prothorace serie dorsali 5-punctato; elytris sat fortiter punctatis. Long. corp. 3¼ lin.

Mas: fere sine notis sexualibus externis.

Antennæ moderately long, with the three basal joints yellow, the rest black; they are slightly thickened towards the extremity; the 4th joint is a little longer than broad, the 10th hardly so long as broad. Mandibles and palpi yellow. Head brassy, rather broader than the thorax, with a deep impression on the middle in front, and two punctures on each side between the eyes, with a few other punctures behind these, and some at the vertex on each side; the middle part smooth and shining. Thorax yellow, a good deal narrower than the elytra, longer than broad, a little narrowed behind, with a series of five coarse punctures on each side the middle, and outside this on each side with about six other coarse punctures. Scutellum rather large, concolorous with elytra, distinctly punctured. Elytra of a tawny-yellow colour, scarcely longer than the thorax, rather deeply and distinctly but not coarsely nor closely punctured, shining, very finely and scantily pubescent. Hind body yellow, with the 5th segment black, except at the extreme base; the 6th black, with the extremity yellow, the 7th yellow; the basal segments are very sparingly punctured; the 5th and 6th segments are more closely punctured, and bear a coarse black pubescence. The legs are yellow. The underside of the head is without punctures.

In the male the front femora are almost without spines; the ventral plate of the 7th segment has its hind part slightly produced in the middle; the produced part is nearly entire, there being only an extremely slight emargination of its hind margin.

Ananá; a single male captured by Dr. Traíl on the 6th September, 1874.

9. *Philonthus capitalis*, n. sp. Capite thoracequo aeneis, nitidis; elytris rufis, angulo apicali nigro; abdomine basi obscure rufo, apice testaceo, segmentis 4, 5, 6 nigris; antennarum basi et articulo ultimo, cum pedibus testaceis. Long. corp. 3½ lin.

Antennae nearly as long as head and thorax; the three basal joints as well as the 11th yellowish, the rest infuscated; 2nd and 3rd joints about equal; 4—10 differing but little from one another in length, each slightly stouter than its predecessor, none of them transverse. Palpi and mandibles yellowish. Head broader than the thorax, the front distinctly produced between the insertion of the antennae and with a deep longitudinal impression in front, with two large punctures on each side between the eyes, and with several other large punctures on each side at the back; it is of a shining-brassy colour. Thorax much narrower than the elytra, rather longer than broad, distinctly narrowed behind, the sides a little sinuate; it is similar in colour to the head, and has on each side the middle a row of five punctures, and also on each side near the front six or seven other punctures; all these punctures very large. Scutellum nearly black, densely and finely punctured. Elytra about as long as the thorax, reddish, broadly black towards the extremity, finely and rather closely punctured. The hind body is of a brownish colour at the base, then with the 4th, 5th and 6th segments black, the hind margin of the latter and the 7th yellow; it is on the upper side rather closely punctured, especially on the 4th, 5th and 6th segments, where also there is a coarse, depressed, dense, black pubescence. The legs are yellowish, with the anterior and middle coxae infuscated. The breast pitchy, the intermediate coxae distant.

Ega; one specimen. This I believe to be a male on account of its large head; the front femora have a few very short spines, but the 7th segment of the hind body is rounded at the extremity, so that the individual may possibly be a female.

10. *Philonthus lustrator*, n. sp. Fulvus, capite thoraceque æneis, elytris apicem versus, abdomineque ante apicem nigris, ano pedibusque testaceis; prothorace serie dorsali 5-punctato. Long. corp. 3½ lin.

Mas: abdomine segmento 7° ventrali margine apicali leviter emarginato, tarsis anticis simplicibus.

Antennæ nearly as long as head and thorax, the three basal joints yellow, the rest infuscated; 3rd joint hardly so long as 2nd, 4—10 each a little shorter and a little stouter than its predecessor, the 4th longer than broad, 10th scarcely so long as broad. Mandibles and palpi yellowish. Head quite as broad as the thorax, with the front distinctly produced between the insertion of the antennæ, with a deep longitudinal impression in front, with a line of four deep punctures between the eyes, and with numerous others on each side at the back; it is of a brassy colour, but through this the original tawny colour is perceptible. Thorax much narrower than the elytra, longer than broad, a little narrowed behind, a little sinuate at the sides; it is similar in colour to the head, and has on each side the middle a rather curved row of five coarse punctures, and has besides six other punctures near the front angles. The scutellum is densely and finely punctured. The elytra are broad, scarcely so long as the thorax, of a shining-tawny colour, broadly black towards the extremity, rather coarsely and moderately closely punctured. The hind body is of a tawny colour, with the 5th and 6th segments black; the extremity of the 6th and the whole of the 7th segments yellow; segments 2—4 sparingly punctured, 5 and 6 more closely punctured and with a coarse depressed black pubescence. Legs yellow.

Pará; one individual, ♂.

11. *Philonthus ænicollis*, n. sp. Rufus, thorace æneo, elytris maculâ laterali ante apicem, abdomineque ante apicem nigris, ano pedibusque testaceis; antennis fuscis, basi cum articulo ultimo testaceis, thorace serie dorsali 5-punctato. Long. corp. 3½ lin.

Mas latet.

Antennæ rather shorter than head and thorax, a little thickened towards the extremity; the three basal joints yellow, the following ones blackish, the last joint again yellow; 2nd and 3rd joints about equal, the penultimate joints not quite so long as broad. Mandibles and palpi yellow. Head bright yellowish-red, slightly broader than the thorax (in the ♀), the front distinctly produced between the antennæ, with a deep longitudinal impression in front, with two punctures on each side in a line near the front part of the eyes, and with several other large punctures on each side at the back. Thorax much narrower than the elytra, about as long as broad, a little narrowed behind, the sides slightly sinuate, the front angles distinct, and nearly right angles; it is of a brassy colour, and has on each side the middle a row of five very coarse punctures, and with six other very large punctures near the front angles. Scutellum densely and finely punctured. Elytra broad, about as long as the thorax, bright tawny, with a large lateral spot on each side near the extremity blackish; they are rather deeply and rather closely but not coarsely punctured. Hind body tawny, with the 5th and 6th segments black, the extremity of the latter and the 7th yellow; it is distinctly and moderately closely punctured, the punctuation being denser on the 5th and 6th segments than elsewhere, these segments also with a coarse depressed black pubescence. Legs yellow; under surface tawny, intermediate coxæ distant.

Ega and St. Paulo; two specimens; one of them I have proved by dissection to be a female, and the other does not differ from it by any external character.

12. *Philonthus palpalis*, n. sp. Rufo-testaceus, nitidus, antennis medio obscurioribus, abdomine ante apicem nigro; prothorace serie dorsali 5-punctato, punctis grossis; palpis labialibus articulo ultimo clavato. Long. corp. 3¼ lin.

Antennæ moderately long, reddish, the three basal joints yellow, and the two apical ones a little paler than the preceding ones; 3rd joint about equal to 2nd; 4th joint rather longer than broad, 10th not quite so long as broad. Second joint of maxillary palpi a good deal broader than the others; the last joint of the labial palpi slender at the base, but dilated towards the extremity, which, however, is not truncate, but almost acuminate. Head yellowish, broader than the thorax, without distinct impression on the middle of the front, but with two punctures on each side between the eyes, and a few other coarse punctures towards the rounded hind angles. Thorax longer than broad, yellowish, with a slight brassy reflection; it is a good deal narrower than the elytra, and distinctly narrowed behind; on each side the middle it has a series of five extremely coarse punctures, and outside these about six other very coarse punctures on each side. Elytra yellow, about as long as the thorax, coarsely and rather closely punctured. Hind body yellow; the basal segments sparingly punctured; the 5th and 6th segments blackish, except the hind margin of the latter; the 7th yellow, anal styles nearly black. Legs yellow.

The male has the hind margin of the ventral plate of the 7th segment of the hind body very slightly emarginate.

Amazons; a single specimen, without more special locality, from Mr. Bates.

Obs.—I had supposed this specimen to be a female, until I dissected the apical segments; it is just possible that the remarkable form of the last joint of the labial palpi may be peculiar to the male.

13. *Philonthus aberrans*, n. sp. Angustus, nitidus, capite suborbiculato, æneo; prothorace rufo, serie dorsali 6-punctato; elytris fuscis, basi fulvis; abdomine fulvo, ante apicem nigro, segmento 7°, cum seg. sexti margine apicali, testaceo; antennis pedibusque testaceis. Long. corp. 3¼ lin.

Mas: tarsis anticis simplicibus, abdomine segmento 7° ventrali apice medio triangulariter exciso.

Antennæ about as long as head and thorax, slender, yellowish; 3rd joint longer than 2nd, 4—10 each is a little shorter but scarcely stouter than its predecessor, the 4th quite twice as long as broad, and even the 10th longer than broad. The mandibles and palpi are yellowish. The head is slightly broader than the thorax; the front distinctly produced between the antennæ, with a deep, longitudinal impression along the front, with two punctures on each side in a line near the inner margin of the eyes, and with numerous other coarse punctures on each side at the back; it is of a shining-brassy colour. Thorax much narrower than the elytra, its length considerably (one and a-half times) greater than its width; it is narrowed behind and sinuate at the sides, the front angles deflexed and rounded; it is of a shining-reddish colour, has a dorsal row of six coarse punctures on each side the middle, and has also six or seven other coarse punctures on each side near the front part. The scutellum is smoky, closely and finely punctured. The elytra are scarcely so long as the thorax, tawny at the base, the other part infuscate; they are coarsely and moderately closely punctured. The hind body is rather slender; segments 2—4 are tawny in colour, 5 and 6 black; the hind margin of the 6th and the base of the 7th yellow; segments 2—4 are each at the base sparingly and finely punctured, the 5th is more closely, and the black part of the 6th still more closely punctured; the extremity of the 7th segment, as well as the anal styles, is fuscous. The legs are yellow, long and slender; the intermediate coxæ distant.

Ega; two specimens, ♂, ♀.

Obs.—Besides these individuals, I have three other specimens which I believe represent two distinct but very closely allied species, but as they are all females I cannot feel quite sure about it.

14. *Philonthus conformis*, n. sp. Nitidus, rufus, pedibus testaceis, capite æneo, elytris basi et summo apice rufis, abdomine ante apicem nigricante, apice testaceo; prothorace serie dorsali 6-punctato; elytris crebre (basi dense) punctatis. Long. corp. 3½ lin.

Mas: abdomine segmento 6° ventrali apice leviter emarginato, 7° late minus profunde exciso.

Antennæ elongate and slender, red, the middle joints rather more obscure in colour than the basal and apical ones; 3rd joint longer than 2nd, 10th longer than broad. Thorax red, elongate and narrow, on each side of the middle with a row of six punctures; the punctures are only moderately large, each being separated from the neighbouring one by a perfectly distinct interval; and also with about seven other coarse punctures on each side. Elytra hardly so long as the thorax, red at the base, then smoky, with the extreme margin again reddish; the punctuation at the base is dense and only moderately coarse, it becomes more sparing towards the hind margin. Hind body reddish, with the 5th and basal portion of the 6th segments reddish, the red segments very sparingly punctured, the black not densely punctured; the 7th segment and hind part of the 6th yellow, anal styles nearly black. Legs yellow.

The male has the hind margin of the 6th segment beneath broadly but slightly emarginate; the following segment has a rather shallow, rounded emargination at the extremity.

Amazons; a single male, without special locality.

Obs.—Though very closely allied to *P. aberrans*, this species has the basal portions of the 5th and 6th segments of the hind body more sparingly punctured, and the male has the hind margin of the 6th segment quite distinctly emarginate.

15. *Philonthus propinquus*, n. sp. Angustus, nitidus, capite suborbiculato, æneo; prothorace rufo-testaceo, serie dorsali grosse 6-punctato; elytris fulvis, apice late fusco-ænescentibus; abdomine fulvo, ante apicem nigro, parce punctato. Long. corp. 3½ lin.

This species is extremely close to the preceding one, and differs from it only as follows. The thorax is narrower, the elytra are a little brassy towards the extremity, the hind body is more sparingly punctured, the extreme base of each of the 5th and 6th segments being much more sparingly punctured than in *P. aberrans*; the punctures on the thorax are larger, and the elytra are rather more sparingly punctured.

Ega; one specimen, ♀.

Obs.—I have also another female, found by Mr. Bates, which I believe belongs to a closely allied but distinct species; the antennæ have joints 4—9 distinctly infuscate, and the two apical ones yellow, and the femora and tibiæ are a little infuscate at their apices.

16. *Philonthus regillus*, n. sp. Fulvus, capite æneo, elytris apicem versus, abdomineque ante apicem nigricantibus; antennis basi, pedibus, anoque testaceis; prothorace grosse punctato, punctorum numerus utrinque circiter sexdecim. Long. corp. 4 lin.

Allied to *P. aberrans*, but easily distinguished by the different punctuation of the thorax. The antennæ are rather longer than head and thorax, the three basal joints yellow, the rest infuscate; joint 3 longer than 2nd; 4—10 differing but little from one another, even the 10th considerably longer than broad. Head brassy, a little broader than the thorax, rather long, a little narrowed behind, the front produced between the insertion of the antennae, and with a deep longitudinal impression, with two punctures on each side between the eyes, the back part coarsely and rather numerously punctured. Thorax much narrower than the elytra, considerably longer than broad, and considerably narrowed behind, the sides sinuate, front angles seen from above not much rounded; it is of a reddish-yellow colour, very shining, and has on each side the middle a dorsal row of seven very coarse punctures, and between these, near the base on each side, three or four accessory punctures, so placed as to render the dorsal row confused, also with five or six other coarse punctures on each side, near the front angles. Scutellum densely and finely punctured. Elytra about as long as the thorax, tawny at the base, smoky towards the extremity, very coarsely and rather sparingly punctured. Hind body tawny, with the 5th and 6th segments blackish, the hinder portion of the latter and the 7th pale yellow, anal styles fuscous; it is sparingly punctured, except on the black parts of the 5th and 6th segments, where the punctuation is much closer. The legs are yellow, the intermediate coxæ distant.

St. Paulo; two specimens, both, I think, ♀.

17. *Philonthus abactus*, n. sp. Angustus, nitidus, rufo-testaceus, capite æneo, elytris apicem versus infuscatis; abdomine ante apicem nigro, ano pedibusque pallidis; prothorace omnium grosse punctato, punctorum numerus utrinque circiter viginti, lineâ mediâ lævi fere nullâ. Long. corp. 3½ lin.

Mas: abdomine segmento 7° ventrali margine apicali minus profunde triangulariter emarginato.

A remarkable Stilicoid species allied in structure to *P. aberrans*, but with the punctuation of the thorax quite different. The antennæ are long, rather longer than head and thorax; they are rather slender and scarcely at all thickened at the extremity, they are of a yellowish colour, joints 4—11 being duskier than the three basal joints; 2nd and 3rd joints about equal, 4th nearly as long as 3rd; from the 5th to the 10th each joint is a little shorter than its predecessor, the 11th joint about as long as the 9th. Head brassy, broad and short, much broader than the thorax, nearly as broad as the elytra, the front much produced between the antennae—this part with a deep impression, a line of four punctures between the eyes, two on each side, and between them a fifth, a continuation of the front longitudinal impression; all the back part and behind the eyes rather closely punctured, the punctures being very coarse and deep. Thorax reddish-yellow, shining, nearly twice as long as broad, distinctly narrowed behind, the sides sinuate, the front angles greatly depressed, the front part somewhat produced in the middle; it is covered with extremely coarse and deep punctures, only the posterior angles and a space between the two hinder punctures being distinctly free from them. Scutellum narrow, closely punctured. Elytra much narrower than the thorax, of a shining-tawny colour, the back portion infuscated and slightly metallic; they are very coarsely and moderately closely punctured. The hind body is slender, it is sparingly but rather coarsely punctured, with erect long black setæ; it is shining and of a tawny colour at the base, the 5th and 6th segments being blackish; the hind margin of the latter and the 7th pale yellow, tip of the 7th and anal styles fuscous. Legs long and slender, pale yellow, the pubescence at the extremity of the tibiæ (especially the intermediate ones) rather darker.

Ega; one specimen, ♂.

18. *Philonthus longipes*, n. sp. Elongatus, capite suborbiculato, æneo; prothorace basin versus angustato, fulvo, disco ænescente; elytris dense punctatis, fulvis, fasciâ latâ mediâ violaceâ; abdomine fulvo, segmentis 2—4 late nigricantibus, 5 et 6 nigris, ano flavo; pedibus et antennis basi articuloque ultimo fulvo-testaceis; prothorace serie dorsali 5-punctato. Long. corp. 4½ lin.

Mas latet.

Antennæ slightly longer than head and thorax; three basal joints reddish-yellow, the rest infuscated, the terminal joint being again paler; 2nd and 3rd joints about equal, 4—10 differing little in length, each slightly stouter than its predecessor, even the 10th considerably longer than broad; 11th joint rather long, obliquely acuminate. Mandibles and palpi reddish-yellow. Head slightly broader than the thorax, the hinder angles rounded, the front distinctly produced between the insertion of the antennæ, with a deep longitudinal impression on the front part, with four large punctures in a line between the eyes, with several other large punctures at the back, and between and behind these with numerous finer punctures; it is of a shining-brassy colour. Thorax rather longer than broad, distinctly narrowed behind, the front angles greatly deflexed and rounded, the sides sinuate; it is of a reddish colour, with the disc indistinctly brassy; it has a dorsal row of five coarse punctures on each side the middle, and has three or four other coarse punctures on each side, and besides this has the extreme sides and front angles densely and rugosely punctured. The scutellum is densely and finely rugose-punctate. The elytra are greatly broader than the thorax, and about as long; they are of a tawny colour, with a broad violet band across the middle; they are densely and rather finely punctured, and not shining. The hind body is tawny; segments 2—4 black across the middle, 5 and 6 nearly entirely black, the hind margin of the latter and whole of the 7th segment pale yellow; the black parts are closely and rather coarsely punctured; besides the very coarse, depressed, black pubescence of segments 4—6, there are numerous other long, outstanding, black setæ. The legs are long, of a yellowish colour. The breast tawny.

Obydos; two individuals, both, I believe, females.

19. *Philonthus serraticornis*, n. sp. Fulvus, nitidus, capite nigro; antennis fuscis articulo ultimo testaceo, articulis 4—10 intus productis; prothorace serie dorsali subtiliter 6-punctato. Long. corp. 4½ lin.

Mas: tarsis anticis simplicibus.

This is another aberrant species which will probably ultimately be separated generically from *Philonthus*, the antennæ being distinctly serrate and the lateral lines of the under surface of the prothorax not joined till quite at the front part of the coxal cavities. It is broad and rather depressed. The antennæ are rather shorter than head and thorax, of an obscure colour, each joint a little red at the base; the 11th joint yellowish; joints 5—10 distinctly produced and pointed on the inner side, at their broadest part broader than long; 3rd joint long, being considerably longer than the 2nd. Mandibles and palpi reddish, the mandibles very short. Head black and shining, orbiculate, the eyes large, the clypeus convex, so as to render the insertion of the antennæ a little different to what is usual in *Philonthus*; the middle parts of the head smooth, the back part and the sides of the eyes coarsely punctured. Thorax a little broader than the head, a little narrower than the elytra, straight at the sides, and not distinctly narrowed either before or behind, the front angles but little rounded, the hind angles nearly absent; it is of a shining-tawny colour, with a dorsal row of six small punctures on each side the middle, and with four or five others placed in an irregular row between this and the side. The scutellum is yellowish, finely and rather sparingly punctured. The elytra are about as long as the thorax, of a yellowish colour, finely and sparingly punctured. The hind body is very broad; it is of a yellowish or tawny colour, extremely finely, moderately closely punctured. The legs are yellow, the middle coxæ distant, the lateral margins of thorax without a spiracular membrane.

Ega; one specimen. I have ascertained by dissection that it is a male, though there is nothing external to indicate this.

HOLISUS.

This peculiar genus at present consists of half a dozen species, to which I now add five new ones. All the species known are South American. The genus was placed by Erichson in the *Xantholinini*, but is assigned by Kraatz to the *Staphylinini*. I have not made a sufficient investigation to enable me to pronounce an opinion as to its nearest allies, but it appears to me to be as yet a remarkably isolated form. *H. discedens* here described differs from the other species in the structure of its hind body, which is more convex, so as to be subcylindric.

1. ***Holisus depressus***, n. sp. Niger, nitidus, depressus, ano rufo, pedibus piceis; capite thoraceque parce fortiter punctatis; elytris crebre sat subtiliter punctatis. Long. corp. 3¾ lin.

Antennæ nearly black; 3rd joint a good deal longer than 2nd, 4th and 5th each about as long as broad, 6—10 very similar to one another, each a little transverse, 11th pale at the extremity. Palpi reddish. Head large, rather broader than the thorax, oblong, quite straight at the sides, on the front part with coarse punctures, forming on each side an irregular patch placed in a slight depression; also on each side of the middle with a patch of about eleven coarse punctures, also punctured at the sides and at the vertex; it is black and shining. Thorax about as broad as the elytra, rather broader than long, a little sinuate at the sides, and a good deal narrowed behind, black and shining, and with a slight opalescent reflection, bearing a few irregularly-placed, rather coarse punctures. Scutellum rather large, coarsely punctured. Elytra pitchy black, much longer than the thorax, rather coarsely and closely but yet not densely punctured, distinctly shining. Hind body rather broad and depressed, coarsely and rather closely punctured, the 6th segment more sparingly so, and the yellowish 7th segment still more sparingly. Legs short and stout, pitchy.

Ega; a single individual.

Obs. I.—A second individual, brought by Mr. Bates from the same locality, may be either the other sex of *H. depressus* or a closely allied but distinct species; it is a little smaller, and has the head rather shorter and the hind body not quite so closely punctured; it has the hind margin of the ventral plate of the 7th segment of the hind body rounded, while this part is more truncate in *H. depressus*.

Obs. II.—*H. depressus* is very closely allied to *H. analis*, Er., but is rather larger and broader, has the thorax shorter in proportion to its length, and the elytra less densely and rather more coarsely punctured.

2. *Holisus picipes*, n. sp. Parallelus, minus latus, nitidus, niger, pedibus piceis; capite thoraceque vage, elytris subtiliter minus crebre, abdomine fortiter, punctatis. Long. corp. vix 3 lin.; lat. ½ lin.

Allied (judging from description) to *H. humilis*, Er., but larger, and with the elytra more finely punctured. Antennæ rather longer than the head, not stout; 3rd joint distinctly longer than 2nd; 5th joint about as long as broad, 6—10 each a little transverse. Mandibles pitchy; palpi dusky yellowish. Head oblong, straight at the sides, the hind angles not much rounded, the length from clypeus to vertex a little greater than the width; it is black and shining, rather strongly and coarsely punctured, the punctures being disposed as follows: a patch of about seven on each side placed in an obscure depression behind the antennæ, and between these two patches two or three other punctures; behind these a broad, longitudinal space is impunctate, and at each side behind is a large patch of fifteen or sixteen punctures, each patch consisting of three indistinct oblique rows. Thorax just about as broad as the head and elytra, its width distinctly greater than its length; it is a little narrowed behind, and has on each side sixteen or eighteen punctures irregularly scattered; the middle space between these punctures much narrower at the back part than in front. Elytra longer than the thorax, pitchy black, finely and not closely punctured. Hind body parallel, rather coarsely but not densely punctured. Legs pitchy.

Ega; two specimens.

3. *Holisus excavatus*, n. sp. Piceo-testaceus, nitidus, antennis pedibusque testaceis, thorace concavo, elytris dense subtiliter punctulatis. Long. corp. 1¾ lin.

Antennæ considerably longer than the head, yellow; 3rd joint scarcely longer than 2nd, 6—10 a little transverse. Mandibles and palpi yellow. Head pitchy, broad, even a little broader than thorax or elytra; the length about as great as the width; the hind angles considerably rounded; the upper surface sparingly and irregularly punctured, the punctures leaving scarcely any distinct smooth space in the middle. Thorax as broad as the elytra, about as long as broad, narrowed behind, the sides rounded; it is of a pitchy-yellow colour; the whole of the upper surface concave, sparingly and irregularly punctured. The elytra are longer than the thorax, finely and closely punctured, of a pitchy-yellow colour, their disc concave. Hind body pitchy yellow, rather finely, moderately closely punctured. Legs yellow.

Ega; one specimen.

Obs.—This individual is a little immature, so that the colour of the species may be somewhat darker than is here described, and it is probable that when fully developed the elytra may be without impression, but I expect the thorax is naturally concave on the upper surface.

4. *Holisus umbra*, n. sp. Omnium perdepressus, angustus, nitidus, piceus, antennarum basi, pedibus, anoque testaceis; thorace concavo, elytris crebre subtiliter punctulatis. Long corp. 1½ lin.

Antennæ rather slender, a little longer than the head, dusky yellow, paler at the base; 3rd joint small, shorter than the 2nd, 6—10 rather transverse. Palpi yellow.

Head rather large, fully as broad as thorax or elytra, oblong-quadrate, the sides straight, the hind angles a little rounded; it is on the upper side rather finely and sparingly punctured. Thorax as long as broad, narrowed behind and rounded at the sides, the upper surface deeply impressed, finely and sparingly punctured, an oblong space on the disc impunctate. Elytra quite as broad as, and longer than the thorax; like it, and the head, of a pitchy colour, closely and finely punctured. Hind body parallel, pitchy, with the extremity yellow, evenly, moderately finely, and moderately closely punctured. Legs yellow.

Ega; one specimen.

5. *Holisus discedens*, n. sp. Niger, nitidus, pedibus piceis; capite thoraceque parce fortiter punctatis; elytris crebre subtiliter punctatis; abdomine minus depresso, sat crebre punctato. Long. corp. 2½ lin.

Palpi pitchy; 1st and 2nd joints of antennæ pitchy (the rest wanting). Head black, quite depressed, quite straight at the sides and vertex, the hind angles rounded; it is rather broader than the thorax or elytra, and bears coarse and distinct but not dense punctures. Thorax short, about long as broad, rounded at the sides, and a good deal narrowed behind; the front much rounded, so that the front angles are quite rounded, with a series of punctures on each side of the middle, leaving between them a broad, impunctate space, and outside these with some punctures, which are most numerous behind the middle; it is black and shining, and shows at the base, in the middle, a very short and fine channel. Scutellum rather coarsely punctured, the punctures disappearing towards the apex. Elytra much longer than the thorax, of a pitchy-black colour, rather finely and closely punctured, the punctures rather closer at the extremity than at the base. Hind body elongate and narrow, less depressed than the front parts; the dorsal plates are convex, so that it is subcylindric, and the lateral margins are extremely fine; it is a good deal narrowed towards the extremity, and also has the basal segment slightly narrower than the following ones; it is quite black in colour, and is rather coarsely but not altogether densely punctured. The legs are pitchy.

In the individual described, which I believe to be a

female, the apical segment has a lateral style on each side, and two more slender ones between them.

A single specimen was found by Mr. Bates, but it bears no special indication of locality.

Diochus.

This genus consists of seven described species, found in widely separated parts of the world, and I now add four new species from the Amazons; these call for no special remark, as they appear to be extremely closely allied to one another.

The genus is one of considerable importance, notwithstanding the insignificant and unattractive appearance of the species which compose it. Some years ago, I examined it, and came to the conclusion that it could not be satisfactorily classed with the *Xantholinini*, with which it is usually associated. On glancing at some of its points of structure again, I am inclined, however, to suspect that it may prove to be one of the earliest and least specialized forms of the *Xantholinini*, and that a careful study of its peculiarities may throw considerable light on the nature of the modifications distinctive of that group, as well as suggest the species of the *Staphylinini*, with which the *Xantholinini* are most directly connected.

1. *Diochus longicornis*, n. sp. Obscure rufo-testaceus, nitidus, capite elytrisque circa scutellum infuscatis. Long. corp. 2 lin.

At once distinguished from *D. flavicans* by the much longer antennæ. These are slender and reach not quite to the extremity of the thorax; they are of a yellow colour; 3rd joint is long and slender, longer than the 2nd; from 4—10 each is a little shorter than its predecessor, even the 10th considerably longer than broad. Head pitchy-red, slightly narrower than the thorax, very little narrowed in front, the front part indistinctly punctured, the punctures placed in irregular lines. Thorax a little narrower than the elytra, longer than broad, very slightly narrowed in front, dusky reddish-yellow, with four punctures near one another on the middle, behind; in front of these with two others farther apart, and with four or five others on each side. Elytra hardly so long as the thorax, yellow, darker about the scutellum, sparingly and very indistinctly punctured. Hind body dusky yellow, extremely finely and densely punctured. Legs yellow.

Tapajos; seven specimens.

2. *Diochus vicinus*, n. sp. Rufescens, capite elytrisque infuscatis, his apice dilutioribus; antennis tenuibus, sat elongatis. Long. corp. 2 lin.

Antennæ reddish, not thickened towards the extremity, moderately long; 3rd joint elongate and slender; 4—10 each is a little shorter, but scarcely visibly broader, than its predecessor, the 10th quite as long as broad; 11th joint much acuminate. Head small, infuscate-red. Thorax bright reddish-yellow. Elytra about as long as the thorax, infuscate-red, with the hind margin paler, shining, with a few indistinct punctures. Hind body dark red, densely and finely punctured. Legs yellow.

Tapajos; three specimens.

Obs.—This species is extremely similar to *D. longicornis*, but the antennæ are less elongate.

3. *Diochus tarsalis*, n. sp. Rufo-fuscus, antennis pedibusque testaceis, illis sat elongatis; tarsis anterioribus dilatatis. Long. corp. 2 lin.

Antennæ moderately slender, reddish; 3rd joint slender and elongate; 4—10 each a little shorter, but scarcely visibly broader, than its predecessor; 10th about as long as broad, 11th much acuminate. Palpi yellow. Head infuscate-red, impunctate along the middle, with a few punctures at the sides. Thorax infuscate-red, rather brighter in colour than the head, with the usual punctuation. Elytra about as long as the thorax, infuscated-red, shining, with a few impressed punctures. Hind body densely punctured, blackish-red, with the hind margins of the segments red, the hind part of the two slender and elongate apical segments broadly red. Legs yellow, front tarsi rather strongly dilated.

Tapajos; three specimens.

Obs.—This species is extremely similar to *D. longicornis* and *D. vicinus*, but is rather darker in its colouration, and has the front tarsi a good deal dilated, while in the two species named they are nearly simple.

4. *Diochus flavicans*, n. sp. Rufo-testaceus, nitidus,

antennis articulis 3—10, elytrisque basi infuscatis. Long. corp. 1½ lin.

Antennæ considerably longer than the head, the two basal joints and the last joint yellow, the others infuscate; 3rd joint about as long as 2nd, 6—10 rather transverse. Head narrower than the thorax, narrowed in front, reddish-yellow, convex, the front part sparingly and indistinctly punctured, the vertex smooth. Thorax a little narrower than the elytra, considerably longer than broad, distinctly narrowed in front, the sides slightly curved; it is reddish-yellow, and has four punctures placed near one another on the middle, behind; in front of these, two others much farther apart, and on each side five or six other punctures near the lateral margin. Elytra scarcely so long as the thorax, yellowish, indistinctly darker near the base, sparingly and indistinctly punctured, the punctures disposed in lines. Hind body dusky yellowish, very finely and densely punctured. Legs pale yellow. Front tarsi not dilated.

Tapajos; two specimens.

STERCULIA.

This genus consists at present of six species, all peculiar to South America, and I here describe seven new ones. The genus contains two apparently distinct groups: the first of these consists of large and brilliant species, which are amongst the most splendid of the *Staphylinidæ*; the species, however, bear such an extreme resemblance to one another, that, although I have twenty different forms belonging to it separated in my collection, I am by no means sure how many species they represent: the second group consists of black and rather smaller species, which have the mandibles much less elongate than the metallic species. *S. amazonica*, *S. pauloensis* and *S. discolor* belong to the group of metallic species, while the other four here described are black species with short mandibles.

1. *Sterculia amazonica*, n. sp. Cyanea, nitida, thorace elytris breviore, capite oblongo, subtus lateribus parce punctato, medio lævi, mandibulis capite brevioribus. Long. corp. 10 lin.; lat. elytrorum apice 2½ lin.; prothoracis basi 1½ lin.

Antennæ rather stout, not thickened after the 4th joint; 3rd joint about twice as long as 2nd, 4—10 transverse;

the three basal joints blue, the rest obscure. Palpi stout. Mandibles about two-thirds of the length of the head. Head quite as broad as the greatest width of the thorax, its width about as great as from the front of the eyes to the back of the head; the whole of the upper surface densely, evenly and coarsely rugose-punctate, beneath very shining, the punctures there about twenty-four on each side; those at the back angles small, the others very large, the middle part quite free. Thorax considerably longer than broad, a little narrowed from the base to the middle, more narrowed from the middle to the front; a deep oblique impression on each side, near the back; the sides in front of the impression rather sparingly punctured, a broad middle line impunctate. Elytra longer than the thorax, finely and sparingly punctured. Hind body very shining, very finely and very sparingly punctured. Legs violet. Hind body beneath less sparingly and finely punctured than above.

Ega; six individuals.

Obs.—These specimens vary somewhat in their colour, which is sometimes purple, sometimes of a bluer or greener tint; the punctures also on the underside of the head are more numerous and distinct in some individuals than in others. Two specimens have the 7th segment beneath a little depressed in front of the hind margin, and more finely punctured and pubescent, the hind margin itself being very slightly emarginate; I think they are probably males.

2. *Sterculia pauloensis*, n. sp. Cyanea, nitida, thorace elytris breviore, capite oblongo, subtus basin versus parcius punctato, mandibulis capite brevioribus. Long. corp. 12 lin.; lat. elytrorum apice $2\frac{3}{8}$ lin.; prothoracis basi $1\frac{5}{8}$ lin.

Very closely allied to the preceding species, but larger, and with the mandibles a little longer, the head and thorax broader, the head being distinctly broader than the thorax, while the thorax is less slender in front; the punctures on the under side of the head are only eleven in number, smaller punctures near the hind angles being quite absent.

St. Paulo; a single individual.

Obs.—It is quite possible this may not prove to be a distinct species from the *S. amazonica*, which, as I have above noticed, appears to be somewhat variable.

3. *Sterculia discolor*, n. sp. Supra violacea, abdomine aureo-purpureo, subtus viridi-cærulea, nitida, capite oblongo, mandibulis hoc brevioribus, thorace elytris vix breviore. Long. corp. 9—11 lin.

Readily distinguished from *S. amazonica* by the discordant colour of the hind body above, as well as by the much less transverse penultimate joint of the antennæ. Antennæ rather long, reaching more than half-way to the back of the thorax, not thickened after the 4th joint; the three basal joints violet, the others dusky; from 6—10 each joint is a little longer than the preceding one, so that the 10th is considerably longer than the 6th. Mandibles much shorter than the head. Head rather narrow, considerably longer than broad, slightly broader than the thorax; the eyes distinctly prominent; it is of a violet colour above, densely and coarsely rugose-punctate; beneath it is shining blue, sparingly punctured on each side with about fourteen punctures. Thorax one and a half times as long as broad, gradually narrowed from the base to the front, with an oblique impression on each side before the base, a space along the middle smooth; the sides rather sparingly punctured, more closely at the front than elsewhere. Elytra much broader and slightly longer than the thorax, of a dark-violet colour, finely and moderately closely punctured. Hind body above of a brilliant golden purple, or copper colour, sparingly and finely punctured, beneath brilliant green. Legs violet; pubescence of the tibiæ conspicuous, being nearly white.

In the male the hind margin of the 7th segment beneath is less rounded than in the female, and the punctuation in front of this is a little denser and finer.

Ega; six specimens.

Obs.—This species is very closely allied to the Bolivian *S. splendens*, but is rather more slender, and has the mandibles shorter; the antennæ are considerably shorter, and the sculpture of the head at the vertex is less dense and rugose.

4. *Sterculia funebris*, n. sp. Nigro-subænea, opaca, densius subtiliusque punctata, thorace lineâ mediâ lævi, elytris thorace brevioribus. Long. corp. 9 lin.

Antennæ black, rather stout, reaching about half-way to the back of the thorax, a little thickened from the 4th to the 10th joint; 3rd joint considerably longer than 2nd, this latter red at the extreme base; joints 4—10 strongly transverse, scarcely differing in length. Mandibles about half as long as the head. Head slightly narrower than the thorax, considerably longer than broad, oblong, the hind angles much rounded, but the sides straight; above black, and quite dull, with a very faint brassy tinge, extremely densely and finely rugulose-punctate, with a very dense, short and fine erect pubescence; lateral grooves broad, shallow and densely punctured, these limited on the underside by a smooth, rather elevated line, the rest of the under surface coarsely and numerously punctured; punctuation at the hinder part fine and dense, passing gradually into that of the upper surface. Thorax about two-thirds of the width of the elytra; its length quite one and a half times its width, slightly broader from the base to a little in front of the middle, thence much narrowed and rounded to the front, extremely densely and finely punctured, with an extremely short and fine, erect pubescence, quite dull, with a middle line smooth, shining and impunctate; it is also slightly transversely impressed some distance in front of the base, and in this transverse impression the smooth middle line is very nearly interrupted; it is similar in colour to the head. Elytra with their greatest length (*i. e.*, measured from the humeral angle to the outer apical angle) just equal to that of the thorax, extremely densely and finely punctured and pubescent, opaque black. Hind body rather shining, brassy black, extremely finely and rather closely punctured, and delicately pubescent. Legs blackish, the tarsi pitchy; the inside of the tibiæ with a dense-grey pubescence.

Ega; four specimens; sex unknown.

Obs.—This species is clearly closely allied to *S. formicaria*, Er., but contradicts his description in several important points. *S. formicaria* was originally described by Laporte, but his description and figure are quite worthless, and it is impossible to decide whether they relate to Erichson's species or not. Erichson places as synonyms of *S. formicaria*, the *flagellicornis* and *pubescens* of Nordmann; these two descriptions are very carefully drawn up by Nordmann, and it appears to me clear that they refer to two distinct species, and are erroneously united by Erichson under the name of *formicaria*. Hence I consider the name *formicaria* should be entirely dropped, and Nordmann's *flagellicornis* be used instead, leaving it to a comparison of the types (in the Museum at Berlin, see.

Nordmann) to determine whether my opinion as to *pubescens*, Nord., being a distinct species be correct. A specimen from Laporte's collection is in my possession, belonging to a closely allied but distinct species from *S. funebris*, but whether it be the individual from which he drew his description of *S. formicaria* I cannot say, though it bore that name in his collection.

5. *Sterculia fimetaria*, n. sp. Nigro-ænea, nitida, capite oblongo-ovali, opaco; thorace crebre punctato lineâ latâ impunctatâ, elytris hoc brevioribus. Long. corp. 6—7 lin.

Antennæ stout, much thickened to the extremity, the scape pitchy, the flagellum obscure ferruginous; 3rd joint not quite one and a half times so long as the 2nd; 4—10 transverse, differing but little in length, but the 10th quite twice the width of the 4th; 11th joint stout and rather short, notched on one side at the extremity. Mandibles pitchy, very short; palpi reddish. Head narrow, scarcely so broad as the thorax, nearly twice as long as broad, coarsely and closely punctured, a little smooth in the middle near the front; beneath it is shining and irregularly punctured, the punctures there moderately coarse, with some coarser ones towards the front in the middle; the lateral grooves narrow and punctured, the smooth space limiting them beneath broad and distinct. Thorax elongate, nearly twice as long as broad, about half the width of the elytra, the sides nearly straight from the base to in front of the middle, then gently narrowed; it is brassy black, the middle broadly smooth, very shining and impunctate, the sides evenly, moderately closely, and moderately coarsely punctured; the pubescence scanty, the oblique impressions quite obsolete. Elytra rather shorter than the thorax, brassy, rather shining, finely and rather closely punctured, with a fine erect pubescence. Hind body brassy, finely and closely punctured and pubescent. Legs pitchy black; tarsi paler.

In the male the hind margin of the 7th segment beneath is a little emarginate, and just in front of this the segment is rather more closely and finely punctured.

Ega; three individuals, two ♂, one ♀.

Obs.—One of these specimens is labelled by Mr. Bates as found under dung.

6. *Sterculia clavicornis*, n. sp. Nigro-ænea, nitida,

capite opaco, crebre subtiliter punctulata, thorace lineâ
mediâ lævi; antennis pedibusque obscure ferrugineis.
Long. corp. fere 6 lin.

Allied to *S. fimetaria*, but very distinct by its smaller and
shorter head, which is very differently punctured beneath.
The antennæ are stout, much thickened towards the extremity, dark reddish, the basal joints pitchy; 3rd joint
longer than 2nd; 4—10 very short and broad, not differing
from one another in length, 10th nearly twice as wide as
the 4th; 11th joint short, stout and pointed, not oblique
at the extremity. Mandibles pitchy, very short; palpi
yellowish. Head rather longer than broad, nearly straight
at the sides, but greatly rounded at the back, about as
broad as the thorax; above quite dull, densely rugulose-punctate, the punctuation at the back very fine, the
pubescence dense and rather long; beneath closely punctured at the sides, more sparingly and coarsely towards
the middle, the lateral grooves closely punctured, not deep),
and not separated from the part beneath by any distinct
smooth space. Thorax much narrower than the elytra,
nearly twice as long as broad, slightly broader from the
base to in front of the middle, then narrowed to the front;
the sides are rather closely and not very finely punctured,
with an impunctate space along the middle, the oblique
impression not well marked, the pubescence rather dense.
The elytra are rather shorter than the thorax, moderately
finely and rather closely punctured, with a rather long
pubescence; like the thorax, of a shining brassy-black
colour. Hind body closely and finely punctured, with a
dense soft pubescence. Legs pitchy red; the tibiæ on the
inside with a grey pubescence.

Obydos; one specimen, probably a female.

7. *Sterculia minor*, n. sp. Nigro-ænea, nitida, capite
opaco, thorace elytrisque crebre subtilissime punctulatis,
illo lineâ mediâ lævi; capite subtus crebrius punctato,
sulcis lateralibus nullis. Long. corp. $5\frac{1}{2}$ lin.

Closely allied to *S. clavicornis*, but with the head and
thorax rather broader, the punctuation of the thorax and
elytra much finer, and the head without any lateral
grooves. The antennæ are rather stout, nearly black, the
extreme base of the 2nd joint red; 3rd joint considerably
longer than the 2nd; 4—10 each a little stouter than the
preceding one, transverse, differing but little in length,

10th not twice as wide as the 4th; 11th joint rather long, pointed and obliquely sinuate at the extremity. Mandibles short, pitchy. Head almost as broad as the thorax, oblong, nearly straight at the sides, the hind angles rounded, above very densely rugose-punctate, quite dull, brassy black, the pubescence dense, especially at the back; beneath it is shining and rather closely punctured, the punctures coarser in the middle than at the sides, a narrow line in the middle smooth. Thorax considerably narrower than the elytra, one and a half times as long as broad, in the middle slightly broader than at the base, then narrowed to the front, oblique impression rather distinct, a space along the middle smooth; the sides closely and very finely punctured, the pubescence rather dense. Elytra about as long as the thorax, closely and very finely punctured, brassy black like the thorax. Hind body finely and moderately closely punctured, brassy black, pubescent. Legs nearly black; tibiæ on the insides with a grey pubescence.

Fonteboa and Ega; two specimens, probably females.

AGRODES.

This genus was established by Nordmann for an elegant South American species, but was united with *Sterculia* by Erichson. The genus appears to me, however, to be so distinct, that I have used Nordmann's name as indicating a separate genus. The differences in the trophi from *Sterculia* (*Arœvememis*, Nordmann), are accompanied by a marked distinction in the form of the head, and by a greater development of the prosternum in *Agrodes*, and these appear to me sufficiently important and constant to justify the acceptation of Nordmann's two genera. The species of *Agrodes* appear to be excessively rare; so much so, that I have never seen any other specimens of it than the two individuals here described as two distinct new species.

1. *Agrodes conicicollis*, n. sp. Cyanea, nitida, thorace elytrisque parce subtiliter punctulatis, illo elongato; capite subtus sat crebre punctato, mandibulis hoc duplo brevioribus. Long. corp. 8 lin.

Antennæ rather stout, the three basal joints black, with a bluish tinge, the 2nd red at the extreme base, 4—10 pitchy; 3rd joint nearly one and a-half times longer than

2nd; 4—10 very similar to one another in length, and each scarcely stouter than the preceding; the last joint sinuate at the extremity. Mandibles nearly one-half as long as the head. Head as broad as the thorax, the hind part gradually narrowed into the neck; it is scarcely shorter than the thorax, considerably longer than broad; above it is densely and coarsely rugose-punctate, and with a rather long pubescence, beneath it is very shining and rather sparingly punctured; the lateral grooves are absent, but a raised, smooth space, proceeding from the base of the mandible, indicates what should be the boundary of the groove beneath. The thorax is elongate, about twice as long as broad, about half as broad as the elytra; it is nearly straight at the sides from the base to the middle, thence narrowed to the front; it is of a shining-blue colour, finely and rather sparingly punctured, with an impunctate space along the middle. Elytra scarcely so long as the thorax, and darker in colour, rather finely and sparingly punctured. Hind body shining blue, finely and rather sparingly punctured. Legs bluish, tibiæ with a grey pubescence on the inside.

Ega; one specimen, probably female.

2. *Agrodes longiceps*, n. sp. Cyanea, nitida, thorace elytrisque subtiliter punctulatis, illo elongato; capite subtus dense grosseque punctato, mandibulis hoc plus duplo brevioribus. Long. corp. 8½ lin.

Allied to the preceding species, but with the head differently formed, and closely and coarsely punctured beneath. The antennæ are nearly black, with the three basal joints bluish; the extreme base of the 2nd joint red; they are not thickened after the 4th joint, joints 4—10 scarcely differing in length and breadth; last joint sinuate at the extremity, ferruginous at the tip. Mandibles not half the length of the head. Head long and narrow, with the eyes very prominent; it is about as long and about as broad as the thorax, the hinder part much narrowed to the neck; it is above densely rugose-punctate, of a violet colour, not shining, with a rather dense pubescence; beneath it is very shining, closely and very coarsely punctured, without any trace of lateral grooves. Thorax considerably more than half the width of the elytra, nearly twice as long as broad, nearly straight at the sides from the base to the middle, thence narrowed to the front; it has a smooth space along the middle, and is finely punctured at the sides, the transverse impression on each side well marked. Elytra not so long as the thorax, darker in colour, finely and sparingly punctured. Hind body shining blue, finely and rather sparingly punctured. Legs blue.

Tunantins; one specimen, sex unknown.

Tesba.

Antennæ fractæ.

Labrum totum corneum, sexdentatum.

Palpi maxillares filiformes, labiales articulo ultimo oblongo.

Prothoracis linea marginalis externa integra.

Prosternum inter coxas anticas processu verticali acuminato munitum.

Corpus magnum, robustum. Frons inter antennas fortiter carinato-compressa. Labrum porrectum, sexdentatum, margine longe setoso. Mandibulæ validiores, medio dentatæ. Maxillæ malâ exteriore corneâ, apice extrorsum setoso, introrsum barbato. Palpi maxillares articulo primo minuto, secundo basi angustato, tertio hoc paulo breviore, quarto tertio longiore. Palpi labiales articulo secundo basi angustato, tertio oblongo, præcedente longiore. Antennæ basi valde approximatæ, articulo primo leviter curvato. Prothorax lineâ marginali superiore integrâ, in lineam marginalem anticam continuatâ, lineâ inferiore paulo ante marginem anticam desinente, superiore haud conjunctâ. Prosternum medio processu acuminato munitum. Tibiæ omnes fortiter spinulosæ, anticæ leviter incrassatæ; tarsis anticis subdilatatis, articulo primo sequentibus multo longiore.

Genus *Xantholininorum* insignis, *Scytalino* affinis.

This genus differs from *Scytalinus* by the presence of the upper line of the side piece of the thorax, by the thick antennæ, which are more approximate at their insertion and separated by a compressed carina-like space, as well as by its much more robust build. I have a third species found by Mr. Belt in Central America, and the nearest allies known to me are some undescribed species from Natal and Madagascar; in the Natal species the antennæ are extraordinarily thick, and separated at their insertion only by a thin lamina.

1. *Tesba gigas*, n. sp. Nigra, sat nitida, capite crebre punctata, antennis articulis 4—10 leviter transversis. Long. corp. 15 lin.; lat. (abdominis) 2⅔ lin.

A giant among the *Xantholinini*. Head broader than the thorax, quite as broad as the elytra, not quite so long as broad, narrowed to the front, at the back part with numerous coarse round punctures, in front of these with finer oblong ones, the very front part without punctures; beneath coarsely and rather sparingly punctured, at the hind angles with a sinuate suture. Palpi reddish. Antennæ black, the first three joints shining, the rest pubescent; the 1st joint long and stout, 3rd rather longer than 2nd; 4—10 not at all thicker from the 4th joint to the extremity, and with little difference in the length of these joints; last joint about twice as long as the 10th, obtusely pointed. Thorax narrower than the elytra, narrowed behind, considerably longer than broad, black and shining, with a very fine and short impressed line in front of the scutellum, with a coarse puncture on each side near the front angles, and at the front angles with several other finer punctures; also on each side the disc, a little behind the middle, is a single puncture, and between this and the front are two or three others on each side placed close together, and numerous others close to the lateral margin. Scutellum with three or four punctures in the middle. Elytra about as long as the thorax, black, rather coarsely and sparingly punctured, the punctures closer near the suture than elsewhere, with a few fine setæ. Hind body broad, black, rather closely and sparingly punctured, with long rufescent setæ, especially distinct at the sides; 8th segment pitchy. Legs black, with the tarsi pitchy.

St. Paulo; one specimen.

2. *Tesba laticornis*, n. sp. Nigra, nitida, capite dense rugoso-punctato, antennis articulis 4—10 valde transversis. Long. corp. 9 lin.; lat. (abdominis) 1¾ lin.

Shining black with the exception of the head, this being dull black from its coarse rugose sculpture. Antennæ short and very stout, slightly thickened from the 4th to the 10th joint; 1st joint stout and distinctly curved; 2nd and 3rd joints short, 3rd not so long as broad, shorter than 2nd; joints 4—10 strongly transverse; 11th joint rather short, obtusely pointed. Palpi dark yellowish. Mandibles shining black, stout. Head quite as broad as the thorax, the upper surface covered with a dense and coarse longitudinally rugose sculpture, the carina between the antennæ smooth and shining, the under surface sparingly but extremely closely punctured, the raised line at the posterior angles continued quite half-way to the base of the mandibles. Thorax nearly as broad as the elytra, a little narrowed behind, about one and a half times as long as broad, smooth and shining, black, with a very coarse puncture on each side near the front angle, with several others quite on the front part, and behind the large puncture with two others on each side placed very near one another; also with several others close to the lateral margin, and with a very fine short line in front of the scutellum. Scutellum very indistinctly punctured. Elytra quite as long as the thorax, shining black, sparingly and rather coarsely punctured, the turned-down part more closely punctured. Hind body shining black, not narrowed till the 5th segment, sparingly and rather coarsely punctured, sparingly furnished with dark outstanding setæ. Legs black; tarsi rufescent.

Tunantins; one individual.

Linidius.

Antennæ fractæ.

Labrum totum corneum, transversum, quadridentatum, vel indistincte sexdentatum.

Palpi omnes filiformes.

Prothoracis linea marginalis externa integra; linea interna cum externâ conjuncta.

Genus corporis habitu *Scytalino* affinis. Mandibulæ validæ, medio unidentatæ, basi lobo membraneo instructæ. Labrum breve, transversum, totum corneum, quadridentatum, dentibus intermediis latis, minus distinctis, vix separatis. Maxillæ lobo superiore angusto, apice barbato, angulo externo vix setoso: palpi maxillares articulo tertio secundo paulo breviore, quarto tertio longiore. Palpi labiales articulo ultimo præcedente longiore, apice subacuminato. Antennæ basi sat approximatæ. Frons sulcis duobus antennariis. Prothorax lineâ marginali externâ, ante angulum anticum valde deflexâ, pleuris abbreviatis, linea interna cum externâ conjuncta. Coxæ intermediæ distantes. Tarsi antici articulo basali haud elongato. Generis typus *L. recticollis.*

This genus appears to be intermediate between *Scyta-*

linus and *Xantholinus*: the three species I have placed in it will probably be ultimately separated generically; indeed the *L. tenuipes* and *extremus* appear to be congeneric with *Thyreocephalus Jekeli*, Guér. ; but as Guérin's genus is not accepted at present, and as the *Linidius recticollis* does not agree therewith, I have associated all the three species under the name *Linidius*, leaving the union of *L. tenuipes* and *L. extremus* with *T. Jekeli*, until the genus *Thyreocephalus* is again rehabilitated. The limits of the genera of *Xantholinini* are, in fact, at present about as uncertain as possible.

1. *Linidius recticollis*, n. sp. Niger, nitidus, elytris cyaneis, fortiter punctatis, capite fortiter minus crebre punctato, prothorace lateribus parallelis. Long. corp. 6 lin.

Antennæ black, the basal joint very slightly curved; 2nd joint short, 3rd considerably longer than 2nd; joints 4—10 differing but little from one another in width or length, transverse ; 11th joint short. Palpi pitchy yellow, last joint of the labial a little thickened on the inside. Mandibles short and stout, black. Head about as broad as the thorax, longer than broad, slightly narrowed in front, shining black, coarsely and sparingly punctured, the punctures wanting on the front part; beneath also sparingly and coarsely punctured, the punctures wanting in the middle. Thorax narrower than the elytra, straight at the sides, nearly twice as long as broad, shining black, and only punctured at the sides and front angles. Scutellum with four or five coarse punctures. Elytra scarcely so long as the thorax, dark shining blue, coarsely and moderately closely punctured. Hind body shining black, sparingly pubescent, rather narrowed towards the extremity, sparingly and rather coarsely punctured. Legs pitchy black.

Ega ; one specimen.

2. *Linidius tenuipes*, n. sp. Niger, nitidus, elytris cyaneis, fortiter punctatis, ano ferrugineo, capite lateribus et vertice punctatis, utrinque pone oculos punctis duobus majoribus. Long. corp. 7 lin.

Antennæ but little longer than the head, not thickened after the 4th joint ; 3rd joint much longer than 2nd ; 4 − 10 transverse, differing very little from one another; last joint rather small, obtusely pointed. Palpi dusky yellow. Mandibles moderately long, shining black. Head broad, rather broader than the thorax, about as broad as the elytra, shining black, the whole of the middle part impunctate, a row of coarse punctures at the back, two other coarse punctures near the inner hind angle of the eye, and another on each side between the eyes, the sides behind the eyes with finer punctures; the under surface is also impunctate, except at the sides. Thorax a little narrower than the elytra, slightly narrowed behind, not twice as long as broad; black, shining and impunctate, except a few punctures near the front angles. Scutellum with four or five coarse punctures. Elytra not quite so long as the thorax, dark shining blue, rather coarsely punctured. Hind body shining black, hind portion of the 6th and the whole of the 7th segment dark reddish, sparingly punctured. Legs pitchy black; tarsi slender.

Ega; one specimen.

3. *Linidius extremus*, n. sp. Niger, nitidus, elytris cyaneis fortiter punctatis, ano rufo-testaceo; capite lato, fere circulari, lateribus et vertice punctatis, utrinque pone oculos punctis tribus majoribus. Long. corp. 7 lin.

This insect so greatly resembles the *L. tenuipes*, that a reiterated description is useless. In *L. extremus* the head is broader, more curved at the sides, and so more circular in form; the punctures at the side of the head are more closely packed, and on each side, just behind and internal to the eye, are three larger punctures, placed near to one another, so as to form a triangle. The extreme vertex, as in *L. tenuipes*, bears sparing coarse punctures. The basal portion of the mandibles is more slender than in *L. tenuipes*, and the legs and tarsi are even a little more slender than in that species.

Upper Amazons; a single individual found by Mr. Bates.

XANTHOLINUS.

Under this generic name there are at present placed something more than one hundred species, found in all parts of the world. I enumerate here ten Amazonian species, seven of which are new; but I have no doubt the species to be found in this rich valley are very much more numerous than this. The genus at present contains a number of very different forms, some of which have been considered by some authors as distinct genera, but have not been generally received as such at present. About thirty species from South America have been as yet described, and it is in this part of the world that the largest and most brilliant of the forms included in the genus are found.

1. *Staphylinus rutilus*, Perty.

Pará, Ega, Tapajos.

The fine series brought back by Mr. Bates of this species show that it varies much in size, large individuals being 9 lin. in length, and the smallest only about 5 lin.; the yellow colour of the extremity of the hind body is in the larger individuals nearly entirely absent.

2. *Eulissus Mannerheimii*, Lap.

Pará, Tapajos, Ega.

This species, recorded hitherto only from Cayenne, appears to be not uncommon in the Amazon valley.

3. *Xantholinus bicolor*, n. sp. Fulvus, nitidus, capite nigerrimo, minus crebre punctato, medio lævi; prothorace serie dorsali subtiliter bi- vel tri-punctato, propeque angulos anteriores parce punctato; elytris subtiliter punctatis. Long. corp. 7 lin.

Slightly larger than *X. glabratus*, of a shining-tawny colour, elytra rather paler and the head black. Antennæ with the three basal joints dark red, the rest pitchy; the 11th joint yellowish at the extremity; 3rd joint nearly twice as long as 2nd; joints 4—10 scarcely differing from one another in length, and only very slightly in width, transverse; 11th joint pointed, about twice as long as the 10th. Palpi reddish. Mandibles short, stout, black, pitchy at the base; labrum considerably advanced, rounded in front, with a deep narrow division in the middle, apparently entirely horny. Head rather broader than the thorax, about as broad as the elytra, a little narrowed in front, black and shining, the disc impunctate; the sides and the under surface sparingly and rather finely punctured. Thorax rather narrower than the elytra, considerably longer than broad, narrowed behind, very shining reddish-tawny; a little in front of the middle on each side is a single puncture, and behind this 1—3 others; there are also five or six larger punctures near the front angle on each side. Scutellum obsoletely punctured. Elytra as long as the thorax, yellow, rather sparingly and finely punctured, the punctures arranged in indistinct lines.

Hind body dark tawny, sparingly punctured, and with a scanty but rather long concolorous pubescence. Legs yellowish; tibiæ darker than the femora.

Ega: five specimens.

4. *Xantholinus anticus*, n. sp. Subdepressus, nitidus, rufus, capite, prothorace anterius abdomineque nigris, hoc apice rufo; capite parce fortiter punctato; thorace versus angulos anteriores utrinque 4- vel 5-punctato, serie dorsali nullâ; elytris obsolete punctatis. Long. corp. 5 lin.

Antennæ dark red, short and stout, basal joint curved, and towards the extremity thickened, as long as the four or five following ones together; 2nd and 3rd joints short, the latter slightly the longer, 4th joint transverse, 5—10 differing little from one another, each strongly transverse; 11th joint stout and obtusely pointed, rather paler at its extremity. Mandibles pitchy red; palpi red, the labial ones scarcely thickened. Head black, rather broader than the thorax, with the frontal furrows rather curved, and each at its extremity bearing punctures; along the sides are placed distant coarse punctures, and at the extreme vertex there are also some punctures, the hind angles are rounded; there is a longitudinal channel extending from the vertex to near the frontal furrows; beneath there is a strigose sculpture on each side. Thorax red, with the anterior part black; it is longer than broad, a good deal narrowed behind, and it is impunctate except for four or five punctures on each side near the front angles. Scutellum red, large, bearing only three or four not very distinct punctures. Elytra red, as long as the thorax, each bearing two or three not very distinct series of longitudinal punctures, and also obsoletely rugulose. Hind body black, with the 7th segment and the 6th, except at its base, red; the segments only very sparingly punctured. Legs red.

Rio Purus; a single individual, found by Dr. Trail on the 24th September, 1874.

5. *Xantholinus pygialis*, n. sp. Niger, nitidus, capite, thorace elytrisque læte violaceis, ano testaceo, pedibus piceis, capite thoraceque lævissimis. Long. corp. 6 lin.

Antennæ short, but little longer than the head, pitchy, the basal joint about as long as the six following; 3rd joint longer than 2nd; joints 4—10 transverse, similar in length, 10th considerably broader than the 4th; 11th joint yellowish at the extremity, obtusely pointed. Palpi yellowish; mandibles rather long and slender at the extremity. Head as broad as the thorax, much narrowed in front, the hind angles rounded, with an elongate oblique puncture on each side between the eyes, a similar one behind the eye, and one or two other punctures near the back; otherwise impunctate, very shining, and of a beautiful violet-blue colour. Thorax a little narrower than the elytra, longer than broad, straight at the sides, similar in colour to the head, very shining, with three or four coarse punctures on each side close to the front; otherwise impunctate. Scutellum with three or four punctures. Elytra shining dark blue, quite as long as the thorax, sparingly but rather coarsely punctured. Hind body shining black, the hind margin of the 6th and the whole of the 7th segment bright yellow, very sparingly punctured. Legs and prosternum pitchy.

Ega; one specimen.

6. *Xantholinus temporalis*, n. sp. Niger, nitidus, capite thoraceque violaceis, elytris nigro-cyaneis, ano flavo; thorace ad angulos anteriores parce punctato. Long. corp. 7½ lin.

This insect is peculiar by the very thick hind part of the head and by a peculiar form of the mandibles. Antennæ rather longer than the head, nearly black; 3rd joint nearly twice as long as the 2nd; joints 4—10 transverse, scarcely differing in length, but the 10th broader than the 4th; 11th joint obtusely pointed, yellow at the extremity. Palpi slender and elongate, yellowish. Mandibles pitchy black, elongate, the left one longer than the right and much curved towards the extremity, and near the extremity sinuate or emarginate on the upper edge. Head large, broader than the thorax, narrowed in front, the hind angles rounded, the upper surface rather uneven, a puncture on each side between the eyes, a few others on each side near the hind angle and back margin, and behind the eye with a curved broad impression; it is of a shining violet-blue colour, on the under surface but little shining, and there with shallow longitudinally subrugose sculpture. Thorax scarcely narrower than the elytra, rather longer than broad, slightly narrowed behind, similar in colour to the head, impunctate with the exception of five or six coarse punctures on each side close to the front angle. Scutellum obsoletely punctured. Elytra of a dark-greenish or bluish colour, slightly longer than the thorax, sparingly punctured. Hind body black, the hind margin of the 6th and the whole of the 7th segment yellow, very sparingly punctured. Legs nearly black; tarsi long and slender.

Ega; one specimen.

7. *Xantholinus æneiceps*, n. sp. Piceus, nitidus, capite æneo, lateribus parce punctato, medio lævi, elytris pedibusque testaceis; prothorace serie dorsali irregulari 6—7 punctato, angulosque anteriores versus 7—8 punctato. Long. corp. 3¾ lin.

Antennæ dull yellowish, rather stout, 3rd joint longer than 2nd; 4—10 transverse, 10th broader than 4th; last joint obtusely rounded, paler at the extremity. Mandibles pitchy yellow. Head rather narrow, scarcely so broad as the thorax, shining brassy; the middle part smooth, the sides sparingly, moderately coarsely punctured. Thorax obscure reddish, with a brassy reflection, very slightly narrowed behind; on each side the middle with a very irregular row of about six punctures, and with six or seven other punctures near the front angle. Elytra yellow, about as long as the thorax, rather sparingly and moderately finely punctured. Hind body pitchy, with a brassy tinge; hind portion of the 6th segment paler; sparingly and finely punctured. Legs yellow.

Ega; one specimen.

8. *Xantholinus Batesi*, n. sp. Niger, nitidus, elytris, ano, pedibusque rufis; capite crebre fortiterque punctato; thorace lateribus sat crebre punctato, seriebus dorsalibus irregularibus 10—12 punctatis, lineis marginalibus ad angulum anteriorem conjunctis. Long. corp. 4 lin.

Antennæ with the three basal joints dark red, the following ones obscure; 3rd joint longer than 2nd; joints 4—10 scarcely differing in length, the 4th narrower than the 5th, and much narrower than the 10th; last joint obtusely pointed, paler at the extremity. Mandibles pitchy. Head about as broad as the thorax, narrowed to the front, the upper surface rather coarsely but not closely punctured, the front part impunctate; under surface moderately closely and coarsely punctured. Thorax, like the head, shining black, much longer than broad, the front and hind angles rounded, the sides nearly straight, a broad space in the middle smooth; on each side with two series of punctures, an irregular, somewhat double series internally, and another irregular series about the outside; beneath, the marginal lines are not joined till the front margin of the prosternum. Scutellum indistinctly punctured. Elytra bright red, about as long as the thorax, rather finely and not closely punctured, the punctures arranged in lines. Hind body black, with the whole of the 6th and 7th segments red, finely and sparingly punctured. Legs reddish-yellow.

9. *Xantholinus amazonicus*, n. sp. Depressus, nigerrimus, nitidus, abdomine segmentis duobus ultimis totis rufo-testaceis; capite subquadrato, antice minus angustato, canaliculato, utrinque sulcato, angulis posterioribus acutis. Long. corp. 7 lin.

It is possible that this insect is a local form of *X. canaliculatus*, Er., from which it differs in the shape of the head, and by the whole of the 6th and 7th segments of the hind body being bright reddish-yellow.

Ega; one individual.

10. *Xantholinus attenuatus*, Er.

Pará; a series of individuals.

This species appears to be one of the most widely distributed and abundant of the South American *Staphylinidæ*.

LEPTACINUS.

This genus at present consists of about twenty-five species, distributed over most parts of the globe; only two, however, have as yet been indicated from South America. As regards the single Amazonian species here described, I may remark that it differs much from the ordinary form of *Leptacinus* by its long and slender legs and scarcely spinulose tibiæ, as well as by the greater development of the prosternum; I had at first intended giving it a new generic name, but as the limits of the neighbouring genera are at present uncertain, and as I cannot make a sufficiently complete examination of the only individual I possess, I have decided on calling it *Leptacinus nitidus*.

1. *Leptacinus nitidus*, n. sp. Rufo-piceus, nitidus, pedibus testaceis, capite subtiliter parcissime punctato; prothorace lateribus subtiliter sat crebre punctatis, medio lævi; elytris parce subtilissime punctatis. Long. corp. 2 lin.

Antennæ dull yellowish, stout, thickened towards the extremity; 3rd joint shorter than 2nd; 4—10 very short, and strongly transverse; 11th joint pointed. Head long, straight at the sides, not narrowed in front, the hind angles rounded, the antennal grooves short, indistinct and diverging behind; it is rather convex above, of a very shining-pitchy colour, finely and sparingly punctured, with scanty, fine, but rather long, exserted setæ. Thorax twice as long as broad, narrower than the elytra, nearly straight at the sides, being only very slightly broader in front, all the angles rounded; of a very shining-pitchy colour, the sides evenly and finely but not closely punctured; the middle smooth, with a few long, exserted setæ at the sides. Scutellum shining, depressed in the centre, impunctate. Elytra scarcely so long as the thorax, of a shining pitchy-yellow colour, very finely and sparingly punctured. Hind body cylindric, pitchy, shining, rather sparingly and very finely punctured. Legs yellow; tarsi slender.

Ega; one specimen.

LITHOCHARODES.

Antennæ geniculatæ.

Labrum medio profunde triangulariter excisum.

Palpi articulo ultimo subulato, præcedente multo breviore.

Elytra suturâ imbricatâ.

Coxæ intermediæ distantes.

Tarsi antici simplices.

Prothoracis linea marginalis superior caret.

Mandibulæ validæ, breviores. Palpi labiales articulis duobus primis crassiusculis, subæqualibus; articulo ultimo tenuissimo, præcedente breviore. Antennæ crassiusculæ, sat elongatæ. Pedes elongati, femoribus linearibus. Abdomen apicem versus paulo latius.

This genus seems to be most allied to *Leptolinus*, from which it differs by the simple undilated front tarsi, and by its much shorter maxillary palpi; the single species I refer to it differs much in appearance from *L. nothus*, on account of its shining and sparingly punctured surface. In these respects it approaches *Typhlodes*, but it would not be proper to associate it at present in the same genus with the eyeless *T. italicus*.

1. ***Lithocharodes fuscipennis*, n. sp.** Rufo-testaceus, nitidus, elytris fuscis, apice summo pedibusque testaceis; capite prothoraceque subtiliter punctatis, hoc lineâ mediâ impunctatâ. Long. corp. 2¼ lin.

Antennæ about as long as the head and half the thorax, stout, thickened towards the extremity, reddish; 3rd joint shorter than 2nd; 4—10 transverse, not differing in length, but the 10th twice as broad as the 4th; 11th joint stout and pointed. Head rather broader than the thorax, longer than broad, a little narrowed to the front, the hind angles rounded, the antennal grooves very fine; it is of a shining-reddish colour, convex above, rather finely and moderately closely punctured; the back, and a line along the middle, smooth; on the under surface it is sparingly and finely punctured in front, impunctate behind. Thorax rather narrower than the elytra, twice as long as broad, rather dilated in front, the front angles very rounded; it is of a shining reddish-yellow colour, at the sides finely and rather sparingly punctured, a broad line down the middle smooth. Scutellum with one or two indistinct punctures on each side. Elytra not so long as the thorax, pitchy, with the extremity yellow, very finely and sparingly punctured. Hind body yellow; the 6th segment much longer than the others, and a little infuscate, extremely finely and rather sparingly punctured. Legs yellow.

Tapajos; one specimen.

Metoponcus.

This generic name is applied by Kraatz to designate the species forming Family I, of Erichson's genus *Leptacinus*. It at present covers only seven species found in Eastern Europe, tropical Asia, and South America. I refer three Amazonian species to the genus, one of which, however, viz. *M. holisoides*, is very different in its appearance from the others, and will almost undoubtedly be ultimately considered a distinct genus; but I have not been able in my examination of the only individual I have seen of the species to detect characters that would justify me in making a new generic name at present for it.

1. *Leptacinus filarius*, Er.

Ega; Bates; a single individual; a second specimen, from the Amazons, has also been given me by Dr. Trail.

2. *Metoponcus basiventris*, n. sp. Elongatus, angustus, subdepressus, nitidus, piceus, abdomine basi pedibusque testaceis; abdomine segmentis duobus ultimis obscure rufis, segmentis singulis subtus medio flavescentibus. Long. corp. 2½ lin.

Antennæ short and stout, scarcely so long as the head, obscure reddish; 3rd joint very small, smaller than the 2nd, not longer than broad; 4th joint much narrower than 5th, 5—10 strongly transverse; last joint paler than the rest. Head elongate, quite as broad as the thorax, the sides straight; smooth and shining, on the under surface with longitudinal channel along each side. Thorax nearly as broad as the elytra, much longer than broad, nearly straight at the sides; like the head, of a pitchy colour, and with only one or two fine punctures; superior marginal line wanting. Scutellum impunctate. Elytra pitchy yellow, about as long as the thorax, scarcely punctured; suture indistinctly imbricate. Hind body elongate and parallel; the first visible segment pale yellow, the next three pitchy, the two last dark reddish; it is shining and impunctate; below each segment is pale in the middle. The legs are pale yellow, the femora stout; the front tarsi simple, the four hinder ones very slender; the hind tibiæ a little curved at the base.

St. Paulo; one specimen.

3. *Metoponcus holisoides*, n. sp. Depressus, nigro-piceus, nitidus, antennis obscure rufis, pedibus piceis, elytris piceo-testaceis. Long. corp. 2½ lin.

Very depressed and rather narrow. Antennæ about as long as the head, obscure red, not very stout; 3rd joint smaller and shorter than 2nd, 4th about equal to 3rd, 5—10 transverse; 11th joint paler than the rest. Head slightly broader than the thorax; at the sides with two elongate punctures in a line behind the eyes, otherwise impunctate, but extremely finely longitudinally strigose, both above and below; the middle grooves between the antennæ short, but distinct: no channel on the under surface at the side. Thorax scarcely narrower than the elytra, scarcely narrowed behind; one and a half times as long as broad, with four punctures in a transverse line across the middle; exterior marginal line short and indistinct, terminating at the corner of the coxal cavity. Elytra about as long as the thorax, sordid testaceous, extremely finely punctured. Hind body impunctate. Legs pitchy yellow; femora very stout.

Ega; one specimen.

OPHITES.

This genus at present contains three very remarkable species, described by Erichson, from South America; it appears to me to approach very closely to *Cryptobium*, though Erichson interpolates many genera between the two. The single remarkable Amazonian species I here describe has the head more abruptly narrowed, to form a slender neck, than is the case in the species described by Erichson, and its general appearance is rather that of a *Stilicus* than a *Cryptobium*; it also has the antennæ and palpi shorter and stouter than in *O. velitaris* and *Raphidioides*.

1. *Ophites stilicoides*, n. sp. Piceus, opacus, antennis pedibusque rufis, omnium dense subtilissimeque punctulatus. Long. corp. 3⅓ lin.

This insect has very much the form of an elongate *Stilicus*. It is quite dull above, and everywhere extremely densely and finely punctured. The antennæ are reddish, about as long as head and thorax, very slightly thickened towards the extremity, the 1st joint as long as the four or five following together; after the 1st, each joint is a little shorter than its predecessor, the 10th joint only about as long as broad; the last joint small and rounded. The head is broader than the thorax, nearly as broad as the elytra, the front part much produced; the eyes placed about the middle of the sides, the grooves for the antennæ very distinct. The thorax is only about half as broad as the elytra; about twice as long as broad, very slightly broader from the base till two-thirds towards the front, then narrowed to the apex; it is obscurely elevated in the middle at the base, and slightly depressed on each side of this elevation; the elytra are about as long as the thorax. The margins of the 6th and 7th segments of the hind body paler than the rest. The legs elongate, yellowish. The basal joint of the hind tarsi twice as long as the second.

Ega; one specimen, ♀.

SCOPÆODES.

Labrum transversum, medio profunde emarginatum.
Caput pedunculo brevi tenui.
Antennæ fractæ.
Tarsi articulo quarto simplice, posteriores articulo primo secundo longiore.

Caput ante oculos elongatum, pedunculo brevi tenui thoracis apici affixum; oculis parvis, rotundatis. Labrum transversum, medio profunde emarginatum. Maxillæ malâ superiore brevi, apice barbatâ. Palpi maxillares articulo primo minuto, secundo tertioque longitudine subæqualibus, illo clavato, hoc apice incrassato, quarto a tertio occulto. Ligula biloba, late emarginata, paraglossæ ei longitudine æquales. Palpi labiales articulo primo secundo breviore, hoc elongato, cylindrico; tertio angusto, præcedente fere duplo breviore. Antennæ fractæ, articulo primo elongato. Thorax angustus, apice attenuatus. Pedes sat elongati, tibiis intermediis leviter spinulosis, ceteris fere inermibus; tarsis omnibus simplicibus, posterioribus articulo primo secundo longiore.

Genus intermedium inter *Scopæum* et *Cryptobium*; ab illo antennis fractis, et ligulâ bilobâ, ab hoc capitis collo angusto, thoracis apice attenuato, distinctum.

I give the above generic name to two new species allied to *Cryptobium* with some reluctance, because that genus contains a variety of forms, several of which, in one or more respects, approach these insects; the very narrow neck by which the head is articulated with the thorax is, however, not met with in *Cryptobium*.

1. *Scopæodes gracilis*, n. sp. Elongatus, testaceus, subtilissime punctulatus; capite vertice fere lævi; thorace basi minus distincte bi-impresso. Long. corp. $2\frac{1}{2}$ lin.

Mas: abdomine segmento 7° ventrali medio exciso.

Antennæ rather shorter than head and thorax, distinctly

thickened towards the extremity; 1st joint rather stout, about as long as the four following together: 3rd joint slightly shorter than 2nd, 4—10 each slightly shorter and thicker than its predecessor, the 10th slightly transverse; last joint about equal to the 10th. Head about as broad as the elytra, the front part much produced, the eyes placed at the middle of the sides: behind these it is narrowed and rounded, so that the hinder angles have entirely disappeared; the front part is quite dull, from a very dense and obsolete punctuation, the hind part almost impunctate and shining. Thorax not quite so long as the head, hardly half as broad as the elytra, quite twice as long as broad, the front third much narrowed, extremely finely, scarcely visibly punctured at the base, with an obscure, broad, double impression. Elytra longer than the thorax, darker in colour at the base than at the apex, closely and finely punctured, but more distinctly so than the rest of the body. Hind body extremely densely and finely punctured. Legs rather long; 1st joint of hind tarsi about twice as long as the 2nd.

Tapajos; a series of specimens.

2. *Scopæodes fusciceps*, n. sp. Gracilis, testaceus, capite elytrisque infuscatis; capite oblongo, opaco, dense subtiliter rugoso-punctato. Long. corp. 1¾ lin.

Mas: abdomine segmento 1° ventrali apice anguste profundeque exciso, 6° late longitudinaliter impresso.

Much smaller than *S. gracilis*; the head more strongly punctured, and less rounded behind the eyes. Antennæ distinctly thickened towards the apex; 1st joint as long as the three or four following together, 3rd joint much shorter than 2nd, 4—10 each a little shorter and broader than its predecessor, the penultimate two or three slightly transverse; 11th joint rather less than the 10th. Head broader than the thorax, narrower than the elytra, elongate, the front part much produced, the eyes placed even a little behind the middle, the sides straight behind the eyes, the hind angles a little rounded; it is quite dull, of a smoky colour, the front part paler; it is very finely and densely, but yet distinctly sculptured. Thorax about as long as the head, not much more than half as broad as the elytra, the front third much narrowed; it is scarcely visibly punctured, and has a double impression at the base. The elytra are rather longer than the thorax, smoky yellow, very finely punctured. The legs are shorter, the hind tarsi much shorter than in *S. gracilis*. The male has a deep, narrow notch in the 7th segment of the hind body beneath, and the 6th segment has along the middle a broad, longitudinal impression, the sides of which are more densely pubescent than the other part of the segment.

Tapajos; one specimen.

CRYPTOBIUM.

This genus is known to be much more richly represented in species in the New World than it is in the other hemisphere; about twenty-two species have been hitherto described from the southern half of America, and I here add twenty species,—eighteen discovered by Mr. Bates and two by Dr. Trail. These twenty species show a remarkable diversity in form. The broad, flat, and ferocious-looking *Cryptobium gigas* offers, indeed, a most striking contrast in its appearance to the completely cylindric *Cryptobium cylindricum*. One of the most interesting peculiarities of the genus is, that in the males of some of the species the ventral plate of the fourth segment of the hind body is furnished with projections or appendages of size and form differing according to the species. I have ascertained that in one of the species here described (*Cryptobium alternans*) this lobe varies in its development in different males, in a manner similar to that which occurs in the male projections and processes found on the more anterior parts of the body in various *Coleoptera*. It appears probable to me that these developments in *Cryptobium* are of a similar character and origin to those secondary sexual characters considered by Darwin to play an important part as influencing sexual selection; and that they exercise little or no influence on the direction of the movements of the abdominal segments, as the notches and processes so common in *Staphylinidæ* on the more apical segments appear to me undoubtedly to do.

1. *Cryptobium gigas*, n. sp. Latum, nigrum, capite thoraceque opacis, parce obsolete punctatis; elytris subopacis, dense fortiterque punctatis. Long. corp. 7 lin.

Mas: abdomine segmento 7° ventrali medio profunde exciso, excisione ad apicem impressionis majoris locato.

Fem.: abdomine segmento 7° ventrali apice medio leviter emarginato.

A very robust species for the genus. Antennæ black, rusty towards the apex, slender, nearly as long as head and thorax; 1st joint as long as the five following, 3rd joint longer than 2nd, 4—11 each a little shorter than its predecessor, each longer than broad. Head broader than the thorax, nearly as broad as the elytra, suborbiculate, being about as broad as long, and narrowed and rounded behind the eyes, the front part but little produced, elevation at base of the antennæ very marked; it is black and quite dull, obsoletely and sparingly punctured; close to the inner margin of the eye is a peculiar ocellated puncture, and there is another similar one at the side, some distance behind the eye. Thorax not much more than half as broad as the elytra, subcylindric, about as long as the head, one and a half times as long as broad, dull black, like the head, and obsoletely and sparingly punctured, a little in front of the base, with a longitudinal smooth elevation, and on each side of this slightly depressed. Elytra scarcely so long as the thorax, their common width rather greater than their length, densely and rather coarsely punctured; the punctuation rugose. Hind body broad, slightly contracted at the base, black, with the extreme apex rusty, moderately closely and finely punctured; the side margins much elevated. Legs black, tibiæ and tarsi a little paler; lower part of front tibiæ clothed with a fulvous pubescence, the front femora in the middle with a tubercle beneath.

In the male the 7th segment beneath is furnished in the middle with a longitudinal impression, pointed at the front part and there serrate at the margin; the impressed portion is yellowish, and has at its extremity a rather deep notch, the sides of which are a little sinuate.

In the female the hind margin of the 7th segment beneath is slightly emarginate at the extremity.

Ega; three specimens, 1 ♂, 2 ♀.

Cryptobium maxillosum, Guérin, is closely allied to the *C. gigas*, and is even a little larger; it has the upper surface a little more shining, and the 4th and 5th segments of the hind body have, on the upper surface, a longitudinal plica at the base in the middle, of which there is no trace in *C. gigas*.

2. *Cryptobium plagipenne*, n. sp. Latum, nigrum, elytris plagâ laterali, ano, femoribusque ex parte rufescen-

tibus; capite thoraceque opacis, elytris crebre fortiter rugoso-punctatis. Long. corp. 5 lin.

Mas latet.

Femina: abdomine segmento 7° ventrali apice obsolete emarginato.

Closely allied in structure to the *C. gigas*. Antennæ pitchy at the base, then nearly black, obscure reddish at the extremity; 1st joint as long as the five following. Mandibles pitchy. Head subcircular, much rounded behind the eyes, quite as broad as long, as broad as the elytra, dull black, very sparingly and rather coarsely but not deeply punctured, with a large ocellated puncture close to the inner margin of the eye, and a similar puncture at some distance behind the eye. Thorax much narrower than the elytra, quite one and a half times as long as broad, opaque black, sparingly and indistinctly punctured, a little in front of the base a slightly elevated raised smooth line. Elytra not quite so long as the thorax, their common width rather greater than their length, a little shining, black, near the outer margin broadly ferruginous, coarsely and closely rugose-punctate. Hind body broad, rather narrowed at the base, the lateral margins much developed; it is black, with the 7th segment reddish, and is rather finely and sparingly punctured. The legs are pitchy black, the femora paler at the base, the front femora nearly entirely yellow, and with a tubercle beneath; the front tibiæ towards the apex with a fulvous pubescence.

Ega; one specimen.

3. *Cryptobium opacum*, n. sp. Nigrum, peropacum, antennis pedibusque rufo-testaceis; capite elytrisque omnium dense subtilissimeque rugoso-punctatis; prothorace dense punctato, lineâ mediâ impunctatâ. Long. corp. 4½ lin.

Mas: abdomine segmento 7° ventrali apice minus profunde et late exciso.

Antennæ nearly as long as head and thorax, yellowish; 3rd joint considerably longer than 2nd, 9th joint much longer than 10th. Palpi yellowish. Head broad, rather broader than the thorax, very nearly as broad as the elytra; the front part much produced, so that the eyes are placed a little behind the middle; it is extremely dull, being very densely and indistinctly punctured. The thorax is about three-fourths the width of the elytra, longer than broad, nearly straight at the sides, closely punctured, the back part more coarsely than the front part, with a line down the middle impunctate. Elytra one-third longer than the thorax, quite dull, densely and finely punctured. Hind body opaque, finely and obscurely punctured, the margins of the segments rufescent; the setæ numerous and distinct. Legs reddish-yellow.

The male is only distinguished from the female by a very small notch at the extremity of the 7th segment beneath.

Tapajos; numerous specimens, ♂ and ♀. Besides these, I have also an imperfect female from Ega, agreeing very closely with *C. opacum*, except that the head is rather longer and narrower (as in the *C. opacifrons*), and the legs longer. It is only by the examination of a male individual that I could decide whether it belongs to a distinct species or not.

4. *Cryptobium opacifrons*, n. sp. Nigro-fuscum, opacum, antennis pedibusque rufo-testaceis; capite omnium dense subtilissimeque rugoso-punctato; prothorace dense punctato, lineâ mediâ impunctatâ; elytris dense minus subtiliter rugoso-punctatis. Long. corp. 4½ lin.

Mas: abdomine segmento 7° ventrali medio longitudinaliter impresso, apice late minus profunde exciso.

At first sight exactly similar to the *C. opacum*, but differing therefrom by its more coarsely punctured elytra, by the different male characters, and also by its head being a little longer and narrower. Antennæ nearly as long as head and thorax, yellowish; 3rd joint one and a half times as long as second, 4—9 differing little from one another; 10th and 11th joints markedly shorter. Palpi yellowish. Head elongate, quite as broad as the thorax, the front part much produced, the eyes placed about the middle; it is extremely dull, being very finely and densely rugose-punctate. The thorax is about three-fourths of the width of the elytra; it is longer than broad, nearly straight at the sides, with a line along the middle smooth and shining, on each side of this coarsely and closely punctured; the punctuation at the sides towards the front part much finer than the rest. The elytra are longer than the thorax, dull, but densely and not altogether finely rugose-punctate. The hind body is finely and indistinctly punctured; the setæ distinct; the margins of the segments very slightly reddish. The legs are dark yellow.

The male has a very broad, but not deep, notch at the extremity of the 7th segment beneath, in front of which the segment is distinctly channelled, and is, moreover, furnished at the sides with dense black pubescence.

Ega; one specimen, ♂.

5. *Cryptobium longiceps*, n. sp. Elongatum, angustum, subdepressum, piceum, capite opaco, dense obsolete punctato; thorace nitido, crebre minus profunde punctato, lineâ mediâ lævi; elytris dense obsoleteque punctatis, opacis. Long. corp. 3½ lin.

Mas: abdomine segmento 7° ventrali apice minus profunde exciso, 6° apice emarginato.

Allied to *C. fracticorne*, but very differently punctured, and with the front of the head much more produced. The antennæ are long and slender, yellow; 2nd joint about equal to the 3rd, very little difference from joints 2—8, the three last each a little shorter than the preceding one. Head elongate, about as broad as the thorax, narrowed to the front, the front part so much produced that the eyes are placed behind the middle. It is quite dull, densely and indistinctly punctured. Thorax narrower than the elytra, longer than broad, nearly straight at the sides, rather closely but very shallowly punctured; a line down the middle impunctate. Elytra slightly longer than the thorax, densely and indistinctly punctured, opaque. Hind body finely and indistinctly punctured, with numerous distinct outstanding setæ. Legs yellowish.

Ega; one specimen, ♂. I have also another very immature individual from the same locality, which is probably the female of this species; it has a very slight notch in the last segment beneath.

6. *Cryptobium ruficorne*, n. sp. Nigro-fuscum, opacum, abdomine segmentorum marginibus antennisque rufescentibus, pedibus testaceis; capite omnium dense subtilissime rugoso-punctato; prothorace dense minus profunde punctato, lineâ mediâ lævi. Long. corp. 3½ lin.

Mas latet.

Allied to *C. longiceps*, but readily distinguished by its more robust build, and its especially broader head and thorax. The antennæ are yellowish; 2nd joint about equal to 3rd, but little difference in joints 2—8; 9—11 each a little shorter than the preceding. Mandibles and palpi yellowish. Head quite as broad as the thorax; the front part broad, much produced, so that the eyes are placed a little behind the middle; it is quite dull, being very densely and finely rugose-punctate. Thorax narrower than the elytra, longer than broad; very slightly narrowed behind, closely and rather coarsely but not deeply punctured, with a narrow line along the middle, smooth. Elytra longer than the thorax, quite dull, closely and indistinctly punctured. Hind body with the margins of the segments red, finely and indistinctly punctured, the outstanding setæ distinct. Legs reddish-yellow.

Ega; one ♀ individual, in very bad condition.

7. *Cryptobium subfractum*, n. sp. Subdepressum, subnitidum, piceum, antennis elytrisque rufescentibus, pedibus testaceis; capite elytris fere latiore, dense fortiterque punctato. Long. corp. $3\frac{1}{2}$ lin.

Mas latet.

Antennæ reddish, nearly as long as head and thorax; 1st joint slightly longer than 2nd and 3rd together, 3rd nearly twice as long as 2nd, 4—10 each a little shorter than its predecessor. Mandibles and palpi reddish. Head broad and short, scarcely produced in front, subquadrate, the hind angles rounded; it is coarsely and at the back very closely punctured, more sparingly so in the front, and has a small transverse space in the middle impunctate. Thorax narrower than the elytra, longer than broad; a little narrowed behind, very coarsely and moderately closely punctured, a line down the middle smooth. Elytra a little longer than the thorax, reddish, coarsely and closely punctured, rather shining. Hind body dull reddish, rather sparingly punctured. Legs yellowish.

Ega; one specimen, ♀.

8. *Cryptobium longicorne*, n. sp. Subdepressum, nigrum, antennis rufescentibus, pedibus testaceis; capite subopaco, dense punctato, subtiliterque pubescente; thorace elytrisque sat nitidis, dense fortiter punctatis, illo lineâ mediâ lævi. Long. corp. 4 lin.

Mas: abdomine segmento 7° ventrali margine posteriore late obsolete emarginato.

A rather narrow species, with long, slender antennae; these are rather longer than head and thorax, pitchy red, rather paler at the extremity; 1st joint scarcely so long as the three following together, 3rd not twice as long as 2nd, joints 4—8 differing little from one another, 9—11 each shorter than its predecessor. Mandibles and palpi pitchy red. Head broader than the thorax, nearly as broad as the elytra, long behind the eyes, the part in front of the eyes not long; it is densely and rather coarsely rugose-punctate, only the extreme front being free from punctures, and is clothed with a fine depressed pubescence. Thorax narrower than the elytra, longer than broad, nearly straight at the sides behind, and only a little narrowed in front; it is closely, and rather coarsely punctured, with a line along the middle, smooth. Elytra rather longer than the thorax, densely and rather coarsely punctured, rather shining, the extreme apex very narrowly yellow. Hind body rather long, finely punctured, and pubescent. Legs yellow, the coxae and knees slightly darker.

The male is only distinguished by the broad but very shallow emargination of the extremity of the ventral plate of the 7th segment.

Tapajos; six individuals.

9. *Cryptobium scutigerum*, n. sp. Antennis, pedibus, elytrorumque apice testaceis, capite opaco, dense obsolete punctato, subtiliterque pubescente; thorace sat nitido, dense fortiterque punctato, lineâ mediâ angustâ lævi; elytris thorace longioribus, dense minus fortiter punctatis. Long. corp. 3¾ lin.

Mas: abdomine segmento 3° ventrali apicem versus foveâ parvâ transversâ impressâ, quarto in lobo lato, apice rotundato, dense longeque setigero, producto, basin versus foveâ parvâ impresso.

Antennae nearly as long as head and thorax, yellow; 3rd joint one and a half times as long as 2nd, 8th joint about as long as 4th, 9—11 each a little shorter than its predecessor. Head longer than broad, about as broad as the thorax, quite dull and opaque, rather closely but indistinctly punctured, finely pubescent. Thorax two-thirds the width of the elytra, nearly one and a half times as long as broad, nearly straight at the sides, only very slightly narrowed in front; it is coarsely and closely punctured, with a narrow line down the middle, smooth. Elytra longer than the thorax, densely, rather finely punctured. Hind body quite dull, finely and closely punctured. Legs yellow.

In the male the 3rd segment of the hind body beneath has a transverse fovea in the middle, near the extremity; the 4th segment has a similar but smaller fovea, and is produced into a broad shield, the margin of which is rounded, and densely fringed with long hairs.

Tapajos; two specimens, ♂ and ♀.

Obs.—As my individuals of this species are evidently immature, I have not alluded to their general colour.

10. *Cryptobium alternans*, n. sp. Rufescens, capite, elytris, abdominisque apice nigris, pedibus testaceis; thorace nitido, crebre fortiter punctato, lineâ mediâ lævi. Long. corp. 3 lin.

Mas: abdomine segmento 3° ventrali apicem versus puncto setigero instructo, segmento quarto medio producto, cumque puncto setigero, apice longe denseque setoso.

Var. abdomine segmento ultimo rufescente.

This species is remarkable by its alternate colouration. Antennæ as long as head and thorax, the base yellow, the rest dusky reddish; 3rd joint longer than 2nd, 8th quite as long as 4th, 9—11 each a little shorter than its predecessor. Palpi yellow; mandibles red. Head rather broader than the thorax, about as broad as the elytra, dull slaty-black, all the back part densely and distinctly, the front sparingly punctured. Thorax rather narrower than the elytra, straight at the sides, of a rather shining-red colour, coarsely and rather closely punctured, with a line down the middle, smooth. Elytra rather longer than the thorax, dull bluish-black, densely and rather coarsely punctured. Hind body red at the base, the three apical segments blackish; it is dull, and finely punctured. Legs yellow.

In the male the 3rd segment of the hind body has beneath, near the extremity, a setigerous puncture; the next segment has a similar puncture, and is moreover in the middle more or less backwards, and has the hind margin fringed with very long hairs; this lobe is, however, variable in its development, and may be entirely absent, in which case the long hairs fringing the hind margin are also entirely absent; on dissecting an individual in which the lobe is largely developed, and another in which it is in the intermediate condition, I find no difference in the ædeagus.

Tapajos; several specimens.

11. *Cryptobium punctipenne*, n. sp. Sat angustum, piceum, subopacum, antennis pedibusque testaceis; capite opaco, obsolete punctato; thorace opaco, subcylindrico, dorso utrinque lineâ punctorum irregulari, lateribusque sat crebre punctatis; elytris sat nitidis, crebre fortiterque punctatis. Long. corp. $3\frac{1}{2}$ lin.

Mas: segmento 7° ventrali apice profunde triangulariter exciso.

Allied to *C. fracticorne*, but larger, and not shining. Antennæ long and slender, quite as long as head and thorax, yellow, a little infuscate in the middle; 1st joint long and slender, about as long as the five following together, 3rd a little longer than 2nd, 4—10 each a little shorter than the preceding. Mandibles and palpi yellowish. Head long, much produced in front, so that the eyes are placed at the middle of the sides, of an opaque-pitchy colour, scarcely visibly punctured, the front part impunctate. Thorax about as broad as the head, and two-thirds as broad as the elytra, subcylindric, similar in colour to the head; a broad space down the middle impunctate, on each side of this a rather irregular line of punctures, and besides this with the sides rather sparingly and not deeply punctured. Elytra rather longer than the thorax, closely and rather coarsely punctured, rather shining. Hind body dull, very finely punctured. Legs yellow.

In the male the 7th ventral segment has a deep, narrow notch in the middle of the hind part.

Tapajos; one specimen.

12. *Cryptobium scrobiculatum*, n. sp. Castaneum, subopacum, elytris apice dilutioribus, antennis pedibusque testaceis, illis medio infuscatis; capite obsolete punctato; thorace crebre subtiliter punctato, medio lineâ latâ impunctatâ; elytris prothorace longioribus, dense subtiliterque subrugoso-punctatis. Long. corp. 3 lin.

Mas: abdomine segmento 7° ventrali apice profunde triangulariter exciso, seg. 6° apicem versus medio late minus distincte impresso.

Of a dull reddish or chestnut colour, with the elytra

infuscate; moderately broad. Antennæ long and slender, rather longer than head and thorax, yellowish; joints 3—7 darker than the rest, 1st joint nearly as long as the five following together, 2nd and 3rd joints about equal in length, 4—11 each a little shorter than the preceding. Head rather broad, quite as broad as the thorax, the front part much produced, so that the eyes are placed at the middle of the side; it is of a dull-reddish colour, the back part closely but obsoletely punctured, and finely pubescent, the front part very sparingly but more distinctly punctured. Thorax two-thirds of the width of the elytra, longer than broad, similar in colour to the head; a broad, longitudinal space along the middle smooth, the sides very finely punctured. Elytra longer than the thorax, darker in colour; the apex paler, densely but very finely rugosely punctured. Hind body reddish, quite dull, very finely and closely punctured. Legs yellow.

In the male the 7th segment of the hind body has a deep triangular notch at the posterior part, and the 6th segment has an indistinct broad impression at the extremity.

Amazons; without particular locality; two male individuals.

13. *Cryptobium fuscipenne*, n. sp. Angustum, rufescens, antennis pedibusque testaceis, illis medio, capite (plus minusve), elytris, abdomineque apice infuscatis; capite opaco; thorace elytrisque dense fortiter punctatis, his apice testaceo, illo lineâ mediâ lævi. Long. corp. 2½ lin.

Mas: abdomine segmento 7° ventrali apice medio anguste triangulariter exciso, seg. 6° apice medio late profundeque semicirculariter impresso.

Antennæ long and slender, rather longer than head and thorax, yellow, infuscated in the middle; 1st joint as long as the five or six following together, 3rd joint about equal to 2nd, 4—11 each a little shorter than the preceding one. Mandibles and palpi yellow. Head about as broad as the thorax, eyes prominent, and placed at the middle of the sides; it is quite dull, densely and finely rugose-punctate, the punctuation on the front part more distinct than at the back. Thorax shining reddish, about two-thirds the width of the elytra, longer than broad, coarsely and closely punctured, with an indistinctly raised line along the middle, smooth. Elytra considerably longer than thorax, fuscous, the extremity paler, densely and distinctly punctured, rather dull. Hind body reddish, with segments 6 and 7 infuscated, except at their hind margins, extremely finely punctured. Legs pale yellow.

In the male the 7th segment of the hind body has a narrow and rather deep notch, and the 6th segment has the hind margin broadly and very distinctly impressed in the middle.

Two specimens, ♂ and ♀. The ♂ (without further locality than Amazons) is described as above. The female is from Pará, and differs a little from the ♂, in being slightly broader, and having the head reddish.

14. *Cryptobium angustum*, n. sp. Elongatum, subcylindricum, nigro-fuscum, elytrorum apice, antennis, pedibusque testaceis; capite opaco, dense subtiliter rugoso-punctato; prothorace opaco, obsoletissime punctulato, basi lineâ elevatâ nitidâ; elytris prothorace vix longioribus, dense fortiterque punctatis. Long. corp. 3¼ lin.

Mas: abdomine segmento 4° ventrali in spinam elongatam producto, trochanteribus posticis spinoso-elongatis.

Antennæ scarcely so long as head and thorax, yellowish; joints 3—6 infuscate, first joint as long as the five or six following together, 3rd joint longer than 2nd, 4—10 each a little shorter than its predecessor. Head narrow, but rather broader than the thorax, the front part much produced; the pterygia very broad, the eyes prominent, placed at the middle of the sides; it is of a blackish colour, quite dull, densely and finely rugose-punctate, the extreme front not punctate; the mandibles and palpi yellow. Thorax quite twice as long as broad, only about half as broad as the elytra, dull, with an elevated line in front of the base, shining, scarcely punctured. Elytra slightly longer than the thorax, black, the extremity yellow, closely and rather coarsely punctured, a little shining. Hind body cylindric, closely and rather strongly punctured. Legs pale yellow.

In the male the fourth segment of the hind body beneath is produced into a long, stout tooth or spine, reaching quite to the extremity of the next segment; the hinder trochanters are produced into a long, slender spine.

Ega, Tapajos: three specimens, 1 ♂, 2 ♀.

15. *Cryptobium cylindricum*, n. sp. Elongatum, peran-

gustum, nigro-fuscum, elytrorum apice, antennis pedibusque testaceis; capite opaco, dense subtiliter rugosopunctato; prothorace opaco, basi lineâ elevatâ nitidâ; elytris crebre fortiterque punctatis. Long. corp. 3½ lin.

Mas: abdomine segmento 4° ventrali in spinam breviorem producto, trochanteribus posticis spinoso-elongatis, femoribusque posterioribus medio obtuse dentatis.

Extremely closely allied to *C. angustum*, but even narrower than that species; the spine on the 4th ventral segment of the male much shorter, the femora on the other hand distinctly angularly dilated on the underside; in other respects nearly the same as *angustum*.

Ega; one specimen, ♂.

16. *Cryptobium laticolle*, n. sp. Nigrum, nitidum, antennis pedibusque rufis; capite crebre fortiterque punctato, thoracis latitudinis; hoc subquadrato, elytris paulo angustiore, crebre punctato, lineâ mediâ lævi; elytris crebre fortiterque substriato-punctatis. Long. corp. 3 lin.

Mas latet.

Very different from the other species here described by the shorter middle joints of the antennæ. These are yellowish and rather short, not reaching half-way back the thorax; 1st joint about as long as the four following together, 3rd joint scarcely longer than second, 4th joint rather longer and narrower than 5th, joints 5—10 each about as broad as long, 11th joint slightly longer than 10th. Palpi yellowish; mandibles pitchy. Head rather short and broad, quite as broad as the thorax, the antennæ inserted not far from the eyes; it is closely and coarsely punctured. Thorax rather longer than broad, nearly as broad as the elytra, straight at the sides, closely and rather coarsely punctured, a line down the middle smooth. Elytra rather longer than the thorax, blackish, a little paler at the base and shoulders, rather closely and coarsely punctured, the punctures distinctly arranged in lines. Hind body rather closely and not altogether finely punctured. Legs yellowish.

Ega; one specimen, ♀.

17. *Cryptobium angustifrons*, n. sp. Rufo-piceum, nitidum, antennis pedibusque rufo-testaceis; capite elytris duplo angustiore, subopaco, subtiliter punctato, vertice elongato; prothorace utrinque serie dorsali punctorum

minorum, laterilbus parce punctatis; elytris parce obsolete striato-punctatis. Long. corp. 4—4¼ lin.

Mas: abdomine segmento 7° ventrali apice anguste sat profunde exciso, seg. 6° apice emarginato.

This species has a peculiar facies, arising from its narrow head, with long vertex and its broad thorax. The antennæ are yellow and about as long as head and thorax; 1st joint not much longer than the two following together, 3rd joint one and a half times as long as the 2nd, the following joints slender and elongate, each a little shorter than its predecessor. Palpi yellow. Head longer than broad, quite one-half narrower than the base of the thorax, the eyes placed in front of the middle; it is dull, sparingly and rather finely punctured, and with a fine rather long scanty pubescence. Thorax shining reddish, at the base slightly narrower than the elytra, distinctly narrowed to the front, rather longer than its breadth at the base; on each side the middle with a row of fine punctures, and with other fine punctures at the sides. Elytra rather longer than the thorax, shining, red at the base, the rest infuscate, each with four or five rows of very fine punctures; these, with the exception of the row close to the suture, being very indistinct. Hind body distinctly and not altogether sparingly punctured. Legs yellowish.

In the male the hind margin of the sixth segment beneath is distinctly emarginate in the middle, and the seventh segment has a rather deep and narrow notch.

Tapajos; numerous specimens.

18. *Cryptobium alienum*, n. sp. Nitidum, rufescens, elytris infuscatis; capite crebre fortiter punctato; thorace elytris vix angustiore, his punctato-striatis. Long. corp. 3½ lin.

Mas: abdomine segmento 7° ventrali apice sat profunde triangulariter exciso.

Obs.—Facie, antennarumque structurâ, generi *Dolicao* similis.

Antennæ yellow, stout for this genus; 1st joint nearly as long as the three following together, 3rd longer than 2nd, 4—10 each a little shorter than its predecessor, the 10th about as long as broad. Head pitchy red, a little narrower than the thorax, but little produced in front; the eyes placed before the middle, coarsely but not closely punctured, very sparingly pubescent. Thorax but little narrower than the elytra, straight at the sides, but little longer than broad, of a shining-reddish colour, moderately finely and not closely punctured; a broad space along the middle, smooth. Elytra darker in colour than the thorax, scarcely longer, shining; each with six distinct rows of punctures. Hind body reddish; the intermediate segments infuscate, rather finely but not closely punctured. Legs yellow.

Tapajos; one specimen, ♂.

19. *Cryptobium triste*, n. sp. Angustum, nigro-fuscum, antennis, palpis, pedibusque testaceis, antennis late infuscatis; capite dense punctato, fere opaco; prothorace nitido, crebre sat fortiter punctato, medio lævigato; elytris thorace longioribus, dense punctatis, subnitidis. Long. corp. 2½ lin.

Antennæ moderately long, slender; base of 1st joint yellowish, its apical portion and the following joints infuscate; the apical joints again paler; 3rd joint not so long as 2nd, 10th quite as long as broad. Palpi yellow. Head about as broad as the thorax, constricted in front of the eyes, which are placed about midway at the sides; its surface is densely punctured, the punctures becoming finer and denser towards the vertex, so that the part behind the eyes is quite opaque; it is nearly black in colour, its greatest breadth just in front of the rounded hind angles. Thorax pitchy, shining, much narrower than the elytra, longer than broad, a little rounded and narrowed towards the front; rather coarsely punctured, but with a broad, smooth space along the middle; the punctures bounding this space on each side are closely packed, so as to form an irregular series, which becomes indistinct towards the front. Elytra a little longer than the thorax, densely punctured, the punctures rather deep and moderately coarse, the interstices not dull,—they are nearly black. Hind body very finely punctured, black, and quite dull. Legs yellow; coxæ and under face of the insect obscure reddish in colour.

A single individual, which I believe to be a female, was captured by Dr. Trail on the 5th November, 1874, but he has not sent me the special locality.

20. *Cryptobium Traili*, n. sp. Elongatum, brunneum, antennis pedibusque testaceis, abdomine segmento 6° nigricante; capite angusto, verticem versus attenuato; prothorace subcylindrico, antrorsum leviter angustato; elytris

dense, profunde, fortiter regulariterque punctatis. Long. corp. 5 lin.

Antennæ yellow, slender and elongate; 3rd joint much longer than 2nd, even the 10th slender and elongate, fully twice as long as broad. Palpi yellow, elongate; mandibles yellow. Head elongate and narrow, narrower than the thorax; the eyes convex and prominent, placed far from the vertex and at a considerable distance from the insertion of the antennæ; the antennal elevations very marked, the space between them smooth and shining, and with only three or four punctures; the back part of the head is gradually narrowed from the eyes till the neck; it is opaque, and is coarsely and rugulosely but not densely punctured. Thorax much narrower than the elytra, greatly longer than broad, distinctly narrowed towards the front, and very slightly towards the base; it is shining and of a brownish-yellow colour, coarsely punctured, but with a rather broad, smooth space along the middle. Scutellum coriaceous. Elytra longer than the thorax, of a brownish colour, with the hind margin a little paler; they are densely covered with coarse and deep, rather regularly-arranged punctures, the interstices of which are quite shining. Hind body brownish-yellow, with the 6th segment blackish; it is not shining, is only scantily punctured, and sparingly pubescent. The legs are elongate and yellow.

Rio Madeira; a single female found by Dr. Trail on the 25th May, 1874; it was attracted by light.

Obs.—This peculiar species appears to approach the genus *Ophites* in some of its peculiarities; it is the most remarkable *Cryptobium* known to me, and I have very great pleasure in naming it in honour of its discoverer, Dr. Trail, to whom I am indebted for the only individual known of it.

*Sphæronum, n. gen.

Labrum transversum, late emarginatum.
Palpi maxillares articulo tertio incrassato, basi angusto.
Antennæ crassiusculæ, sub-fractæ.
Tarsi articulo quarto simplice.
Corpus elongatum, angustum, alatum. Caput liberum, collo tenui a verticis prolongatione tricarinatâ, obtecto.

---

* I had written this word *Sphærinum*, and it so stands in the list of species at the commencement of this paper; but as Erichson has used the word *Sphærina* for a genus of Coleoptera, I have thought an alteration necessary.

Labrum breve, medio sinuatum. Mandibulæ robustæ, medio tridentatæ. Maxillæ malâ interiore latâ, barbatâ, exteriore subelongatâ, apice dense longeque barbatâ. Palpi maxillares articulo primo secundo duplo breviore, hoc sat elongato, basi angustiore; tertio secundo longiore, valde incrassato, basi constricto. Labium* paraglossis valde elongatis, acuminatis. Palpi labiales articulo primo secundo fere duplo breviore, hoc elongato, cylindrico; tertio brevi, angustissimo, secundo fere quadruplo breviore et angustiore. Antennæ crassiusculæ, elongatæ, vix fractæ. Thorax angustus, elongatus, basi apiceque attenuatus; prosternum magnum, convexum. Elytra truncata. Abdomen apicem versus leviter incrassatum; penis magnus, latus, oblongo-ovalis. Pedes sat elongati; tibiis anticis basi dilatatis, medio subito constricto; tarsis omnibus simplicibus, articulo primo secundo duplo longiore.

Habitus singularis, capitis forma *Ophitidem* referens, a quo oris partibus, antennis subfractis, tibiarumque anticarum structurâ singulari, discedit.

1. *Sphæronum opacum*, n. sp. Elongatum, opacum, capite, thorace, antennisque nigris, pedibus testaceis, elytris abdomineque obscure rufescentibus; dense subtilissimeque punctulatum, subtilissimeque griseo-pubescens. Long. corp. 5¼ lin.

Mas: abdomine segmento 6º ventrali apice medio depresso, semicirculariterque exciso; segmento 7º late impresso, apice emarginato excisoque.

Antennæ nearly as long as head and thorax, black, not in the least thickened towards the extremity; 1st joint more than twice as long as 2nd, 4—10 differing but little from one another; last joint rather longer than the 10th. Head black, about as broad as the elytra, closely and finely punctured and pubescent, produced behind into a stout, tricarinate neck, the hind margin of which is truncate. Thorax considerably narrower than the elytra, twice as long as broad, much narrowed to the base and to the front, its greatest width at about two-thirds of the

---

* The ligula is distorted in the preparation I have made of the trophi, but as far as I can see it appears to be entirely corneous; if this be the case, it will add another remarkable character to this very distinct genus.

distance from base to front; it is nearly black, very opaque, extremely densely and finely punctured, and very finely pubescent, the hind half with an indistinct carina along the middle. Elytra dull reddish, about as long as the thorax, extremely densely and finely punctured, and extremely delicately pubescent. Hind body obscure reddish, distinctly broader from the base to the extremity, very densely and finely punctured. Legs dull yellowish.

In the male the 6th segment of the hind body beneath is depressed at the extremity, and provided with a rather broad but not deep notch; the 7th segment is broadly flattened or depressed, is emarginate at the extremity, and has in the middle of the emargination a distinct notch, not narrower at the front.

Ega; two specimens, ♂.

2. *Sphæronum depressifrons*, n. sp. Capite thoracequc nigro-piceis, nitidis, fere impunctatis; clytris abdomineque rufescentibus, opacis, dense subtilissimeque punctatis; pedibus testaceis. Long. corp. 5 lin.

Antennæ shorter than head and thorax, pitchy; 1st joint rather shorter than the three following together; joints 2—10 differing but little from one another, the 3rd rather longer than the others; 11th joint longer than the 10th. Head pitchy, as broad as the elytra, with a strong prominence on each side in front over the insertion of the antennæ, and between these prominences depressed and without any carina; it is narrowed behind into a stout, strongly tricarinate neck, and is smooth and shining, towards the sides very sparingly and finely punctured, and very sparingly pubescent. Thorax not much more than half as broad as the elytra, more than twice as long as broad, a little narrowed behind, but more so towards the front; it is shining, it is very sparingly and finely punctured, with a distinct carina along the middle behind. The elytra are distinctly shorter than the thorax, dull reddish, very finely and densely punctured. The hind body is rufescent, a little broader from the base to the extremity, very finely and closely punctured. The legs are yellow.

Ega; a single specimen.

Obs.—I suppose this specimen to be a female. The hind body shows no peculiar structure beneath, but on the upper side the hind part of the 7th segment is distinctly produced in the middle.

3. *Sphæronum carinifrons*, n. sp. Nigro-fuscum, opacum, abdomine rufescente; pedibus testaceis; dense et (fronte exceptâ) subtilissime punctatum. Long. corp. $4\frac{1}{4}$ lin.

Allied to *S. opacum*, but much smaller, the head narrower, and the antennæ shorter. Antennæ nearly as long as head and thorax, dusky red; 1st joint longer than the two following together, 2nd shorter than 3rd, 4—10 scarcely differing from one another. Palpi reddish. Head nearly as broad as the elytra, quite dull, densely and finely punctured, the front part not so densely and finely as the back; it is also densely and finely pubescent; it is produced behind into a short stout tricarinate neck, and in front has besides the prominences over the antennæ,—a distinct elevation between these. The thorax is distinctly narrower than the elytra, about twice as long as broad, narrowed behind, and still more strongly narrowed to the front; its greatest breadth is at fully two-thirds of the distance from base to apex; it is very densely and finely punctured, and has a distinct fine carina in the middle at the base. The elytra are very dark and obscure reddish, very densely and finely punctured, not quite so long as the thorax. The hind body is rather paler than the other parts, obscure dull reddish, very densely and finely punctured, rather incrassated at the extremity. The legs are yellow.

Ega; one specimen.

4. *Sphæronum elongatum*, n. sp. Angustum, nigrofuscum, opacum, elytris abdomineque rufescentibus, pedibus testaceis; capite subopaco, dense subtiliter punctato. Long. corp. $3\frac{3}{4}$ lin.

Mas: abdomine segmento 6° ventrali longitudinaliter impresso, apice minus profunde exciso; segmento 7° basi late impresso, apice sat profunde exciso.

Closely allied to *S. carinifrons*, but smaller and narrower, and with the head less densely and not rugosely punctured, so that it is not altogether opaque. The antennæ are pitchy black, nearly as long as head and thorax, joints 2—10 differing but little from one another. Head about as broad as the elytra, slightly shining, densely and finely punctured and pubescent; the back part more finely punctured than the front, the three frontal eminences very distinct, as are also the three carinæ of the neck; the middle one of these narrow, and strongly elevated. Thorax considerably narrower than the elytra, more than twice as long as broad, narrowed in front and behind, and with a distinct central carina visible along quite two-thirds of its length; extremely finely and densely punctured, quite dull. Elytra nearly as long as the thorax, dull red, very densely and finely punctured. Hind body dull reddish, densely and finely punctured, broader towards the extremity. Legs yellowish.

In the male the 6th segment of the hind body beneath is distinctly impressed along the middle, and notched at the extremity; the 7th segment is very broadly impressed at the base, and its hind margin rather deeply notched.

Ega; one specimen, ♀.

5. *Sphæronum carinicolle*, n. sp. Rufescens, capite piceo, crebre minus subtiliter punctato, sub-nitido; prothorace per totam longitudinem carinato, dense subtilissime punctato; pedibus testaceis. Long. corp. 2⅜ lin.

Mas: abdomine segmentis 6° et 7° apicibus excisis.

Of a dull-reddish colour, with the head and antennæ darker than the other parts. Antennæ not so long as head and thorax, dull red; joints 2—10 differing little from one another, the penultimate quite as long as broad. Head about as broad as the elytra, pitchy, closely and not finely punctured; the extreme base impunctate, the three frontal eminences large, the central carina of the neck elongate, and continued quite to the front as a very narrow impunctate line. Thorax quite twice as long as broad, broader from the base to near the front, then narrowed to the front; it is dull reddish, very densely and finely punctured, with a raised central carina through its whole length, which is, however, but little distinct at the front part. Elytra rather shorter than the thorax, dull red, densely and finely punctured. Hind body red, densely and finely punctured. Legs pale yellow.

In the male the 6th segment of the hind body beneath has a very slight impression along the middle, and a broad shallow notch at the extremity; the 7th segment has a rather deep triangular notch at the extremity.

Ega; one male specimen.

6. *Sphæronum pallidum*, n. sp. Nitidulum, testaceum, capite picescente, lateribus parce sat fortiter punctato. Long. corp. 2⅓ lin.

Mas: segmento 6° ventrali apice medio emarginato, segmento 7° sat profunde triangulariter exciso.

The small size, very narrow form, very pale colour, and sparing punctuation, render this a very distinct species. The antennæ are reddish, rather shorter than head and thorax. The head is about as broad as the elytra, dark reddish, or pitchy colour, shining, the middle and back part impunctate; the sides sparingly but not finely punctured, the three frontal eminences very distinct. Thorax yellowish, shining, very sparingly and finely punctured, at the back part with a distinct elevation along the middle. Elytra yellow, shining, sparingly and finely punctured, rather shorter than the thorax. Hind body yellow, scarcely shining, but indistinctly punctured. Legs pale yellow.

In the male the 6th segment of the hind body beneath is impressed along the middle, and a little emarginate at the extremity of the impression; the 7th segment is flattened at the base, and has a rather deep triangular notch at the extremity.

Tapajos; ten specimens.

LATHROBIUM.

The genus *Lathrobium*, consisting of about one hundred described species, is distributed throughout the world, although comparatively few species are yet known from the tropics and subtropical regions. South America is the part of the world in which hitherto it might have been, with apparent reason, surmised that the genus is represented by fewer species than elsewhere; only two or three species having been described from these parts, and but few others existing, so far as I know, in collections. I am enabled here, however, to distinguish no less than twenty-five Amazonian species of the genus, so that it becomes evident that the want of South American species in collections is not really indicative of anything more than our very limited acquaintance with the tropical *Staphylinidæ*.

Of these twenty-five species the first, *L. macrocephalum*, is about the largest and most peculiar species of *Lathrobium* I am acquainted with, and will probably be ultimately considered a distinct genus. Then follow eight species bearing an extraordinary resemblance to one another in appearance and general characters, but distinguished nevertheless by striking and highly important primary and secondary sexual characters. In the case of some of these species (*L. puncticeps* and *L. decisum*, for instance), after a very careful examination, I am unable to see any satisfactory distinctive characters except the sexual ones; and an examination of the male intromittent organ has convinced me that it is extremely doubtful whether fertilization could be effected by the sexes of different species, even if attempted. By this I mean that if, for example, the male and female organs in *L. opalescens* be mutually adapted for the facilitation of fecundation, as it is only reasonable to suppose is the case, then from the great difference we find to exist in the intromittent organs of the males of the exactly similar *L. puncticeps*, we are fairly entitled to conclude that fecundation of the female of *L. opalescens* by it would be difficult. It may, perhaps, not be out of place to state here my conviction that these modifications of sexual characters will be found to be very directly in relation with those "laws of variation," a knowledge of which is so much to be desired for the further elucidation of the question of the differentiation of species.

1. **Lathrobium macrocephalum**, n. sp. Robustum, nigrum, nitidum, elytris abdomineque nigro-piceis, pedibus piceis; capite magno, crebre fortiter punctato; prothorace elytrisque parcius punctatis, illo tenuiter canaliculato. Long. corp. 6¼ lin.; lat. (capitis) 1⅛ lin.

Mas: abdomine segmento 7° ventrali apice profunde triangulariter exciso; segmento 6° late triangulariter impresso, apice emarginato.

The massive head of this species distinguishes it from all others of the genus. The antennæ are stout, shorter than head and thorax, slightly more slender at the extremity than the base; 1st joint about as long as the three following together, 3rd longer than 2nd, 4—10 differing little from one another in length; 11th joint slender and pointed, longer than the 10th, rusty at the extremity. The head is rather broader than the thorax; it is a little narrowed towards the front, is coarsely and rather closely punctured; the punctuation rugose behind the eyes, a narrow space along the middle, smooth. Thorax fully as broad as the elytra, slightly broader than long, a little narrowed behind, rather coarsely and irregularly punctured, a narrow space along the middle smooth, and in the centre of this a fine channel; it is black and shining. Elytra distinctly longer than thorax, shining pitchy black, rather finely and not closely punctured. Hind body pitchy, dull, very finely and not closely punctured. Legs pitchy, the hind ones reddish; first joint of hind tarsi very short, quite hidden by the tibia. In the male the 7th segment of the hind body has on the underside, at its extremity, a rather deep triangular notch. The 6th segment has the hind margin broadly emarginate; in front of this it has a broad triangular impression, the middle part of which is smooth, and the sides furnished with short, coarse, black hairs.

Ega; one specimen.

2. *Lathrobium opalescens*, n. sp. Piceum, nitidum, antennis pedibusque testaceis; capite, thorace elytrisque subtiliter opalescentibus; capite vertice angulisque posterioribus dense, subtilissime rugoso-punctatis, opacis et pubescentibus, disco lævi, fronte fortiter parcius punctatâ; prothorace crebre punctato, lineâ mediâ impunctatâ; elytris crebre punctatis. Long. corp. 4½ lin.

Mas: segmento 7° ventrali apice medio late, sat profunde semicirculariter exciso, ante excisionem leviter impresso; segmento 6° leviter emarginato.

Allied to *L. brunnipes*, but greatly broader, and less cylindric. Antennæ slender and elongate, rather longer than head and thorax, yellow; the 3rd joint longer than 2nd, the penultimate joints shorter than the intermediate ones. Head nearly as broad as the thorax, the back and the hind angles very densely and finely punctured, quite opaque, the front part with an opalescent reflection; a broad space in the middle smooth, in front of which it is sparingly punctured. Thorax quadrate, about as broad as the elytra, its length equal to its breadth, regularly but not closely punctured, a line along the middle smooth; it is of a pitchy colour, with opalescent reflection. Scutellum impunctate. Elytra rather longer than the thorax, moderately closely, and not altogether finely punctured. The hind body is obscurely rufescent, very finely and closely punctured, the 7th segment sparingly punctured. Legs yellow.

The male has a very broad but not deep notch at the

extremity of the 7th segment, beneath; this notch is of a peculiar shape, being somewhat contracted at its entrance, in front of it the segment is a little impressed; the 6th segment has the hind margin broadly but slightly emarginate.

Ega; three male specimens; also one from Santarem, but I am not sure that this indication of locality is correct.

I have also from Ega two female individuals of a *Lathrobium*, which I had at first described under the name of *L. quadraticolle*, but on re-examination I think it highly probable that they may be females of *L. opalescens*, from which they differ by being much larger and broader, their length being 5½ lin. I have ascertained by dissection of one of them that it is a female; the ventral plate of the 7th segment is not rounded in the middle, but is very slightly emarginate; the dorsal lobe of the 8th segment is very compressed at the extremity, so that it appears to form a sharp longitudinal carina.

3. *Lathrobium decisum*, n. sp. *L. opalescenti* omnino similis, notis sexualibus tantum differt.

Mas: abdomine segmento 7° ventrali apice medio profunde, sat late semicirculariter exciso: segmento 6" late profundeque longitudinaliter impresso, apice medio emarginato, utrinque angulatim producto.

Tapajos; a single male.

I am unable to find any characters except the sexual ones to distinguish this species from *L. opalescens*, but these are very marked and important. The notch of the 7th segment is narrower and deeper, not contracted at the entry. The 6th segment has a broad and deep longitudinal impression along the middle; the sides of this impression are remarkably abruptly defined, and project beyond the hind margin, so as to form an acute angle.

I have also, from the same locality, a female individual, which I believe to be the other sex of this species; it is about the same size as the male; it has the hind margin of the 7th segment beneath slightly emarginate in the middle; it differs from the female described as that of *L. opalescens* by its much smaller size, and by the shorter and much less laterally compressed dorsal lobe of the 8th segment.

4. *Lathrobium puncticeps*, n. sp. Piceo-rufum, nitidum, antennis pedibusque rufo-testaceis; capite subopaco,

dense punctato, medio parce fortius punctato; prothorace crebre fortiter punctato, lineâ mediâ lævi. Long. corp. 4½ lin.

Mas: abdomine segmento 7° ventrali apice profunde exciso; sexto medio longitudinaliter impresso, 5" obsolete impresso.

Rather broader and more parallel than *L. geminum.* Closely allied to *L. opalescens,* but slightly narrower, and the male characters different. The antennæ are reddish-yellow, long and slender, rather longer than head and thorax; the 3rd joint longer than the 2nd; the penultimate joints shorter than the intermediate ones. The head is as broad as the thorax, pitchy, dull, the sides and back part very densely and finely punctured, the middle and front part sparingly and more coarsely punctured. The thorax is quadrate, about as broad as the elytra, straight at the sides, quite as long as broad, the whole of the sides rather coarsely and closely punctured, a line along the middle smooth; it is of a reddish colour and rather shining. The elytra are rather longer than the thorax, reddish, moderately closely punctured. The hind body is reddish, closely and finely punctured. The legs are yellow.

Tapajos; numerous specimens.

This species is so closely allied to the two preceding that it is scarcely distinguishable from them except by the sexual characters. The male has a rather deep and narrow notch, not contracted at the entrance, on the hind margin of the 7th segment; the 6th segment has a distinct but ill-defined longitudinal impression along the middle, and there are indications of a very slight depression on the 5th segment, the sides of this having, in fresh specimens, some rough black pile, which appears very easily removed. The female has the hind margin of the 7th ventral segment rounded and entire, and the dorsal lobe of the 8th short, and only very slightly laterally compressed towards the apex.

5. *Lathrobium parallelum,* n. sp. Piceum, nitidum, antennis pedibusque rufo-testaceis; capite dense subtiliter punctato, disco lævi, fronte parce fortiter punctatâ; prothorace crebre sat fortiter punctato, lineâ mediâ impunctatâ. Long. corp. 4 lin.

Mas: abdomine segmento 7° ventrali apice profunde exciso, segmento 6° late profundeque longitudinaliter impresso, ad impressionis apicem exciso.

Rather smaller than *L. puncticeps*, and closely allied; the head shorter and rather differently punctured, the structure of the 6th segment of the hind body in the male different. About as large as *L. brunnipes*, but more parallel and rather broader. Antennæ longer than head and thorax, reddish-yellow; 3rd joint longer than 2nd, 4—10 each a little shorter than the preceding one. Head almost as broad as the thorax, the hind angles densely punctured; a small space in the middle impunctate and shining, the punctures in front of and those surrounding this space coarse and not so close. Thorax about as broad as the elytra, quite as long as broad, quadrate, of a pitchy colour, shining, rather coarsely and closely punctured, with a line along the middle smooth. The elytra are longer than the thorax, rather closely and distinctly punctured. The hind body is pitchy, with the extremity reddish, finely and rather closely punctured. The legs are yellow.

The male has a large deep notch at the extremity of the 7th ventral segment; the 8th segment has a broad and deep longitudinal impression along the middle, at the extremity it is deeply emarginate; the sides of the longitudinal impression are not abruptly defined (as they are in *L. decisum*), nor produced beyond the hind margin, and their extremity forms a rounded right angle. In the female the dorsal lobe of the 8th segment is quite simple, not at all laterally compressed; both the dorsal and ventral plates of the 7th segment are a little produced and rounded at the extremity.

Tapajos; several specimens.

6. *Lathrobium mendax*, n. sp. Piceum, nitidulum, antennis pedibusque rufo-testaceis; capite dense punctato, disco lævi, fronte parce fortiter punctatâ; prothorace crebre punctato, lineâ mediâ impunctatâ. Long. corp. vix 4 lin.

Mas: abdomine segmento 7° ventrali apice profunde exciso; 6° medio indistincte longitudinaliter impresso; 5° medio depresso, basi utrinque impressione transversâ profundâ.

This species resembles exactly the *L. parallelum*; it is scarcely smaller, and has the antennæ, thorax and elytra slightly shorter; but the male characters are remarkable, and very different. In this sex the 7th ventral segment has a notch similar to that of *L. parallelum*, but it is not quite so deep, and rather broader at its opening; the 6th segment is only indistinctly impressed along the middle, but the 5th has a broad and deep impression in the middle, and on each side of this, at the extreme base, is a deep, abruptly-defined, curved, transverse impression.

Tapajos; a single specimen.

7. *Lathrobium certum*, n. sp. Piceum, nitidulum, antennis pedibusque rufo-testaceis; capite dense subtiliter, anterius parce fortiter, punctato, disco lævi; prothorace crebre punctato, lineâ mediâ impunctatâ; elytris crebre fortiter punctatis. Long. corp. vix 4 lin.

Mas: abdomine segmento 7° ventrali apice medio profunde exciso; 6° medio leviter emarginato.

Antennæ reddish, about as long as head and thorax. Head as broad as the thorax; the posterior parts densely rugosely punctured, the anterior part sparingly punctured, the central part free from punctures. Thorax quadrate, just about as long as broad, a broad line along the middle impunctate, the sides rather coarsely punctured. Elytra slightly longer than the thorax, their punctuation quite as coarse as, and similar to, that of the thorax. Hind body dull; densely, extremely finely punctured, reddish towards the extremity. Legs yellowish.

The male has at the hinder part of the 7th ventral segment a deep notch, the entry of it being broad and quite rounded at the sides; the 6th segment is flattened along the middle, and distinctly emarginate at the extremity in the middle: in these characters it approaches considerably *L. puncticeps*, but the sides of the notch of the 7th segment are more cut away, so that it is much broader at its entry than in *L. puncticeps*; the 6th segment is less impressed along the middle, but more deeply emarginate at the extremity.

Amazons; a single male individual; without special locality.

This species, extremely closely allied to the five preceding, has the elytra more coarsely punctured than any of them.

8. *Lathrobium rufulum*, n. sp. Rufo-testaceum, nitidulum, parallelum, abdomine segmentis 2—5 infuscatis, capite angulis posterioribus dense subtiliter punctatis, disco lævi; prothorace lateribus sat crebre punctato, lineâ latâ impunctatâ. Long. corp. 3 lin.

Mas: abdomine segmento 7° ventrali apice sat profunde exciso; segmento 6° medio leviter longitudinaliter impresso, apice late emarginato.

About the size of *L. terminatum*, but of totally different colour, and with the head larger and differently shaped. Antennae yellow, about as long as head and thorax, formed much as in *L. terminatum*, but slightly stouter, the penultimate joints a little shorter; 3rd joint distinctly longer than 2nd. Head about as broad as the thorax, the hinder angles not rounded, the front part sparingly and rather coarsely punctured; the disc shining and impunctate; the hind angles densely and finely punctured. Thorax subquadrate, straight at the sides, quite as long as broad, a little narrower than the elytra; a broad line along the middle impunctate; the sides moderately coarsely and not closely punctured; the medial punctures at the hind part separated from the others by a narrow, smooth space. Elytra rather longer than the thorax, finely and not closely punctured, the punctures indistinctly arranged in lines. Hind body with the basal segments pitchy, the hind part reddish-yellow, very finely punctured. Legs yellow.

In the male the 6th segment of the hind body beneath has a longitudinal impression along the middle; its hind margin is broadly but very shallowly emarginate, the emargination limited on each side by a slight projection; the 7th segment has a rather broad and deep notch, the front part of which is rounded, and not notched.

Tapajos; one specimen, ♂.

9. *Lathrobium proximum*, n. sp. Rufo-testaceum, nitidulum, parallelum, abdomine piceo, apice rufo-testaceo; capite angulis posterioribus dense subtiliter punctatis, disco lævi; prothorace lateribus crebre punctato, lineâ latâ impunctatâ. Long. corp. 3 lin.

Mas: abdomine segmento 7° ventrali apice sat profunde exciso; segmento 6° late impresso (impressionis apice impunctato), margine posteriore leviter emarginato.

Extremely close to *L. rufulum*, and differing only as follows: the antennæ are rather longer, the 3rd joint considerably longer than the 2nd, the thorax is more closely punctured; and in the male the 6th segment of the hind body beneath has a broader, ill-defined impression, at the extremity of which is a triangular impunctate (as it were, membranous) space; the hind margin is slightly emarginate, and there is not the least trace of any projection at the outside limits of this emargination; the 7th segment has a notch similar to *L. rufulum*.

Tapajos; about a dozen individuals.

10. *Lathrobium amazonicum*, n. sp. Angustum, piceo-rufum, nitidulum, capite piceo, crebre fortiter punctato; antennis pedibusque testaceis; prothorace crebre fortiter punctato, lineâ mediâ impunctatâ; elytris punctato-striatis. Long. corp. 2½—3 lin.

Mas: abdomine segmento 7º ventrali apice profunde minus late exciso.

A narrow and parallel species. Antennæ yellow, rather longer than head and thorax, moderately stout; 3rd joint longer than 2nd. Head pitchy, rather long, about as broad as the elytra, closely and coarsely punctured; the disc more sparingly punctured, the punctuation at the hind angles rugulose. Thorax slightly narrower than the elytra, straight at the sides, distinctly longer than broad, shining reddish, a line along the middle smooth, the sides coarsely and closely punctured. The elytra are longer and more finely punctured than the thorax, the punctures (rather indistinctly) arranged in lines; they are of a reddish colour, a little infuscated towards the extremity. The hind body is elongate and narrow, finely punctured. The yellow legs are rather short and stout.

The male has a rather deep but narrow notch at the extremity of the 7th ventral segment; the 6th segment has the hind margin slightly projecting in the middle, and in the middle of this is an extremely small emargination or notch.

Tapajos; numerous specimens.

11. *Lathrobium tardum*, n. sp. Rufescens, capite fusco-rufescente, pedibus testaceis; antennis elongatis; capite dense punctato, fere opaco; prothorace crebre punctato, lineâ mediâ lævigatâ; elytris hoc longioribus, sat crebre minus fortiter punctatis. Long. corp. 3½ lin.

Mas: abdomine segmento 7º ventrali apice medio exciso.

Antennæ red, slender, quite 1 line in length; 3rd joint a good deal longer than 2nd. Head slightly broader than the thorax, of an infuscate or somewhat purplish-red colour, coarsely and densely punctured, a small space on the disc, smooth. Thorax a little narrower than the elytra, longer than broad, nearly straight at the sides and slightly narrowed behind, closely and rather coarsely punctured, with a very straight line along the middle smooth; it is of a red colour and a little shining. Elytra a good deal longer than thorax, reddish with a slight purplish obscuration, a little shining, rather finely and not densely punctured. Hind body elongate and narrow, yellow, very finely punctured, dull except towards the apex, where it is a little shining. Legs yellow.

The male has a moderately large notch at the extremity of the ventral plate of the 7th segment of the hind body.

Manaos; two individuals, ♂ and ♀, captured at light by Dr. Trail in August, 1874.

Obs.—This species much resembles *L. amazonicum* in form and colour, but it is larger and has the sculpture of the upper surface denser and finer.

12. *Lathrobium tenuicorne*, n. sp. Elongatum, angustum, parallelum, piceo-rufum, antennis pedibusque testaceis; capite piceo, dense subtiliter rugoso-punctato, fronte parce fortiter punctatâ, disco anguste impunctato; prothorace crebre sat fortiter punctato, lineâ mediâ impunctatâ. Long. corp. 2½ lin.

Mas: abdomine segmento 7° ventrali profunde exciso; segmento 6° late profundeque impresso.

Allied to *L. amazonicum*, but with the antennæ longer and more slender and the head and elytra differently punctured. The antennæ are yellow, very slender and elongate, considerably longer than head and thorax; 3rd joint much longer than 2nd. Head rather long, its sides parallel, quite as broad as the thorax, the sides and back densely and finely rugosely punctured, the front more sparingly and coarsely punctured, a narrow part in the middle impunctate. Thorax reddish, much longer than broad, moderately closely and rather coarsely punctured, a line down the middle impunctate. The elytra are about as long as the thorax, finely and not closely punctured. The hind body is dusky red, with the 5th and 6th segments obscurely darker. The legs are yellow.

The male has a rather deep notch at the extremity of the underside of the 7th segment of the hind body; the 6th segment has a broad and deep impression; this is very deep and well defined at the front part, and the bottom of it is smooth and membranous.

Tapajos; one ♂ specimen.

13. *Lathrobium Batesi*, n. sp. Parallelum, castaneum, nitidulum, antennis pedibusque testaceis; capite piceo, vertice et angulis posterioribus dense subtiliter, disco et fronte parcius fortiter punctatis; prothorace crebre subtiliter punctato, lineâ mediâ impunctatâ. Long. corp. 2 lin.

Mas: abdomine segmento 7° ventrali apice sat profunde lateque exciso; 6° utrinque impressione profundâ.

Allied to *L. amazonicum*, but much smaller and more finely punctured. Antennæ yellow, longer than head and thorax; 3rd joint considerably longer than 2nd, 4th about as long as 2nd. Head quite as broad as the thorax, pitchy, the hind angles and vertex densely and very finely punctured, opaque, the front and middle more sparingly and distinctly punctured, shining. Thorax longer than broad, nearly as broad as the elytra, shining reddish, the sides rather closely and finely punctured; a broad line down the middle impunctate. Elytra rather longer than the thorax, similar to it in colour, finely and not closely punctured, the punctures arranged in lines at the base. Hind body closely and finely punctured. Legs yellow.

The male characters are peculiar; the 7th ventral segment has a rather deep notch in the middle at the extremity; the 6th segment has on each side the middle, near the base, a large deep fovea or impression; the trochanters are peculiarly formed, their hind margin is concave, its apical angle acuminate, and they are externally obliquely truncate.

Tapajos; eight individuals, four of each sex.

14. *Lathrobium minor*, n. sp. Parallelum, testaceum, nitidulum, capite fortiter parcius, angulis posterioribus dense obsolete punctato; thorace crebre subtiliter punctato, lineâ mediâ impunctatâ. Long. corp. 1¾ lin.

Mas: abdomine segmento 7° ventrali apice exciso, segmento 6° emarginato.

Closely allied to *L. Batesi*; paler in colour and rather broader than that species, with the head less densely punctured, and the hind margin of the 6th segment of the hind body emarginate in the male. Antennæ rather longer than head and thorax, yellow; 3rd joint longer than 2nd, 4—10 each a little shorter than the preceding one. Head as broad as the elytra, rather darker than the rest of the insect, the hind angles densely and indistinctly punctured, the disc almost impunctate, the other part more distinctly and sparingly punctured. Thorax rather longer than broad, a little narrower than the elytra; it is of a shining-yellowish colour, with a broad line along the middle impunctate, the sides not altogether finely punctured. The elytra are shining yellow, longer than the thorax, sparingly and finely punctured, the punctures arranged in rows except at the extremity. Legs pale yellow.

The male has a moderately large notch in the middle of the hind margin of the 7th ventral segment, and the hind margin of the 6th segment is also slightly emarginate in the middle.

Tapajos; four individuals, two of either sex.

15. *Lathrobium simplex*, n. sp. Angustulum, testaceum, nitidulum, capite disperse punctato; thorace dorso biseriatim punctato, lateribus sat crebre punctatis; elytris thorace paulo longioribus, seriatim, minus distincte, punctatis; abdomine crebre subtiliter punctato. Long. corp. 2 lin.

Mas latet.

Broader, but only a little longer, than *Lathrobium longipenne*. Antennæ yellow, rather longer than head and thorax, rather stout; 3rd joint a little longer than 2nd. Head dark yellow, shining, the front part sparingly punctured, a space across the middle impunctate, the vertex more closely punctured. Thorax a little longer than broad, nearly straight at the sides, very slightly narrowed behind, on each side the middle with a row of fine punctures, which towards the base are placed in a depression; these rows are separated by a broad impunctate central space, and outside them the surface is sparingly punctured. The elytra are slightly longer than the thorax; they are shining yellow, scarcely lighter in colour than the thorax; their punctuation is indistinct, consisting of four or five rows of obsolete punctures. The hind body is broad in comparison with the front parts; it is reddish in colour, finely and moderately closely punctured, and finely pubescent, scarcely shining; the legs are pale yellow.

Tapajos; three female individuals.

16. *Lathrobium chloroticum*, n. sp. Pallide testaceum, nitidulum, capite parcius fortiter punctato, medio impunc-

tato; thorace subtiliter punctulato; elytris fere impunctatis. Long. corp. 1½ lin.

Smaller than *L. longulum*, parallel, shining yellow. Antennæ about as long as head and thorax. Head quite as broad as the elytra, straight at the sides and rather long, the sides sparingly and rather strongly punctured, the middle part impunctate; it is rather darker in colour than the rest of the insect. Thorax rather narrower than the elytra, longer than broad, shining yellow, with a line of very fine punctures along each side of the middle, and some other extremely fine punctures about the sides. Elytra rather longer than the thorax, pale yellow, scarcely visibly punctured. Hind body scarcely punctured.

Ega; one specimen (I believe a ♀).

The shining, almost impunctate, hind body renders this a very easily distinguished species.

17. *Lathrobium necatum*, n. sp. Pallide testaceum, nitidulum, minus elongatum, antennis brevibus, abdomine subtiliter minus crebre punctato. Long. corp. 1¼ lin.

Mas: abdomine segmento 7º ventrali apice excisione parvâ triangulari.

The smallest species of the genus I have seen. Antennæ short, yellow; 2nd joint scarcely longer than 3rd, the following joints bead-like, little longer than broad. Head shining yellow, sparingly punctured, with the middle part impunctate. Thorax about as long as broad, shining yellow, with two rows of fine punctures along the middle; the sides sparingly punctured. Elytra scarcely longer than the thorax, pale shining yellow, almost impunctate. Hind body very finely, sparingly punctured, the apical segments more sparingly than the basal ones. Legs pale yellow.

The male has a small notch at the extremity of the 7th ventral plate of the hind body; this notch is quite pointed in front.

Tapajos; eight individuals.

This species is smaller than *L. chloroticum*, and is readily distinguished by the much less elongate form of the front parts.

18. *Lathrobium deletum*, n. sp. Rufo-testaceum, elytris

basi infuscatis, anterius nitidulum; capite parcius punctato; thorace oblongo, dorso biseriatim punctato, lateribus antice parce punctatis; elytris thorace paulo longioribus, parcius seriatim minus distincte punctatis; abdomine dense punctato, fere opaco. Long. corp. 1¾ lin.

Mas: abdomine segmento 7° dorsali apice medio excisione parvâ, ante hanc impressione parvâ; 6° medio margine posteriore semicirculariter minus profunde exciso, utrinque angulato.

Allied to *L. rufo-partitum*, Fairm., but smaller, and with the head and hind body pale. Antennæ yellow, about as long as head and thorax; 3rd joint slightly longer than 2nd. Head rather small, reddish-yellow, shining, sprinkled with rather coarse punctures, which are denser on the vertex than elsewhere; the middle part free from punctures. Thorax slightly longer than broad, straight at the sides, not narrowed behind, along the middle with two rows of eight or ten fine punctures, and on each side, near the front, with some other punctures. Elytra a little longer than the thorax, the basal half or more infuscate, but the extreme base a little paler than the middle; along the suture each has a series of fine punctures, and near the side three other series of obsolete punctures. Hind body reddish, densely and finely punctured both on the upper and under sides. Legs pale yellow.

The male has a very small notch at the extremity of the 7th ventral segment, and in front of this a very small impression; the 6th segment has a semicircular notch in the middle of the hind margin, and on each side of this forms a well-marked angle.

Tapajos; one ♂, four ♀ individuals; Ega, a single female.

19. *Lathrobium integrum*, n. sp. Rufo-testaceum, nitidulum, elytris basi obscurioribus; thorace oblongo, dorso biseriatim punctato, lateribus antice parce punctatis; elytris thorace paulo longioribus, seriatim, parce, subtiliter punctatis; abdomine supra dense punctato, apicem versus nitidulo, subtus crebre minus subtiliter punctato, nitidulo. Long. corp. 2 lin.

Mas latet.

This species is extremely closely allied to *L. rufo-partitum*, but has the head and hind body paler, and the antennæ not in the least infuscate in the middle. It is also closely allied to *L. deletum*, but is rather larger, and has the elytra longer, and the underside of the hind body more coarsely punctured and more shining.

Tapajos; a single female.

20. *Lathrobium pictum*, n. sp. Rufescens, nitidulum, capite, elytrorum parte basali, abdominisque apice summo infuscatis, pedibus testaceis; capite parcius punctato, disco lævi. Long. corp. 2 lin.

Mas latet.

Antennæ nearly as long as head and thorax, yellow, with the middle joints darker; 3rd joint slightly longer than 2nd. Head rather short and broad, reddish, but infuscate, sprinkled with rather coarse punctures, which leave a space across the middle free. Thorax shining red, oblong, longer than broad, a little rounded at the sides, with two rows of about seven punctures along the middle, and also with a few punctures on each side, near the front part. Elytra about as long as, but distinctly broader than the thorax; their apex pale yellow, the base reddish, along the middle of each a dark patch; shining, very obsoletely and sparingly punctured, the punctures consisting of a sutural series and some extremely indistinct serial punctures near the sides. Hind body reddish, a little dilated in the middle, closely and finely but a little roughly punctured, a little shining, the apical segment infuscate. Legs pale yellow. Underside of head sparingly and finely punctured, very shining; underside of hind body rather closely and somewhat coarsely punctured, but shining.

Amazons; a single female, without special locality.

This insect bears an extreme resemblance to *L. integrum*, but has the head shorter in proportion to the width, and the hind tarsi considerably shorter.

21. *Lathrobium hilare*, n. sp. Rufo-testaceum, nitidulum, elytrorum parte basali capiteque nigricantibus; capite disperse fortiter punctato, medio absque punctis; elytrorum apice pedibusque flavis. Long. corp. 2 lin.

Mas: abdomine segmento 7° ventrali apice medio excisione minus profundâ, ante excisionem impressione parvâ; 6° medio semicirculariter minus profunde exciso, utrinque angulato.

Antennæ moderately long, reddish-yellow, the middle joints a little more obscure; 3rd joint a little longer than 2nd. Head red, suffused with black, the neck red; it is shining, and has a good many rather coarse punctures on the upper surface, which become more sparing in the middle, so that the disc is free; on the under surface it is also rather coarsely punctured. Thorax shining reddish, a little rounded at the sides, and slightly narrowed behind; along the middle with two rows of about eight punctures,— these rows a little approximate behind; also on each side towards the front with a few punctures. Elytra slightly longer than the thorax, the larger basal half blackish, the smaller apical half pale yellow, shining, with a sutural row of punctures and three rows near the side, which are but indistinct, and become quite obsolete before the extremity. Hind body red-yellow, rather closely and finely punctured, but still a little shining, the extremity a little infuscate. Legs pale yellow. Under face of hind body more coarsely punctured than the upper, and distinctly shining.

Amazons; a single specimen, without special locality.

The species is closely allied to *L. deletum*, but has the head and basal portion of elytra darker; the antennæ are darker in the middle, and have the apical joints less elongate; the thorax is less parallel at the sides; the hind body not quite so densely and finely punctured, and so more shining. The male characters are very similar in the two species. From *L. pictum* it may be distinguished by the coarser punctuation of the underside of the head.

22. *Lathrobium nanum*, n. sp. Rufo-testaceum, elytrorum parte basali infuscatâ, nitidulum; capite disperse, crebre æqualiter punctato; thorace dorso biseriatim punctato, lateribus præsertim anterius punctatis; abdomine dense subtiliter punctato. Long. corp. 2½ lin.
Mas latet.

Antennæ about as long as head and thorax, the intermediate joints scarcely darker; 3rd joint a little longer than 2nd. Head rather large, reddish-yellow, shining, distinctly, moderately closely punctured, the punctures finer on the vertex than in front, and almost as close on the disc as elsewhere; its under surface rather coarsely punctured. Thorax distinctly longer than broad, very little rounded at the sides and scarcely narrowed behind; along the middle with two rows of close punctures, twelve to sixteen in each row,—the punctures in these rows a little irregular or double; the sides rather sparingly punctured, the punctures being almost wanting towards the base.

Elytra slightly longer than the thorax, shining, the apical half yellow; in front of this infuscate, but the extreme base reddish; along the suture with a row of punctures, and towards the side with three or four other indistinct rows, which become obsolete before the extremity. Hind body reddish, densely and very finely punctured, dull, the punctuation of its under face similar to that of the upper. Legs pale yellow.

Tapajos; four females.

This species greatly resembles the four preceding ones (*deletum*, *pictum*, *integrum* and *hilare*), but is a little larger, and may be easily enough distinguished by the more even distribution of the punctures on the head.

23. **Lathrobium glabrum**, n. sp. Nigrum, nitidum, fere lævigatum; antennis, pedibus, elytrorumque apice summo pallidis; capite parce punctato, disco late impunctato; thorace dorso subtiliter biseriatim punctato, abdomine minus crebre punctato, nitido. Long. corp. 2 lin.

Mas latet.

Antennæ rather stout, moderately long, very slightly thickened at the extremity, yellow, with the middle joints obscure; 2nd and 3rd joints about equal to one another. Palpi yellow; mandibles red. Head rather large, a little broader than the thorax, black and shining, the front, sides and vertex with a few punctures. Thorax very shining, nearly black, longer than broad, nearly straight at the sides and hardly narrowed behind; along the middle with two rows of five or six slightly impressed punctures, and with a few other fine punctures at the sides. Elytra about as long as, but distinctly broader than the thorax, shining black, with a small portion at the extremity yellow; along the suture with a row of about eight punctures, and towards the side with a few other remote punctures, forming three indistinct series. Hind body black and shining, only with a short and scanty pubescence; the apical segments sparingly punctured; the basal segments at their base more closely punctured. Legs pale yellow; coxæ pitchy; hind tarsi slender.

Ega; a single specimen, which I believe to be a female.

24. **Lathrobium politum**, n. sp. Nigrum, nitidum, fere lævigatum; antennis, pedibus, elytrorumque apice pallidis;

thorace dorso biseriatim obsolete punctato, lateribus leviter rotundatis; elytris basi punctis paucis. Long. corp. 1⅔ lin.

Mas latet.

Antennæ reddish, rather stout, the middle joints a little obscure, 2nd and 3rd about equal. Palpi and mandibles yellow. Head broad, distinctly broader than the thorax, with a few sparing and fine punctures, which are wanting on the disc. Thorax considerably narrower than the elytra, longer than broad, a good deal rounded at the sides, and a little narrowed behind; on each side the middle with a row of about five indistinct punctures, with a very few indistinct punctures outside these near the front. Elytra a little longer than the thorax, shining black, with the extremity pale yellow; the sutural series of punctures indicated only by one or two at the base, and the lateral series quite as indistinct, or more so. Hind body slender and shining, the segments finely punctured in their basal portion. The under surface pitchy; legs pale yellow, with the coxæ pitchy.

St. Paulo; a single female.

This species is closely allied to *L. glabrum*, but is smaller, has the head shorter, the thorax less parallel at the sides, and the elytra more sparingly and indistinctly punctured.

25. *Lathrobium pumilum*, n. sp. Nitidulum, breviusculum, rufo-testaceum, elytrorum apice nigro, obsolete punctulatum; antennis brevibus. Long. corp. 1 lin.

Antennæ yellow, rather stout, a little thickened towards the extremity; 1st and 2nd joints thick, 2nd shorter than 1st, 3rd much shorter and more slender than 2nd, 4th about as long as broad, 8—10 rather strongly transverse, 11th acuminate. Palpi yellow, 3rd joint of maxillary large. Head as broad as the thorax, shining reddish-yellow, impunctate along the middle, at the sides sparingly and obsoletely punctured. Thorax about as broad as the elytra, about as long as broad, shining yellowish-red, sparingly and scarcely visibly punctured. Elytra about as long as the thorax, shining reddish-yellow at the base, black at the apex, sparingly and very indistinctly punctured. Hind body reddish, with the basal portion of the 6th segment infuscate, scarcely visibly punctured. Legs rather short, yellow; tarsi short, anterior only moderately dilated; basal joint of hind tarsus rather longer than 2nd, 2—4 short and about similar to one another.

Rio Madeira, 25th May, 1874; a single individual, which I believe to be a female, found by Dr. Trail; it was attracted by light.

DOLICAON.

This generic name was first applied by Laporte to a large Staphylinid from the Cape of Good Hope, and Erichson afterwards included under it some European insects very different in appearance from the South African species above alluded to. The genus now comprises over twenty species, most of which are from the Mediterranean area, with one or two from India and Australia. The insect I here describe as *Dolicaon distans* is very different in appearance from any of the forms hitherto included in the genus, though in its structure it appears to be rather similar to the *Dolicaon lathrobioides*, from the Cape of Good Hope. As the genus already contains species very different in appearance, some of which will probably be grouped as distinct genera, there is no harm in my adding to their number a distinct South American form, which appears to offer all the recorded characters of the genus. This insect, as I have above remarked, has a peculiar facies, which at first reminds one of the genus *Œdichirus*, and I should not feel at all surprised if it ultimately prove to mimic or resemble some Amazonian species of that group.

1. *Dolicaon distans*, n. sp. Angustulum, nigrum, thorace piceo, elytrorum apice rufo, pedibus testaceis, antennis fusco-testaceis; thorace biseriatim punctato; elytris hoc brevioribus, fortiter seriatim punctatis; abdomine apicem versus dilatato, crebre subtiliter punctato. Long. corp. 2¾ lin.

Antennæ moderately long, not thickened towards the extremity, reddish at the base, the other joints infuscate; 3rd joint long, rather longer than 2nd, 4—10 each shorter than its predecessor, 10th longer than broad, 11th much acuminate. Palpi reddish. Head black, broader than the thorax or elytra, shining, rather coarsely but not closely punctured, the punctures becoming less numerous towards its middle. Thorax pitchy or dark reddish, rather longer than broad, a little narrowed behind, all the angles rounded and indistinct; along each side of the middle with a series of six or seven punctures, and also outside these sparingly and irregularly punctured. Elytra very small, shorter than the thorax, black and shining, with the hind margin broadly yellowish; on each is three series of coarse punctures, and a few punctures external to these; these series are abbreviated, especially the external ones, and the sutural one is placed in a depression. Hind body a good deal dilated towards the extremity, closely and finely punctured, dull, with a fine greyish pubescence. Legs yellow, with the coxæ pitchy reddish; the front tarsi only moderately dilated, hind tarsi rather long, 1st joint twice as long as 3rd, 2nd intermediate in length between the two.

A single female found by Dr. Trail on the 3rd November, 1874, but no locality mentioned.

Scopæus.

This is another widely distributed genus, and one of which only two or three species have as yet been described from South America. Nevertheless, it is probable that species of it are numerous there, and I here describe seven. Of these seven the three last, viz., *S. distans*, *S. laxus* and *S. lævis*, depart widely in facies from the ordinary species of the genus, and suggest to one greatly, at first sight, our European *Tachyusa ferialis;* the polished surface, elongate and loosely articulated form, and the greatly developed legs, distinguish these species from the ordinary forms of the genus. As, however, they possess the tricuspidate ligula, which is so characteristic a mark of *Scopæus*, as well as all the other characters mentioned in systematic works as distinctive of the genus, I have not thought it advisable to establish a new genus for them. A kindred form has, indeed, been already described by Erichson as a *Scopæus;* at least, I suppose from his description of *S. pulchellus*, from Columbia, that it pertains to the same group as the species in question. I have also some other species allied to these insects from Rio de Janeiro. The *S. chloroticus* is also a very peculiar form, and one which may ultimately give rise to the establishment of a separate genus, which, to judge from facies, would probably be as much allied to *Lathrobium* as to *Scopæus*.

1. *Scopæus tarsalis*, n. sp. Rufescens, sat nitidus, parcius obsolete punctatus, elytris fusco-rufis; prothorace lato, obsolete punctato, medio canaliculâ brevissimâ; abdomine basi angustato, subtilissime punctato; tarsis brevibus, validis. Long. corp. 1¾ lin.

♂. Antennæ elongatæ, longer than head and thorax, reddish-yellow; 1st joint rather stouter than those following, quite as long as 2nd and 3rd together, 3rd joint a little shorter than 2nd, 4—10 each slightly shorter but not broader than its predecessor; 11th joint acutely pointed, a little longer than 10th. Labrum with four almost equidistant teeth in front, a little emarginate between the two middle ones. Mandibles each with three acute teeth in the middle. Head dark reddish, very finely and indistinctly punctured, about as broad as the elytra; the extreme vertex in the middle with a short, deep, fovea-like channel, and on either side slightly emarginate; the front part of the head with two large, ill-defined elevations. Thorax distinctly narrower than the elytra, the greatest width about one-fifth from the front, from thence abruptly narrowed to the neck, and slightly narrowed towards the rounded base; extremely finely and indistinctly punctured, shining, in the middle with a short impression. Elytra a little longer than the thorax, infuscate-red, distinctly impressed at the scutellum, extremely finely and indistinctly, and not densely punctured, a little shining. Hind body reddish-yellow, distinctly dilated towards the extremity, densely, very finely and indistinctly punctured. Legs yellow; tarsi short and stout, the front pair very broad.

In the male, segments 2—5 of the hind body are on the underside distinctly impressed in the middle, the 6th segment is nearly simple, the 7th has a deep narrow notch; and it is probable that the elevations on the front of the head are peculiar to the male sex.

Tapajos; a single individual.

2. *Scopæus ornatus*, n. sp. Dense, subtilissime punctatus, opacus, rufescens; pedibus testaceis; antennis articulis quatuor ultimis albidis; elytris fuscis, apice testaceis; abdomine lato, basi angustato, apicem versus infuscato. Long. corp. 1¾ lin.

Antennæ about as long as head and thorax, reddish, with the four apical joints white; 1st joint distinctly stouter than the others, nearly as long as 2, 3 and 4 together; 3rd scarcely so long as 2nd, 10th quite as long as broad. Labrum with a small notch, in the middle on either side of which is a prominent spine; the inner side of this spine is dilated. Mandibles pale red, elongate, irregularly toothed, the left one with a broad, only little prominent tooth in the middle, and between this and the base with a very minute tooth, and above the middle tooth with a small sharp tooth; the right one with three small approximate teeth in the middle, the upper one of which is very obsolete. Head reddish, broad, rather broader than the elytra, the clypeus in front distinctly impressed on each side; the hind angles much rounded, the surface extremely densely, finely and indistinctly punctured. Thorax rather longer than broad, its greatest breadth in front of the middle, greatly narrowed to the front, and a good deal narrowed towards the base; extremely finely carinate along the middle, the carina being only distinctly visible near the base, in consequence of the surface there being a little flattened or depressed on either side; colour and punctuation similar to the head. Elytra rather longer than the thorax, fuscous with the apex yellow, their punctuation extremely dense and fine. Hind body broad, but a good deal contracted at the base, reddish but infuscate towards the extremity, very densely, finely and indistinctly punctured. Legs yellow; tarsi rather stout but elongate, the hind ones being quite half as long as the tibiae.

In the male, segments 3—6 of the hind body are on the underside impressed along the middle, the hind margin of the 6th segment is broadly emarginate; the hind margin of the 7th segment is also broadly emarginate, and in the middle there is also a small notch.

Tapajos; two males and one female.

3. *Scopæus pauper*, n. sp. Angustulus, subparallelus, rufescens, subtilissime vix perspicue punctulatus, subopacus; vertice emarginato, medio foveolato; pedibus brevibus, validis. Long. corp. 1 lin.

Of narrow form, and almost unicolorous pale-reddish colour. Antennae short; 3rd joint small, a good deal smaller than the small 2nd joint, joints 3—6 differing little from one another; 7—10 each very slightly broader than its predecessor, and shorter than broad; 11th joint short. Head rather long and narrow, slightly broader than the thorax, and about as broad as the elytra, nearly straight at the sides; the vertex distinctly emarginate, and with a small distinct fovea in the middle, the surface extremely finely and densely (but not quite so indistinctly as the head and thorax) punctured; the eyes small. Thorax longer than broad, distinctly narrower than the elytra; its greatest breadth much in front of the middle, much narrowed to the front, but only slightly towards the base; the surface very obsoletely punctured, so as to be a little shining, with faint indications of two foveæ at the base in the middle. Elytra only slightly longer than the thorax, extremely finely and indistinctly punctured, depressed at the scutellum. Hind body very finely and indistinctly punctured, a little dilated towards the extremity. Legs yellow, short and stout; the tarsi short, the anterior ones particularly short and broad.

Tapajos; a single female.

Obs.—I have not been able to examine the mandibles and labrum of this obscure little species; so far as general appearance goes, it may be said to be closely allied to the European *S. minimus*.

4. *Scopæus chloroticus*, n. sp. Pallide testaceus, angustulus, subparallelus, subnitidus, minus pubescens; thorace lateribus parallelis, angulis anterioribus rotundatis, crebre subtiliter punctato, lineâ latâ mediâ lævigatâ; elytris thorace paulo longioribus, albidis, vix perspicue punctulatis, nitidulis. Long. corp. $\frac{7}{8}$ lin.

Antennæ yellow, shorter than head and thorax, slightly thickened towards the extremity; 1st joint stouter than 2nd, rather longer than 2nd and 3rd together; 2nd joint short; 3rd joint rather shorter and narrower than 2nd; 4—7 bead-like, differing little from one another, 8—10 transverse; 11th joint short. Head rather long and narrow, the sides about parallel, the vertex nearly straight, the angles much rounded; the surface obsoletely and not densely punctured, and with a broad longitudinal line along the middle smooth. Thorax a good deal longer than broad, the sides parallel, the front angles rounded in a gentle curve continuous with the front; the surface obsoletely and not densely punctured, with a broad space along the middle smooth. Elytra distinctly broader, and a little longer, than the thorax, very pale yellow, their punctuation scarcely visible. Hind body parallel, densely and indistinctly punctured, more opaque than the front parts. Legs pale yellow, short and stout; the tarsi short, the anterior ones broad. Under surface of head impunctate, with two parallel longitudinal lines along the middle.

Tapajos; a single female.

Obs.—This minute species is peculiar, and probably generically distinct from the ordinary *Scopæi*; the form of the front angles of the thorax is dissimilar from what is usual in *Scopæus*; the general appearance is much that of an extremely minute *Lathrobium*, but, as the structure of the tarsi is like that of *Scopæus*, it may be placed in that genus till its characters can be more fully ascertained.

5. *Scopæus distans*, n. sp. Rufo-testaceus, nitidus, fere impunctatus, parcius setosus; antennis apice pallidioribus, elytris disco abdomineque apice obscurioribus; pedibus elongatis, tarsis gracilibus. Long. corp. 2 lin.

Narrow and elongate in form. Antennæ about as long as head and thorax, a little thickened towards the extremity, reddish, with the four or five apical joints pale yellow; 3rd joint elongate, a good deal longer than 2nd; 10th joint a little longer than broad. Head rather long and narrow, about as broad as the elytra, entirely rounded at the vertex; the surface shining reddish-yellow, impunctate, with some upright black setæ, and with a fine and scanty yellow pubescence. Mandibles each with three large sharp teeth in the middle. Thorax elongate and narrow, much narrower than the elytra, very convex, the greatest width in front of the middle, and thence much narrowed towards the front and a good deal towards the base; impunctate, colour and setæ as on the head. Elytra long and narrow, a little longer than the thorax, shining and impunctate, yellow, but largely dark chestnut about the middle. Hind body narrow at the base, a good deal broader towards the extremity, yellowish, with the 6th segment, except its hind margin, infuscate; its punctuation and pubescence very fine and indistinct. Legs yellow, long, and rather stout; the hind tarsi long and slender, more than half the length of the tibiæ.

In the male the ventral plate of the 7th segment of the hind body has a broad notch or emargination at the extremity.

Tapajos; several individuals.

6. *Scopæus laxus*, n. sp. Rufo-testaceus, nitidus, fere

impunctatus, parcius setosus; antennis apice pallidioribus, elytrorum disco abdominisque apice obscurioribus. Long. corp. 1¾ lin.

This insect so extremely resembles *S. distans*, that to describe it would be in most points to repeat the description of that species; it is, however, rather less elongate in all its parts, so that the 10th joint of the antennæ is hardly as long as broad, and the hind tarsi are distinctly shorter when compared with those of *S. distans*.

In the male the hind margin of the ventral plate of the 6th segment of the hind body is broadly but not deeply emarginate at the extremity, and the 7th has a small notch, and the ædeagus itself is considerably shorter than in *S. distans*.

Tapajos; six male, two female individuals.

7. *Scopæus lævis*, n. sp. Rufo-testaceus, nitidus, fere impunctatus; elytris abdominisque apice nigricantibus, femoribus quatuor posterioribus apicem versus leviter infuscatis. Long. corp. 1½ lin.

Closely allied to *S. distans* and *laxus*, but considerably smaller. Antennæ rather short, a good deal shorter than head and thorax, a little thickened towards the extremity, reddish, with the apical joints a little paler than the middle ones; 3rd joint shorter than 2nd, 8—10 not so long as broad. Head broader than the thorax and as broad as the elytra, rounded at the sides, but with the vertex a little truncate, the surface shining red, without sculpture. Thorax longer than broad, very convex, a good deal narrower than the elytra, much rounded at the sides, being greatly narrowed towards the front and a good deal towards the base, shining red, impunctate. Elytra scarcely longer than the thorax, shining, blackish, without sculpture. Hind body narrow at the base, a good deal dilated towards the extremity, the basal segments reddish, the apical ones blackish; the surface very finely and indistinctly punctured. Legs yellow, the outer portion of the four posterior femora slightly infuscate; the hind tarsi slender and long, a good deal more than half the length of the tibiæ.

Amazons; a single female, without more special locality.

LITHOCHARIS.

The species of this widely-distributed genus are nowhere more numerous than in South America; thirteen species have already been described by Erichson from Columbia, so that it is not surprising that I should here describe twenty-two species from the Amazons. Among these twenty-two species there is sufficient variety in structural points to render it probable that some of them will ultimately be referred to new genera. The *L. munda* bears considerable resemblance in general appearance to a *Scopæus*, and it is probable that this resemblance is indicative of a real affinity. The five species (*L. oculata, quadrata, egena, humilis* and *ardua*) with setose antennae, the two basal joints of which are stout, the others slender, may also perhaps form a distinct genus; indeed, Kraatz has already founded a genus (*Thinocharis*) for some Ceylon species possessing this structure of the antennæ; but, as it is doubtful whether the South American species I am alluding to are really congeneric with the Eastern *Thinocharis*, and as Erichson has already described as members of the genus *Lithocharis* several Columbian species with similarly-formed antennæ, I have preferred associating the new forms here described with the cognate forms from a neighbouring locality.

L. discedens and *L. connexa* are distinguished by a peculiarity of structure of the 4th joint of the hind and middle tarsi, and are probably closely related to the Columbian *L. biseriata*, Er.

The most peculiar of the new species I here describe are the four I have placed at the end of the genus, viz., *L. polita, germana, pagana* and *picta*. These four species I anticipate will be found to be closely allied to *L. macularis* and *L. angularis*, Er., from Venezuela and Columbia. Mr. Solsky, who has in the Hor. Soc. Ent. Ross. (v. p. 142, pl. iv.) described and figured the trophi of *Dacnochilus lætus*, Leconte, has suggested that Erichson's *L. angularis* should be placed in that genus; and in the Munich Catalogue this has been done. On comparing the parts of the mouth of the species here described with Solsky's figures, I find them to be far from agreeing therewith; the labrum in the four species I describe possesses an acute stout tooth on either side of the central notch, while in Solsky's figure the lobes are quite destitute of this; the 3rd joint of the maxillary palpi is more slender than in Solsky's figures, and the 4th joint, which is concealed within the 3rd, appears to be much more slender and acuminate; the last joint of the labial palpi is more slender, quite cylindrical, and only about half as long as the preceding joint. I am, therefore, unable to consider these species as congeneric with the North American *Dacnochilus lætus*, and prefer to place them for the present in the genus *Lithocharis* rather than establish a new genus for them.

Erichson describes the *L. macularis* and *L. angularis* as possessing a labrum similar in structure to the four species I have here described, and I have therefore great doubts whether the *L. angularis* is correctly referred to *Dacnochilus*. I have another closely allied species of the group from Rio de Janeiro.

1. *Lithocharis latro*, n. sp. Lata, depressa, fusco-ferruginea, antennis rufis, pedibus testaceis, dense subtilissime punctata, opaca; elytris dilutioribus, thorace paulo longioribus; tarsis anterioribus dilatatis. Long. corp. 3½ lin.

Antennæ rather slender, reddish, 1½ lin. in length, not at all thickened towards the extremity; 3rd joint distinctly longer than 2nd, 5—10 each slightly shorter than its predecessor, 10th longer than broad. Palpi red, last joint elongate and slender. Labrum red, with a single, short, obscure tooth in the middle, on either side of which it is a little emarginate. Mandibles red, the left with three, the right with four, distinct teeth. Head large, slightly broader than the thorax, the hind angles slightly produced, so that the vertex is a little emarginate, very densely and finely punctured. Thorax very nearly as long as broad, fully half a line in length, the sides a little narrowed behind, very densely and finely punctured, with a very narrow indistinct line along the middle. Elytra ⅝ lin. in length, taken together rather broader than long, paler in colour than the head and thorax, densely and finely punctured. Hind body broad, very densely and finely punctured. Legs yellow; front tarsi moderately dilated.

In the male the ventral plate of the 6th segment of the hind body is rounded in the middle and emarginate on each side; the ventral plate of the following segment has a very broad and deep excision in the middle.

Ega; a single specimen.

Obs.—This species much resembles the Eastern *L. staphylinoides*, but has very different male characters. *L. hepatica*, Er., from Columbia, appears to be an allied species, and I have another closely allied, but distinct, species from Rio de Janeiro.

2. *Lithocharis simplex*, n. sp. Rufescens, capite, elytrorum apice, abdomineque apicem versus infuscatis, pedibus testaceis, dense subtilissime punctata; prothorace quadrato, lineâ mediâ impunctatâ. Long. corp. 1⅔ lin.

Antennæ moderately long, rather stout; 3rd joint about equal to 2nd, 5—9 each slightly broader and shorter than its predecessor, 10th about as long as broad. Mandibles, palpi and labrum reddish. Head infuscate, large, quite as broad as the thorax, densely and extremely finely punctured. Thorax not quite so long as broad, straight at the sides, densely and very finely punctured; rather lighter than the head in colour, with a smooth impunctate line along the middle. Elytra a good deal longer than the thorax and distinctly broader, reddish at the base, infuscate towards the extremity; densely and finely punctured, but the punctuation more distinct than on the head and thorax. Hind body reddish, with the penultimate segments infuscate, densely and finely punctured. Legs yellow.

I do not know the male of this species, but in the female the anterior tarsi are not in the least dilated.

St. Paulo, three individuals; Ega, one individual.

This species is, in form and sculpture, closely allied to our European *L. ochracea*, but it has the labrum quadridenticulate, it being furnished in the middle with four short, stout, not very distinct teeth; the mandibles are very stout at the base, and beyond the stout basal portion, are armed, the left with one, the right with two teeth.

3. *Lithocharis condita*, n. sp. Piceo-testacea, antennis rufis, pedibus testaceis, elytris obscure testaceis, dorso infuscatis, dense subtilissime punctulata, subopaca; prothorace mediâ lineâ obscurâ glabrâ; elytris thorace longioribus; tarsis anterioribus simplicibus. Long. corp. 1½ lin.

Antennæ moderately long, rather stout, reddish; 3rd joint about as long as 2nd; 10th joint distinctly broader than the 4th, about as long as broad. Mandibles red, short and robust, their teeth small. Head pitchy, densely and very finely punctured, very delicately pubescent, scarcely shining. Thorax about as long as broad, scarcely broader than the head, and distinctly narrower than the elytra; densely and extremely finely punctured, so that an impunctate line along the middle is very indistinct, the colour scarcely paler than that of the head. Elytra a good deal longer than the thorax, obscure testaceous, with the disc a little infuscate, closely and finely punctured. Hind body very densely and finely punctured, quite opaque. Legs yellow; front tarsi not in the least dilated, hind tarsi with the basal joint a good deal longer than the 2nd.

The male characters are very slight; the ventral plate of the 7th segment of the hind body being a little emarginate at the extremity in that sex.

St. Paulo; three individuals.

This species appears to be closely allied to the common and widely distributed *L. ochracea*, but is so much smaller that it cannot be confounded therewith. *L. infuscata*, Er., from Columbia, is probably a very close ally of this species, but according to Erichson's description differs in the colour of the hind body.

4. *Lithocharis diffinis*, n. sp. Rufescens, capite, elytrorum apice abdomineque apicem versus plus minusve infuscatis, pedibus testaceis, dense subtilissime punctata; prothorace quadrato, lineâ mediâ impunctatâ. Long. corp. 2 lin.

Similar in size, form and punctuation to *L. ochracea*, but with the head larger, it being almost broader than the thorax; rather larger than *L. simplex*, and with the labrum furnished in the middle with two obscure distant teeth; in other respects it appears to agree exactly with *L. simplex*.

The 7th segment of the hind body has the hind margin of the ventral plate quite simple in the male, so that I am not aware of any external character by which the sexes can be distinguished.

Ega; two specimens.

5. *Lithocharis comes*, n. sp. Rufescens, capite thoraceque obscurioribus, clytris sordide testaceis, apice dilutioribus, pedibus testaceis, dense obsoletequepunctata;

capite thoraceque opacis, hoc lineâ mediâ impunctatâ. Long. corp. 2 lin.

Antennæ red, rather short and stout; 3rd joint scarcely longer than 2nd, 10th not quite so long as broad. Head large, slightly broader than the thorax, pitchy red; the eyes large and rather prominent, the hind angles slightly swollen, so that the vertex appears a little emarginate in the middle; the surface densely, finely and obsoletely punctured, opaque. Thorax scarcely so long as broad, straight at the sides, the front obliquely truncate on each side; the colour slightly paler than that of the head, the surface densely and very obsoletely punctured, opaque, but with a shining impunctate line along the middle. Elytra a little longer and a little broader than the thorax, of a sordid-testaceous colour, but with the apex quite pale, finely and closely punctured. Hind body pointed at the extremity, densely and finely punctured. Legs yellow.

Ega; a single specimen.

Obs.—I do not know the sex of this individual, nor have I been able to examine the labrum and mandibles; it is excessively close to *L. diffinis*, but has the head rather larger and the eyes more prominent, and the vertex more emarginate, and the extremity of the elytra paler; the front tarsi are not in the least dilated. It is similar to *L. ochracea*, but has the head larger and the front of the thorax more oblique on each side.

6. *Lithocharis sobrina*, n. sp. Fusco-rufa, opaca, antennis rufis, pedibus testaceis; capite, thorace, abdomineque subtilissime punctatis; elytris subtiliter sed magis distincte punctatis, fuscis, lateribus margineque apicali testaceis. Long. corp. 2¼ lin.

This species is extremely similar to *L. diffinis* and *L. comes*, but has the elytra differently coloured, their sides being obscurely yellow, while the sutural portion is largely infuscate. This infuscation does not, however, reach to the extremity, which is similar in colour to the lateral margins.

There are no external characters to distinguish the male.

Pará; four individuals, collected by Mr. Smith.

Obs.—Like the *L. diffinis* and *L. comes*, this insect appears to be closely allied to *L. ochracea*. It has the antennæ longer than *L. comes*, joints 4—10 being each distinctly a little longer. Notwithstanding its great resemblance to *L. diffinis*, the aedeagus is different enough in the two to make me feel sure they are quite distinct species, the appendage with which it is furnished being short and hastate in *L. sobrina*, while it is elongate and slender in *L. diffinis*.

7. *Lithocharis crassula*, n. sp. Crassiuscula, castanea, pedibus testaceis; capite thoraceque nitidulis, transversis, crebre fortiter punctatis, hoc lineâ mediâ impunctatâ; elytris thorace longioribus, crebre minus fortiter punctatis, vix nitidis; abdomine dense subtilissime punctato. Long. corp. 2 lin.

Antennæ short, reddish-yellow; 3rd joint about equal to 2nd, 4th shorter than 3rd, 5—9 differing but little from one another; the 9th and 10th, however, slightly stouter than the others, about as long as broad. Labrum with a small notch in the middle, without teeth. Mandibles with the basal part stout, each with three teeth; on the left the two upper ones very small, on the right the upper one small. Head broad and short, about as broad as the thorax, the hind margin distinctly emarginate; the surface rather coarsely and closely punctured, the punctures wanting on a space in the centre. Thorax as broad as the elytra, a good deal broader than long, scarcely narrowed behind, the front a little rounded; the surface coarsely and rather closely punctured, with a smooth line along the middle. Elytra a little longer than the thorax, moderately closely and not coarsely punctured, only a little shining. Hind body a good deal pointed at the extremity, densely and finely punctured, not shining. Legs yellow; front tarsi dilated, hind tarsi slender but not very long; 4th joint simple.

In the male the hind part of the ventral plate of the 7th segment of the hind body has a moderately large notch in the middle.

Tapajos, one male; St. Paulo, one female.

8. *Lithocharis restita*, n. sp. Rufo-castanea, breviter hirsuta, nitidula, antennis brevibus cum pedibus testaceis; capite crebre minus subtiliter punctato, vertice medio profunde impresso; thorace crebre subtiliter punctato; elytris parcius, dorso obsolete biseriatim punctatis. Long. corp. vix 2 lin.

Antennæ yellow, short, rather stout, distinctly thickened

towards the extremity; 3rd joint about equal to 2nd, 4th and 5th about equal to one another, about as long as broad, 6—10 distinctly transverse, 11th stout, pointed, about as long as the two preceding together. Labrum with two distant obscure teeth in the middle. Mandibles each with three teeth in the middle; on the left mandible the centre one is smaller, on the right mandible longer than the others. Head quadrate, rather convex, the vertex with a deep, short longitudinal impression in the middle; the surface shining, at the sides more densely and coarsely punctured than in the middle; the eyes rather small. Thorax distinctly narrower than the elytra, not quite so long as broad, distinctly narrowed behind, the front a little oblique on each side, the front angles rather prominent; the surface shining, rather finely and indistinctly punctured, with a very obsolete line along the middle. Elytra red, shining, a little longer than the thorax, rather sparingly and indistinctly punctured, impressed on each side the suture at the base, and each with two indistinct abbreviated series of punctures on the disc. Hind body rather darker than the front parts, but with the apex paler, a little contracted at the base, closely and finely punctured. Legs yellow, front tarsi only slightly dilated; underside of head coarsely and closely punctured.

Ega; a single individual, of whose sex I am in doubt, the hind body showing nothing peculiar in structure.

9. *Lithocharis integra*, n. sp. Fusco-ferruginea, antennis pedibusque rufo-testaceis, dense subtiliter punctata, opaca; prothorace quadrato. Long. corp. 2 lin.

Antennæ reddish, rather slender; 3rd joint about equal to 2nd, 4th shorter than third, 5—10 differing little from one another, 10th about as long as broad. Labrum reddish, prominent in the middle, and there with three small teeth. Mandibles and palpi red, the former each with three teeth. Head rather long, a little narrowed towards the front, a little broader than the thorax, and quite as broad as the elytra; nearly truncate behind, with the angles rounded, dull reddish in colour, finely and very densely punctured, the eyes small. Thorax a little narrower than the elytra, almost as long as broad, a little narrowed behind, the front distinctly but not greatly oblique on each side; the surface obscure reddish, finely and very densely punctured, without distinct impressions or line along the middle. Elytra distinctly longer than the thorax, and rather darker in colour, very densely, finely and indistinctly punctured, quite dull. Hind body rather slender, very densely and finely punctured. Legs yellow; front tarsi rather short, distinctly dilated; 4th joint of hind tarsi small, simple.

The male has a very slight emargination at the extremity of the ventral plate of the 7th segment of the hind body.

Ega; two individuals.

10. *Lithocharis compressa*, n. sp. Depressa, ferruginea; crebre subtiliter punctulata, subopaca. Long. corp. 1¾ lin.

Antennæ short, red, scarcely at all thickened towards the extremity; 10th joint slightly transverse. Head large and flat, quite as broad as the elytra, reddish, finely and closely punctured, the vertex distinctly emarginate. Thorax rather narrower than the elytra, rather broader than long, distinctly narrowed behind, reddish; extremely finely punctured, with an extremely fine channel near the base, and a very indistinct impression at the base on each side of the middle. Elytra a good deal longer than the thorax, closely and finely punctured, infuscate-reddish. Hind body reddish, densely and finely punctured.

Ananá; a single female taken by Dr. Trail on the 6th September, 1874.

Obs.—This species greatly resembles *L. integra*, and appears structurally closely allied thereto, but it is smaller and more depressed, and has the surface less densely punctured, so that it is less opaque.

11. *Lithocharis discedens*, n. sp. Ferruginea, elytris, antennis, pedibusque rufescentibus, opaca, dense punctata; capite thoraceque dense rugulose punctatis, hoc basi quadripunctato; elytris dense minus distincte, dorso biseriatim punctatis. Long. corp. 2 lin.

Antennæ reddish, fully ½ lin. in length; 2nd, 3rd and 4th joints about equal to one another, 5—10 each a little shorter than its predecessor, 9 and 10 scarcely so long as broad. Labrum slightly emarginate in the middle, without teeth. Mandibles red, each with three teeth; the middle tooth on the left mandible very small, on the right one longer than the others. Head broad and short, very densely and rugosely punctured, the punctures finely ocellated, the interstices very fine. Thorax almost as long as broad, narrowed behind the front, on each side very oblique, the punctuation similar to that of the head; the extreme base with four small foveæ, the middle with a short and excessively fine channel. Elytra brighter red than the head and thorax, distinctly longer than the thorax, closely punctured, each with three series of punctures; one along the suture, two along the middle, these rows not reaching to the apex. Hind body very densely and finely punctured, quite opaque. Legs reddish. Front tarsi short, moderately dilated.

Ega and Tapajos; two individuals.

Obs.—The structure of the 4th joint of the middle and hind tarsi is peculiar in this species; though narrow, this joint is on the upper side deeply bilobed. I do not know the sex of my individuals, but I think them to be males, though they have no external abdominal character to indicate this.

12. *Lithocharis convexa,* n. sp. Ferruginea, elytris rufescentibus, antennis pedibusque rufo-testaceis; capite dense rugulose punctato, opaco; thorace dense haud rugulose punctato, basi quadripunctato, punctis externis obsoletis; elytris crebre, dorso vix distincte biseriatim punctatis. Long. corp. 1¾ lin.

Antennæ reddish-yellow; 3rd joint slightly longer than the adjacent ones. Labrum not emarginate in the middle, without distinct teeth. Head very densely punctured. Thorax closely punctured, the punctures much finer than on the head, and towards the front not so dense as at the base, so that the front part is slightly shining, the extreme base with four puncture-like foveæ, the outer ones being very indistinct. Elytra finely and indistinctly, not very closely punctured; a little shining, with traces of a sutural and two dorsal series of punctures. Front tarsi distinctly dilated.

In the male the ventral plate of the 7th segment of the hind body is slightly emarginate in the middle at the extremity.

Amazons; two male individuals, without special locality.

Obs.—This species is rather smaller than *L. discedens,* to which it is closely allied; but besides some peculiarities of sculpture, which readily enough distinguish it, it has the labrum rather differently formed.

Besides the two individuals above described, I have also

from Ega a specimen rather smaller and more finely punctured and paler, which I believe to be an immature variety of *L. discedens*.

13. *Lithocharis oculata*, n. sp. Nigro-fusca, opaca, antennis, palpis pedibusque testaceis, elytris dilutioribus, angulo externo testaceo; thorace basin versus fortiter angustato, medio canaliculato; oculis permagnis. Long. corp. 1½ lin.

Antennæ yellow, slender, with the two basal joints stout; 3rd joint shorter and greatly more slender than 2nd, joints 3—10 setose. Labrum red, furnished in the middle with two distinct, approximate, stout teeth. Mandibles red, each with two large, sharp teeth. Head short and broad, broader than the thorax, eyes large and convex, reaching to within a short distance of the vertex, which is nearly truncate, the surface densely and finely punctured. Thorax not so long as broad, the front and front angles rounded, the sides much narrowed towards the base; the surface covered with fine, moderately close asperities, and along the middle with a fine channel, which is deepest in its hinder part. Elytra a good deal longer than the thorax and rather paler in colour, each one with the external angle broadly pale yellow; the surface rather closely and finely punctured, the punctuation becoming obsolete towards the extremity. Hind body very densely and finely punctured. Legs yellow; front tarsi not dilated, hind tarsi with the fourth joint small and simple.

Ega; three individuals, of whose sex I am in doubt.

14. *Lithocharis quadrata*, n. sp. Fusca, opaca, dense punctata, pedibus testaceis, antennis elytrisque ferrugineis; capite quadrato, angulis posterioribus rectis; thorace basin versus leviter angustato, medio canaliculato. Long. corp. 1¾ lin.

Antennæ reddish, the two basal joints stout, the others slender; 3rd joint distinctly longer than 4th, 4—10 differing very little from one another. Mandibles red, each with two teeth. Labrum red, with two approximate teeth in the middle. Head large, distinctly broader than the thorax, about straight at the sides, the hind angles not rounded, about right angles, the vertex a little emarginate; the surface finely and very densely punctured, opaque, the eyes reaching about half-way to the hind angles. Thorax not quite so long as broad, dis-

tinctly but not greatly narrowed behind, the front angles
obtuse and not much rounded, the colour pitchy, similar
to that of the head; the surface densely covered with fine
asperities, along the middle with a fine channel, which
does not reach the front, but is continued to the front as a
very fine, smooth line. Elytra a good deal longer than
the thorax, rusty testaceous, their sculpture similar to that
of the thorax; along the middle of each two indistinct
abbreviated lines. Hind body dusky ferruginous, very
densely and finely punctured. Legs yellow; front tarsi a
little dilated, hind tarsi with the 4th joint small and quite
simple.

Ega; a single individual, in which I perceive no external character to indicate the sex.

15. *Lithocharis egena*, n. sp. Ferruginea, dense punctata, leviter nitidula, antennis pedibusque testaceis;
thorace basin versus angustato, medio canaliculato, elytris
thorace paulo longioribus. Long. corp. 1½ lin.

Antennæ moderately long, yellow, setose; the two
basal joints stout, the rest very slender; 3rd joint distinctly longer than 4th. Head rather large, slightly
broader than the thorax, dull fuscous, or obscure red; the
hind angles a little rounded, the vertex about straight, the
surface densely, finely and indistinctly punctured, only
slightly shining. The mandibles have each two teeth in
the middle, and the labrum two distinct, contiguous
denticles in the middle. The thorax is not so long as
broad, the front is rounded, the sides much narrowed
behind, obscure red; the surface closely covered with fine
asperities, but still a little shining, the basal part with a
fine channel along the middle, which, though deep at the
extreme base, is very indistinct on the middle, and does
not reach the front. Elytra only a little longer than the
thorax, obscure reddish, slightly shining, their sculpture
similar to that of the thorax, but not quite so close.
Hind body very densely and finely punctured. Legs
yellow; the front tarsi stout, but not dilated.

Amazons; a single individual, of doubtful sex, without
special locality.

The species is smaller than *L. quadrata*, is not so dull,
and can be readily distinguished by the simple front
tarsi.

16. *Lithocharis humilis*, n. sp. Fusca, dense punc-

tata, antennis pedibusque obscure rufo-testaceis: elytris testaceis, apicem versus infuscatis; prothorace basin versus leviter angustato, medio lineâ glabrâ abbreviatâ, posterius profunde canaliculatâ. Long. corp. 1¾ lin.

Antennæ setose, obscure reddish, the two basal joints stout, the rest slender; 3rd joint distinctly longer than 4th. Labrum reddish, with two approximate denticles in the middle. Mandibles red, each with two sharp teeth in the middle. Head large, broader than the thorax, and quite as broad as the elytra; the vertex straight, the hind angles much rounded; the eyes rather large and prominent, reaching quite half-way to the vertex; the colour obscure pitchy red, the surface densely and indistinctly punctured. Thorax narrower than the elytra, about as long as broad, distinctly narrowed behind, the front and front angles much rounded, the colour obscure reddish, rather paler than the head; the surface densely covered with fine asperities, along the middle a fine glabrous line, the hind part of which is occupied by a deep and distinct channel. Elytra a little longer than the thorax, finely and not very densely punctured, yellowish, infuscate towards the extremity: along the middle of each are indications of two abbreviated impressed lines. Hind body very densely and finely punctured. Legs yellowish; front tarsi distinctly dilated, hind tarsi with the 4th joint quite simple.

Ega; a single individual, which exhibits no certain indication of sex, and is probably a female.

17. *Lithocharis ardua*, n. sp. Fusco-ferruginea, antennis pedibusque testaceis, thorace elytrisque leviter nitidulis; thorace brevi, basin versus angustato, medio lineâ abbreviatâ, glabrâ, canaliculatâ. Long. corp. 1⅓ lin.

Antennæ setose, yellow; 1st and 2nd joints stout, the others slender, 3rd not longer than 4th. Labrum large, with two distinct approximate teeth in the middle, and a little sinuate on each side of these, reddish. Mandibles reddish-yellow, rather slender, each with two rather large, sharp teeth in the middle. Head large, broader than the thorax, and quite as broad as the elytra, quadrate, the hind angles right angles and a little rounded; the vertex almost straight, the colour obscure pitchy, dull; the surface closely and indistinctly, in front obsoletely, punctured.

Thorax a good deal shorter than broad, a little narrower than the elytra, much narrowed towards the base ; the front rather rounded, the surface closely covered with distinct asperities and slightly shining ; along the middle in front of the base is a short, smooth line, in which is a fine channel; the colour is dusky reddish, a little paler than the head. Elytra a good deal longer than the thorax; their colour obscure fuscous, slightly shining, their punctuation indistinct. Hind body obscure ferruginous, very indistinctly punctured. Legs yellow; front tarsi not dilated, hind tarsi slender, with the 4th joint quite simple.

Amazons ; a single individual, without special locality; it is probably a male, as it has the ventral plate of the 7th segment of the hind body a little emarginate in the middle at the extremity.

18. *Lithocharis munda*, n. sp. Castanea, nitidula, antennis breviusculis, apicem versus leviter incrassatis; capitis lateribus posterius profunde strigosis; prothorace sat elongato, crebre minus distincte punctato, lineâ mediâ glabrâ ; elytris thorace paulo longioribus, crebre punctatis. Long. corp. 1 lin.

Antennæ reddish, short and rather stout ; 2nd and 3rd joints about equal, 6—10 each slightly broader than its predecessor, the 9th and 10th distinctly transverse ; 11th joint short, stout, obtusely pointed. Labrum rather small, in the middle with two rather stout, widely-separated teeth, and a little emarginate outside the teeth. Mandibles short, each with two small teeth. Head rather large, almost broader than the elytra, slightly narrowed in front ; the vertex a good deal emarginate, the hind angles much rounded ; on either side, behind the eyes, coarsely longitudinally strigose ; towards the front rather coarsely but indistinctly punctured, the sculpture wanting on the middle. Thorax a little narrower than the elytra, quite as long as broad, a little narrowed to the base: the front angles distinct and not rounded, in front of them abruptly obliquely narrowed to form a rather slender neck ; the surface shining, rather coarsely and closely but indistinctly punctured, with a broad line along the middle smooth. Elytra a little longer than the thorax, rather finely and indistinctly and not densely punctured. Hind body a little contracted towards the base, indistinctly punctured, slightly shining. Legs yellowish; basal joints of front tarsi a little dilated; hind tarsi slender and rather long, 4th joint simple: underside of head coarsely punctured, with only a narrow space along the middle smooth.

Tapajos; six individuals, in which I perceive no indications of external sexual characters.

19. *Lithocharis polita*, n. sp. Rufo-testacea, nitida, parce nigro-setosa, fere impunctata; prothorace subquadrato, basin versus angustato, angulis anterioribus minus rotundatis; elytris thorace paulo longioribus, maculâ longitudinali fuscâ; abdomine parce punctato, segmento sexto utrinque maculato. Long. corp. 3 lin.

Antennae reddish, scarcely longer than head and thorax, rather slender, very slightly thickened towards the extremity; 3rd joint elongate and slender, one and a half times the length of 2nd; 10th joint a good deal longer than broad; 11th joint obtusely pointed, distinctly longer but not broader than 10th, Labrum with a large notch in the middle, reaching almost to the base, the front margin projecting as a tooth on each side the notch. Mandibles short, the basal portion very stout, and at the extremity with three or four fine teeth or serrations; beyond these abruptly contracted. Maxillary palpi with the 3rd joint truncate at the extremity; the 4th joint hidden. Head broad and short, quite as broad as the thorax; the vertex nearly straight, but with a distinct emargination in the middle, shining reddish-yellow; the surface with a few scattered punctures, each bearing a black seta. Thorax in front nearly as broad as the elytra, not quite so long as broad; the sides not curved, but a good deal narrowed towards the base, the front a little rounded, the angles a little rounded; the surface shining reddish-yellow, impunctate, except for some setigerous punctures at the sides. Elytra a little longer than the thorax, quadrate, reddish; each with a dark mark on the middle towards the extremity, smooth and shining, each with a series of fine distant punctures close to the suture, and with two or three other such series of punctures towards the sides. Hind body broad, strongly margined, sparingly punctured, reddish; the 6th segment on either side with a large dark mark. Legs yellow; hind tarsi slender; the basal joint elongate, 2, 3 and 4 each shorter than its predecessor.

In the male the front tarsi are distinctly dilated; the

7th segment of the hind body has on the underside a broad but not deep notch at the extremity, which is continued forwards as a shallow, longitudinal depression. In the female the front tarsi are only slightly dilated, and the 7th segment is simple. In both sexes the 8th segment terminates in two stout, pointed, horny styles of a pitchy-red colour.

Tapajos; one male, three female specimens. In two of the latter the elytra are without the dark mark.

20. *Lithocharis germana*, n. sp. Rufo-testacea, nitida, parce nigro-setosa; prothorace latitudine fere longiore, angulis anterioribus rotundatis; elytris thoracis longitudine, maculâ fuscâ; abdomine sat crebre punctato, segmento 6° utrinque maculato. Long. corp. 3 lin.

Antennæ with the 2nd joint long, 3rd longer than 2nd. Mandibles rather long, each with two distinct teeth in the middle. Thorax very convex, about as long as broad; the sides a little curved and narrowed behind, the front and front angles much rounded. Elytra about as long as the thorax. Hind body finely and moderately closely punctured.

In the male the front tarsi are distinctly dilated; the ventral plate of the 6th segment of the hind body is a little emarginate in the middle at the extremity; on the 7th segment the hind margin is broadly but not very deeply emarginate, and in front of the emargination has a longitudinal smooth space.

Extremely similar to *L. polita*, except in the points mentioned above.

Tapajos; one male, two female specimens. The male is indicated as having been found in an ant's nest.

21. *Lithocharis payana*, n. sp. Rufo-testacea, nitida, parce nigro-setosa; prothorace basin versus vix angustato; elytris thoracis longitudine, suturâ maculâque obliquâ nigricantibus; abdomine subtiliter sat crebre punctato, segmento sexto utrinque maculâ magnâ; mento tuberculo erecto, acuminato. Long. corp. 3 lin.

Antennæ reddish, slender, about as long as head and thorax, not thicker at the extremity; 3rd joint a little longer than 2nd; 10th twice as long as broad. Mandibles red, moderately elongate, each in the middle with two short teeth, which are obtuse or emarginate at the extremity. Thorax convex, about as long as broad, only slightly narrowed behind, and but slightly curved at the sides; the front and front angles moderately rounded. Elytra quadrate, about as long as and as broad as the thorax, shining reddish-yellow; the raised suture and an oblique dash on the middle of each blackish, with fine sutural and two or three lateral rows of fine distant setigerous punctures. Hind body finely but not densely punctured; the 6th segment with a large spot on either side blackish.

In the male the front tarsi are distinctly dilated; the ventral plate of the 6th segment of the hind body has an extremely slight emargination in the middle at the extremity; the ventral plate of the 7th segment has a small emargination at the extremity and in front of this a short smooth depression. In the female the front tarsi are scarcely dilated.

Tapajos: two male, three female specimens.

Obs.—Though excessively similar in general appearance to *L. polita* and *L. germana*, this species is readily to be distinguished from them by the peculiar tubercle on the mentum.

22. *Lithocharis picta*, n. sp. Rufo-testacea, nitida, parce nigro-setosa, nigro-variegata; abdomine crebre subtiliter punctato; mento tuberculato. Long. corp. 3 lin.

Antennæ reddish-yellow, slender, about as long as head and thorax, not thicker at the extremity; 3rd joint distinctly longer than 2nd; 10th quite twice as long as broad. Mandibles moderately long, each with two short teeth in the middle. Head about as broad as the thorax, shining red, the clypeus in front, and an ill-defined transverse mark on the vertex, infuscate. Thorax about as long as broad, scarcely narrowed behind, shining red, the sides and the middle with irregular large dark marks. Elytra about as long as the thorax, shining red, with a large dull mark extending obliquely from the shoulder to the inner angle, the series of punctures along the suture and towards the sides distinct. Hind body finely and rather closely punctured, reddish, with the anterior outer angle of each segment blackish; the 6th segment entirely blackish except towards the hind margin.

St. Paulo.

Obs.—The single individual I have seen of this species

appears to be a female and has the front tarsi scarcely dilated; it comes extremely close to the *L. pagana*, but differs in having the head slightly shorter and the hind body a little more closely punctured as well as in its dark markings. The tuberculate mentum readily distinguishes it from *L. polita* and *L. germana*.

Stilicus.

The species of this genus found by Mr. Bates are only two in number, and are evidently closely allied to the two species *S. jugalis* and *S. carinatus* described by Erichson from Columbia. I have also three or four other closely allied species from Brazil, so that the genus is probably quite as well represented in South America as in any other part of the world; the species, though widely distributed on the globe, being nowhere numerous.

1. *Stilicus amazonicus*, n. sp. Niger, antennis pedibusque testaceis, capite thoraceque opacis, hoc fortiter carinato; elytris nigro-æneis, apice testaceo, minus fortiter punctatis. Long. corp. 2½ lin.

Antennæ rather long, quite yellow; 3rd joint scarcely longer than 2nd; 10th joint quite as long as broad; 11th joint rather long, nearly as long as the two preceding together. Palpi yellow, slightly infuscate. Labrum large. Head dull blackish, rather elongate, very densely punctured; the punctures on a small space on the middle coarser than elsewhere, so that this spot is a little shining, but none of the punctures are wanting, neither are the interstices there any broader. Thorax just ⅓ lin. in length, strongly angulated at the sides and abruptly narrowed towards the front, densely punctured and with a shining conspicuous smooth line along the middle. Elytra just as long as the thorax and about as broad as long, a little shining, of an obscure brassy colour, with the extremity pale yellow; rather sparingly and not coarsely though distinctly punctured. Hind body densely and finely punctured, slightly shining. Legs clear yellow.

Ega; a single female specimen.

This species is rather larger than the European *S. orbiculatus*; its antennæ are longer, the terminal joint of the maxillary palpi is more elongate and more linear, the head behind the eyes is longer and less truncate, the thorax more abruptly angulated at the sides and considerably longer, the hind tarsi more slender and rather longer.

2. *Stilicus punctatus*, n. sp. Niger, antennis pedibusque minus læte testaceis, illis medio palpisque leviter infuscatis; thorace fortiter carinato; elytris nigro-æneis, nitidulis, fortiter profundeque punctatis, apice testaceis. Long. corp. 2¼ lin.

Antennæ short, reddish, the middle joints infuscate. Palpi infuscate-reddish. Labrum large. Head broad and short, very densely punctured, dull, a little shining on a small place in the middle, where the punctures are a little coarser. Thorax black, much angulated at the sides and abruptly narrowed in front; densely punctured, with a conspicuous shining line along the middle. Elytra quite as long as the thorax, distinctly brassy, with the extremity yellow; distinctly shining, deeply and rather coarsely punctured, the punctuation absent on the pale extremity. Hind body densely punctured, a little shining. Legs yellow.

Ega; a single female individual.

This species is about the size of *S. orbiculatus*, and the head is very similar in form thereto, but the palpi are darker, with the apical joint more linear and elongate; the thorax is more angular at the sides, and the punctuation of the elytra is greatly more conspicuous. From *S. amazonicus*, the smaller size, differently-shaped head, shorter antennæ, and more coarsely punctured elytra of *S. punctatus*, readily distinguish it.

MONISTA (nov. gen. *Pæderinorum*).

Labrum medio emarginatum, utrinque rotundatum.
Mandibulæ breviusculæ, robustæ.
Palpi maxillares articulo quarto inconspicuo.
Prosternum post coxas parte corneâ magnâ, sed acetabulis posterius haud occlusis.
Tarsi articulo tertio parvo, quarto bilobo-membranaceo.
Genus ex affinitate *Sunii*, sed facie potius *Lithocharidis*; ab illo mandibulis, labro, et prosterni structurâ, ab hoc tarsorum articulo quarto, facile distinguendum.

Body shining and sparingly punctured. Antennæ short and stout. Mandibles short and robust, toothed in the middle. Labrum large, rounded on either side so as to be emarginate in the middle. Third joint of maxillary palpi a little dilated in the middle on the inner side and pointed at the extremity; 4th joint very minute. Head with a slender neck. Thorax subglobose, the horny portion of the prosternum large, so that in the natural position it extends as far back as the front edge of the mesosternum; but the side pieces of the prothorax are not contiguous with this, so that the anterior coxal cavities are quite open behind. Mesosternum large, forming a well-marked neck. Tarsi apparently only four-jointed, the 3rd joint being short, especially on the upper side; its lower surface longer and hairy, and the 4th joint consisting of a membrane, hairy beneath, enwrapping the sides and undersurface of the fifth joint on the hind foot; the basal joint is as long as the rest of the tarsus. Front tarsi undilated in both sexes.

This genus is allied to *Sunius*, but cannot be amalgamated therewith, owing to the different structure of the prosternum. Its facies and form are also different, and approach *Lithocharis* and *Scopæus*. It forms a connecting link between the *Pæderidæ* with closed anterior coxal cavities and those in which these are widely open, and its position in the usually adopted scheme of classification is between *Lithocharis* and *Sunius*. The above characters are drawn from a Rio de Janeiro species, of which I give a diagnosis in the subjoined note.*

1. *Monista certa*, n. sp. Rufo-castanea, nitidula, antennis, palpis, pedibus, elytrorumque apice testaceis; capite thoraceque fortiter punctatis, hoc subgloboso lineâ mediâ impunctatâ; elytris fere impunctatis. Long. corp. 1¼ lin.

Antennæ yellow, about as long as head and thorax; 1st joint nearly as long as the two following together, 2nd almost as stout as 1st; 3rd quite as long as and rather more slender than 2nd; 4—8 each slightly shorter than its predecessor, bead-like, about as long as broad; 10th quadrate, slightly larger than the intermediate joints; 11th rather stouter than 10th, about as long as 9th and 10th together, pointed. Head rather broader than thorax,

---

* *Monista typica*, n. sp. Castanea, nitidula, pedibus testaceis, antennis basi apiceque testaceis, medio obscurioribus, articulo decimo leviter transverso; capite crebre fortiter punctato; prothorace subgloboso, fortiter sed obsolete punctato; elytris thorace longioribus, obsolete punctatis. Long. corp. 1¼ lin.

Hab.—Rio Janeiro.

Closely allied to *M. certa*, but rather larger, with the thorax more obsoletely punctured and the antennæ clouded in the middle.

narrower than the elytra ; the eyes only moderately large, and placed much nearer the antennæ than the vertex ; the surface rather closely and moderately coarsely punctured, with simple impressed punctures, castaneous in colour, shining, and bearing a fine erect pubescence. Thorax much narrower than the elytra, nearly as long as broad, without angles, rounded in front, narrowed towards the base, and gently curved at the sides ; its colour and sculpture similar to those of the head, but with an impunctate line along the middle, not reaching however to the front. Elytra longer than the thorax, rather inflated, yellowish in colour, but paler at the extremity and slightly clouded about the middle, shining and almost impunctate. Hind body broad and short, strongly margined, the segments very finely and indistinctly punctured; the 6th infuscate, but pale at the extremity. Legs slender, pale yellow.

In the male the ventral plate of the 6th segment of the hind body is emarginate at the extremity; the apical segments are retracted in my only individual.

Ega ; a single male.

2. *Monista longula*, n. sp. Rufo-castanea, nitidula, pedibus elytrorumque apice testaceis, antennis basi apiceque dilutioribus ; capite thoraceque crebre fortiter punctatis, hoc latitudine longiore, lineâ mediâ impunctatâ ; elytris fere impunctatis. Long. corp. 1½ lin.

Antennæ yellowish at the base and extremity, with the middle joints a good deal darker than the others. Thorax longer than broad ; its greatest width in front of the middle, thence much narrowed towards the front, and a good deal towards the base. Hind body with the 6th segment concolorous.

St. Paulo ; a single female.

Obs.—This species is closely allied to *M. certa*, but is readily distinguished by its more elongate thorax.

3. *Monista divisa*, n. sp. Rufo-testacea, nitidula, capite, thorace, elytrisque piceis, his apice late testaceis ; antennis pedibusque testaceis, illis ante apicem infuscatis. Long. corp. 1½ lin.

Antennæ rather shorter than head and thorax ; 3rd joint slightly longer than the contiguous ones ; joints 7—10 infuscate, the others yellowish ; 10th joint transversely quadrate. Head about as broad as the thorax; the sides rounded behind the eyes, towards the narrow neck, pitchy in colour, shining, rather coarsely and closely punctured. Thorax nearly as long as broad, its greatest breadth in front of the middle, greatly narrowed towards the front, and much towards the base; rather coarsely and moderately closely, but obsoletely punctured, with a broad line along the middle smooth. Elytra a good deal broader and longer than the thorax, shining and impunctate, pitchy, with the extremity broadly yellow. Hind body reddish-yellow, very obsoletely punctured. Legs yellow.

In the male the hind margin of the ventral plate of the 7th segment of the hind body is emarginate, and the following segment bears a very broad and deep excision.

Tapajos; a single individual.

ECHIASTER.

Of this remarkable genus only six species have been yet described, viz., three from Northern South America, one from Chili, and two from the United States of North America. I here describe ten (or perhaps only nine) new species, which show a striking variation in form of different parts of the body, and leave no doubt that many other species will be discovered; indeed, I have already two others from Rio Janeiro, very dissimilar to any here described.

A highly important character of the genus appears hitherto to have escaped notice, viz., that the prothorax behind the front coxæ is horny. This character, in conjunction with the others indicated for the genus, gives it an isolated position in the *Pæderidæ*, and renders it probable that it will prove to be one of the most important of the genera of South American *Staphylinidæ*; and also that, notwithstanding the extension of one or two species into Chili and North America, *Echiaster* will be one of the most characteristic genera of the Austro-Columbian *Coleoptera*. Kraatz has described a genus from East India (*Sclerochiton*, Kr. Staph. Faun. von Ostind. pl. ii. f. 8), which appears to possess several points of relationship with *Echiaster*, and to be at present its nearest known ally; it is doubtful, however, to what extent the resemblances between the two genera indicate a real affinity between them. Kraatz indeed in his figure and description gives us no reason to suppose that *Sclerochiton* possesses that peculiar tubular elongation of the apical abdominal segments which is one of the most striking and easily-perceived characters of *Echiaster*.

1. *Echiaster boops*, n. sp. Testaceus, abdominis apice nigricante; capite orbiculato, antennis breviusculis, clavatis, subtus oculis dilatatis. Long. corp. (abdomine extenso) 1½ lin.

Antennæ about as long as head, yellowish; 1st and 2nd joints short and stout; 3rd joint small, rather shorter and much more slender than 2nd; 4—9 each slightly shorter than its predecessor, the penultimate joints rather strongly transverse; 11th joint rather short. Maxillary palpi with the 3rd joint broad and short. Head broader than the thorax, and even slightly broader than the elytra, the eyes occupying a large portion of the side and encroaching greatly on the under surface; the hind angles rounded, the vertex not gradually narrowed but the neck very abrupt; the surface opaque, very densely and indistinctly, though not very finely punctured. Thorax only about half as broad as the elytra, longer than broad, the greatest width in front of the middle, much narrowed towards the front and a good deal towards the base; the surface even, the colour yellowish, the punctuation similar to that of the head. Elytra distinctly longer than the thorax and rather darker in colour, very densely and more roughly and distinctly punctured than the head, quite opaque. Hind body much pointed towards the extremity, similar in colour to the elytra, with the terminal segments blackish, densely punctured, and with a short subsquamose golden pubescence. Legs pale yellow.

Tapajos; nine individuals, which show me no external sexual distinctions.

2. *Echiaster fumatus*, n. sp. Testaceus, abdominis apice nigricante; capite, elytris metasternoque obscurioribus; capite orbiculato, antennis breviusculis, clavatis. Long. corp. 1½ lin.

This species is excessively closely allied to *E. boops*, but it is slightly more elongate; it has the elytra and metasternum infuscate, and the head also is more obscure in colour, and differs a little from that of *E. boops* in its form, it being rather longer in proportion to its breadth, so that the eyes do not occupy so large a portion of the sides; the thorax also is indistinctly carinate along the middle. In other respects it extremely resembles *E. boops*.

Tapajos; a single individual of unknown sex.

3. *Echiaster signatus*, n. sp. Elongatus, angustulus, castaneo-testaceus, elytris fusco-nigrosignatis; crebre fortiter punctatus; capite elongato, vertice angusto. Long. corp. 2½ lin.

Antennæ yellow, rather longer than the head, a little thickened towards the extremity; 3rd joint much narrower but almost longer than 2nd; 4th and 5th joints slender, nearly equal to one another, each rather longer than 3rd; 8—10 bead-like, scarcely broader than long; 11th rather broader, and a good deal longer than 10th, ending in a seta-like spine. Mandibles reddish, very elongate and slender, each with two teeth, of which the upper one is very long. Labrum with two sharp, stout, triangular, approximate teeth in the middle, and on each side these with a smaller sharp projection. Head longer than broad, reddish, gradually narrowed from the eyes to the vertex, densely punctured; the punctures rough and asperate, except on the front part. Thorax elongate and narrow, only half as broad as the elytra, twice as long as broad; its greatest width about the middle, thence a good deal narrowed towards the front and slightly towards the base, dull, reddish in colour, densely punctured. Elytra only slightly longer than thorax, yellowish, but with four large longitudinal marks towards the extremity (often more or less confluent), leaving a basal fascia pale, which, however, is subinterrupted in the middle by the extension forwards of the dark markings; densely and roughly punctured. Hind body elongate, slender and greatly pointed, obscure yellowish; the basal segments rather coarsely and asperately punctured, the apical ones very finely and indistinctly. Legs pale yellow; under surface chestnut-yellow, coarsely punctured.

Tapajos; eleven individuals.

On dissecting one of these specimens I find it to be a male, though there are no external characters to indicate this, and the ædeagus is small and inconspicuous. The black marks on the elytra vary a good deal in their extent.

4. *Echiaster carinatus*, n. sp. Elongatus, angustulus, testaceus, elytris fuscis, basi testaceis; dense punctatus,

opacus; capite elongato, vertice angusto; thorace medio longitudinaliter carinato. Long. corp. 2½ lin.

Antennæ yellow, about as long as head; 1st joint much stouter and longer than the following ones, as long as the three following joints together; the two or three penultimate joints distinctly transverse. Mandibles very long and slender, the upper of the two teeth in the middle very long. Labrum with two widely-separate sharp teeth in the middle, and sinuate and emarginate between them; lateral teeth indistinct. Head reddish, gradually narrowed from the eyes to the vertex; convex in the middle, so as to give an appearance of two obscure anteriorly divergent grooves; the surface dull, densely and intricately punctured, with the interstices extremely fine. Thorax elongate and narrow, hardly half so broad as the elytra, quite twice as long as broad; the greatest width in the middle, thence a good deal narrowed towards the front and slightly towards the base, pale reddish-yellow, closely but indistinctly punctured, dull; along the middle with a broad longitudinal elevation, and depressed on either side of this. Elytra about as long as the thorax, blackish, with the basal portion pale yellow, densely and roughly but not coarsely punctured. Hind body elongate and narrow, very pointed, yellowish, with the extremity darker, and on the side of each segment a small, indistinct, dark mark; the basal segments rather coarsely and roughly punctured. Legs pale yellow.

Tapajos; a single female.

This species at first sight resembles *E. signatus* extremely, but is abundantly distinct by the structure of the labrum and the carinate thorax.

5. *Echiaster latifrons*, n. sp. Latior, piceus, antennis, palpis, pedibusque testaceis; capite lateribus rotundatis, vertice angusto; prothorace elongato, bisulcato; elytris abdomineque fortiter asperato-punctatis. Long. corp. 2¾ lin.

Antennæ reddish-yellow, rather longer than the head; 1st joint very stout, quite as long as the two following together; 3rd joint more slender than 2nd; 4th and 5th slender, each longer than 2nd, differing little from one another; 6th shorter than 5th, 7th shorter than 6th, about as long as broad; 8—10 slightly transverse, 11th acuminate at the extremity. Mandibles reddish. Labrum with four sharp approximate teeth in the middle. Head broad, quite as broad as the elytra, suborbiculate, with the sides evenly curved behind the prominent eyes towards the vertex; the colour is pitchy, the surface quite opaque, densely punctured with umbilicate punctures, more indistinct towards the front; the interstices very fine, the vertex in the middle with an obscure, longitudinal impression. Thorax only half as broad as the elytra, nearly twice as long as broad; the greatest breadth about the middle, thence much narrowed towards the front and slightly towards the base; along the middle with two deep, longitudinal furrows; the colour similar to that of the head, as also the punctuation, but the latter not quite so distinct. Elytra quite as long as the thorax, of an obscure fuscous colour, with some very indistinct paler spaces at the base, densely punctured with a distinct scabrous punctuation. Hind body with the basal segments roughly and coarsely punctured, the apical ones nearly smooth. Legs pale yellow.

Tapajos; a single individual, of doubtful sex.

Obs.—In the form of the head this species is intermediate between *E. boops* and *E. signatus*; the individual described bears a transverse impression on the middle of the head, which I have not mentioned in my description, as I think it is the result of accident.

6. *Echiaster mamillatus*, n. sp. Infuscato-rufescens, elytrorum fasciâ latâ basali, antennis pedibusque testaceis; opacus, dense punctatus; capite angusto, vertice elongato, subtus pone orem bimamillato; thorace minus elongato, latius bisulcato. Long. corp. 2 lin.

Rather narrow. Antennæ reddish-yellow; 1st joint very stout, rather short; 2nd joint stout, bead-like, not longer than broad; 3rd joint small, 5—10 each a little broader than its predecessor, 7—10 transverse; 11th joint rather stout, oblique at the apex. Mandibles red, very slender, only moderately long. Teeth of labrum short and indistinct. Head infuscate-red, about as broad as the thorax; the sides convergent, the vertex forming a neck; the surface densely and obscurely punctured, convex between the eyes. Thorax longer than broad, much narrower than the elytra, rather irregular in form, with a rather obscure elevation along the middle, and another still more obscure on each side of this; colour and sculpture similar to that of the head. Elytra longer than the thorax, blackish at the apex, the basal part yellow, this colour forming an angulated fascia, which occupies nearly half of the elytra; densely asperately punctured. Hind body much pointed towards the extremity, the basal segments densely, moderately coarsely, the apical ones very finely punctured. Legs pale yellow. Front tibiæ slightly dilated towards the apex, and distinctly bisinuate externally. On the underside of the head, at the base of the mouth, are two peculiar fine tubercles.

Tapajos; five specimens.

Obs.—This species varies much in colour, some individuals being nearly black, and others reddish or yellow.

I have dissected the terminal segments of one of these specimens, with the hope of ascertaining it to be a male; but I am unable to say positively whether this is the case or not; for though I have found what may possibly be the ædeagus, yet it is so small and insignificant that I am by no means sure it may not be merely a portion of some dried internal tissue;—this although I have examined it with a very high power.

7. *Echiaster muticus*, n. sp.? Infuscato-rufescens, elytrorum fasciâ basali, antennis pedibusque testaceis; opacus, dense punctatus; capite angusto, vertice elongato, subtus mutico; thorace minus elongato, latius bisulcato. Long. corp. 2 lin.

Tapajos; two individuals.

Obs.—These two specimens present no difference from *E. mamillatus*, except the absence of the tubercles on the underside of the head; it is probable that this may be merely a sexual character, the *E. muticus* being only the other sex of *E. mamillatus*, in which case the species may bear the name *E. mamillatus*.

8. *Echiaster tibialis*, n. sp. Rufus, opacus, dense punctatus, pedibus testaceis; capite elongato, postice angustato; thorace minus elongato, latius bisulcato; tibiis anterioribus extus dilatatis. Long. corp. 1¾ lin.

Antennæ red, stout, as long as the head; joints 5—10 each transverse, and each slightly broader than its predecessor; 11th joint short. Mandibles very slender, rather short. Labrum distinctly quadridentate. Head elongate, the sides narrowed, but not rounded from the eyes to the neck; the surface elevated in the middle, densely and obscurely punctured, quite dull, reddish. Thorax much narrower than the elytra, rather longer than broad, the sides almost rounded; the greatest width about the middle, thence much narrowed to the front, and a little towards the base, with a broad, ill-defined, longitudinal impression along each side of the middle; red, punctuation dense and obscure. Elytra much longer than the thorax, and slightly paler in colour, densely and indistictly punctured. Hind body quite dull, densely punctured, and with a very fine and short, depressed golden pubescence; 6th segment very elongate. Legs pale yellow; anterior tibiæ flattened, so that seen on one face they appear very broad; their tarsi with the basal joint broad, each joint following a little narrower than its predecessor.

Tapajos; three individuals.

Obs.—These three individuals all have the hind margin of the ventral plate of the 6th segment of the hind body distinctly emarginate in the middle, and I suppose them all to be males; the 7th segment is entirely retracted, except in one specimen, in which its hind margin is exposed, and this is emarginate beneath, like the 6th segment.

9. *Echiaster Batesi*, n. sp. Testaceus, dense punctatus, opacus, abdominis apice fuscescente; capite sat elongato, lateribus curvatis, vertice angusto; thorace minus elongato, profunde bisulcato. Long. corp. 1⅜ lin.

Antennæ yellow, short, quite as long as the head; 1st joint rather short, very stout; 2nd almost orbicular; 3, 4, 5 small, subequal to one another; 6—10 each a little broader than its predecessor, 9 and 10 distinctly transverse, 11th moderately long. Mandibles very slender, rather short. Labrum very indistinctly toothed. Head rather long, the sides gradually narrowed in a rounded curve to the narrow vertex, which is not prolonged into a neck, red, densely and very indistinctly punctured, elevated about the middle. Thorax a good deal narrower than the elytra, and almost as broad as the head; rather longer than broad, almost rounded at the sides; the greatest breadth in front of the middle, thence much narrowed towards the front, and distinctly towards the base, deeply longitudinally impressed on each side of the middle, yellow, quite dull, punctuation very indistinct. Elytra rather longer than the thorax, closely and indistinctly punctured. Hind body densely and indistinctly punctured, obscurer in colour towards the extremity.

Legs pale yellow; front tibiæ short, distinctly dilated towards the extremity; their tarsi short and slender.

Tapajos; a single individual, of doubtful sex.

Obs.—The different shape of the head, more slender antennæ, and the differently-formed front tibiæ and tarsi, readily distinguish this species from *E. tibialis* ; the individual described is perhaps somewhat immature.

10. *Echiaster scissus*, n. sp. Testaceus, dense punctatus, opacus, elytrorum apice medio, abdominisque apice fuscescentibus; capite lateribus curvatis, oculis minoribus; prothorace lato, suborbiculato, profunde bisulcato. Long. corp. (vix extenso) 1½ lin.

Antennæ rather slender; 1st joint stout and short, 2nd stout, 3rd and 4th very small, 6—10 each a little broader than its predecessor, 9 and 10 distinctly transverse, 11th rather short and stout. Head slightly broader than thorax, a good deal narrower than the elytra; the eyes small, the sides behind the eyes greatly rounded; the neck very slender, the middle of the surface much elevated, so that the vertical portion appears much depressed; yellow in colour, quite dull, densely and indistinctly punctured. Thorax short, about as long as broad, greatly narrower than the elytra, rounded at the sides, with two deep longitudinal impressions, so as to make it appear tricarinate, densely and indistinctly punctured, quite dull. Elytra much longer than the thorax, rather broad, yellow, with an infuscate patch in the middle at the extremity, densely punctured, quite dull. Hind body rather broad, the basal segments rather coarsely punctured, the 6th segment conical. Legs pale yellow; front tibiæ moderately broad, slightly sinuate.

Tapajos; a single specimen.

Obs.—The small eyes readily distinguish this species from all the others here described.

LINDUS (nov. gen. *Pæderinorum*).

Palpi maxillares articulo tertio magno, subsecuriformi, articulo quarto occulto.

Mandibulæ perelongatæ, tenues, valde curvatæ, edentatæ.

Tarsi anteriores fortiter dilatati, posteriore articulo quarto simplice.

Abdomen immarginatum, stylis duobus rigidis terminatum.

Genus perdistinctum, habitu *Pinophilinorum*, a quibus palporum maxillarium articulo 4° condito, et prothorace post coxas membranaceo differt.

Labrum transverse, with the horny part excessively short, and with two stout triangular teeth in the middle; from the outside of the teeth proceeds a white membrane, which extends all round the exposed part of the labrum, and much increases its size. Mandibles very slender, long, pointed and curved, and without any trace of teeth. First joint of maxillary palpi much shorter than the others, longer than broad; 2nd joint rather slender, more than twice as long as the 1st joint; 3rd joint longer than 2nd, much dilated, especially on the inner side, the extremity rather truncate; 4th joint quite invisible. Ligula broad in front and emarginate, so as to be in fact bilobed; paraglossæ very distinct and greatly developed, extending a good deal beyond the ligula, and distinctly beyond the base of the 2nd joint of labial palpi. First joint of labial palpi concealed in my preparation by the ligula and paraglossæ; 2nd joint cylindrical, quite twice as long as broad; 3rd joint not much more than half as broad as 2nd, about twice as long as broad. Antennæ rather stout, not thickened towards the extremity. Head short and transverse, with a moderately broad neck. Thorax transversely quadrate, with the base rounded, the side pieces broad throughout, and with a long projection near the hinder part; membranous under the coxæ; the coxal cavities thus forming two rather long oblique openings, which are confluent in their hinder part. Mesosternum forming chiefly a horny neck, only forming at the base of the middle coxæ a short angular projection. Middle coxal cavities large and deep, confluent. Hind body subcylindrical, only the basal segment margined, terminated in two stout, pointed, rigid styles. Wings present. Front tarsi with the four basal joints forming a broad patella; 5th joint slender and rather long. Legs moderately long and slender; 1st joint of hind tarsi a little longer than 2nd, 4th not at all lobed.

The curious insect for which I have established this genus has the hind body formed almost as in *Œdichirus*, and gives one, from its facies, the idea of a member of the *Pinophilini* rather than the *Pæderini*; nevertheless its systematic position at present is in the *Pæderini*, in the neighbourhood of *Lithocharis*; but I have a strong impression that it will ultimately prove to be one of the steps of a transition to the *Pinophilini*.

1. *Lindus religans*, n. sp. Piceus, nitidus, subparallelus, antennis, pedibus, palpisque testaceis, elytris fortiter seriatim punctatis; abdomine parcius pubescente, crebre fortiter, profundeque punctato. Long. corp. 3—3¼ lin.

Antennæ reddish, rather shorter than head and thorax; 1st joint elongate, rather longer than 2nd and 3rd together; 3rd joint long, a good deal longer than 2nd, 5—10 each a little shorter, but scarcely broader than its predecessor, 10th about as long as broad; 11th joint short, with a slender spine or seta at its apex. Head short and broad, slightly narrower than the thorax; eyes large and prominent, separated by a narrow space only from the hind margin, with some coarse punctures at the sides and vertex, which are wanting along the middle. Thorax transverse, a little narrower than the elytra, nearly truncate in front, the front angles but little rounded, the base and hind angles rounded, the sides nearly straight and not narrowed behind; on either side of the middle is an irregular, longitudinal series or patch of rather coarse punctures, and between these and the sides are other scattered, pretty numerous punctures. Scutellum small, impunctate. Elytra longer than the thorax, quadrate, their extremity emarginate, depressed on either side of the finely-elevated suture; in the depression a series of rather coarse punctures, and outside this with several other series of coarse, deep punctures. Hind body stout; the basal segment margined at the sides, the others without margins, and each slightly narrower at the base than the extremity, very coarsely, closely and deeply punctured; the 6th and 7th segments more sparingly and more finely than the basal ones; hind margin of upper plate of 7th segment angulate in the middle, and spinous at each side; the connecting membranes of the segments coriaceous, as in the *Pinophilini*. Legs yellow.

In the male the hind margin of the ventral plate of the 6th segment of the hind body is slightly emarginate at the extremity, and is polished and depressed in front of this emargination, and the 7th segment has a large notch or excision in the same place.

Tapajos; four specimens of this interesting species were brought back by Mr. Bates, one of which, however, I have destroyed by an unfortunate accident during examination.

PÆDERUS.

The species of this very widely-distributed genus brought from the Amazons are seven in number; they show nothing remarkable, their colour and appearance being very similar to our European species; they all have the mandibles with a simple bidentate tooth in the middle. Although scarcely a score of species of *Pæderus* have been as yet described from South America and Mexico, the species are really numerous there; the more remarkable of the South American forms of the genus, such as *P. rutilicornis* and *P. ferus*, appear to be unrepresented in the Amazons.

The species of the genus require careful study, and some of them have already given rise to much discussion; the structure of the ædeagus has been hitherto neglected; but when it is considered, it will, I have no doubt, be found to greatly facilitate the recognition and discrimination of the species.

1. *Pæderus solidus*, n. sp. Robustus, alatus, niger, elytris thorace paulo longioribus, cyaneis; abdomine segmentis quatuor primis, thorace, mesosternoque rufis, pedum basi antennarumque basi et apice testaceis, his medio infuscatis. Long. corp. 5—5¼ lin.

Antennæ 1¾ lin. in length, 3 basal joints yellow, the next 6 strongly infuscate, the 10th infuscate-yellow, the last joint yellow. Palpi yellow, terminal joint infuscate at the apex; mandibles red. Head black, with the neck red, broad, a little broader than the thorax, rather sparingly and finely punctured; the punctures wanting towards the middle. Thorax bright red, longer than broad, about ⅞ lin. long and scarcely ⅔ lin. broad, sparingly and finely punctured, with a broad impunctate space along the middle, the punctures bearing fine black hairs. Scutellum red, elytra blue, parallel, rather longer than thorax, from apex of scutellum to extremity of suture being quite ⅞ lin.; their common width also about ⅞ lin., rather finely and moderately closely punctured. Legs rather long; the coxæ and base of femora yellow, apical third of femora and the tibiæ nearly black; the tarsi strongly infuscate. The central notch of labrum shallow, but rather broad.

The notch on the ventral plate of the 7th segment of the hind body in the male is very deep and parallel-sided. In the female the extremity of the same plate forms a large triangle, and on each side of this at the base is a stout, very short tooth.

Ega and Tapajos; several individuals.

This species is allied to *P. æquinoctialis*, Er., to judge from description. I have several closely-allied South American species, but fail to ascertain without doubt which of them is Erichson's species.

2. *Pæderus tridens*, n. sp. Angustulus, niger, elytris humeris angustatis, thorace fere brevioribus, cyaneis; abdominis segmentis quatuor primis, thorace, mesosterno, mandibulisque rufis; palpis, pedum basi, antennarum basi et apice testaceis, his medio fuscis; elytris crebre fortiter ruguloso-punctatis. Long. corp. 4½ lin.

Antennæ slender, 1¾ lin. in length, 3 basal joints yellow, the 3 apical ones reddish, the middle ones deeply infuscate. Palpi yellow, apical joint obscurely infuscate at apex; mandibles and labrum reddish, the latter with the central excision very small. Head black, with the neck reddish, rather elongate, sparingly and finely punctured, the punctures wanting on the middle. Thorax red, not quite ⅞ lin. in length, and not ¾ lin. in width, a good deal narrowed behind; the surface shining and almost impunctate. Elytra narrow, narrower at shoulders, their greatest length about that of the thorax, bluish-green; coarsely and rather deeply but irregularly punctured, so that the interstices are rugulose. Four basal segments of hind body red, the two apical ones blackish. Mesosternum and scutellum red. Legs long and slender, the coxæ and base of femora yellow; apex of femora infuscate, the hinder much more broadly so than the anterior; tibiæ and tarsi also infuscate.

In the female the ventral plate of the 7th segment of the hind body ends in an elongate pointed central spine in the middle, and on either side in a shorter slightly curved spine; the narrow tongue-like process, which forms the dorsal plate of the 8th segment, is much longer than the broader ventral plate.

Tunantins; a single female.

3. *Pæderus lingualis*, n. sp. Angustulus, niger, elytris angustis thoracis latitudine, cyaneis; abdominis segmentis quatuor primis, thorace, mesosterno, mandibulisque rufis; palpis, pedum basi, antennarum basi et apice testaceis, his medio nigris; elytris fortiter minus crebre punctatis. Long. corp. 5 lin.

This species resembles exactly the *P. tridens* above described, and appears to differ chiefly in the form of the labrum, the central notch being larger and much deeper; the elytra also are more sparingly punctured, the punctures being less deep and the interstices not so rugulose.

In the female the structure of the 7th segment of the hind body is the same as in *P. tridens*; but the narrow dorsal process of the 8th segment is less elongate than in *P. tridens*, and projects only a little beyond the ventral plate.

Ega; a single female.

4. *Pæderus mutans*, n. sp. Rufus, elytris viridi-cyanescentibus, antennis medio, pedibus basibus exceptis, metasterno, abdominisque segmentis duobus ultimis infuscatis; elytris thorace paulo longioribus, humeris distinctis, crebre minus fortiter punctatis. Long. corp. $4\frac{1}{2}$ lin.

Antennæ $1\frac{3}{4}$ lin. in length, yellow, with the five middle joints infuscate. Palpi yellow. Labrum with a very small notch in the middle and without lateral emargination. Head red, rather narrow in proportion to the length, finely punctured, the punctures wanting towards the middle. Thorax $\frac{3}{4}$ lin. in length and $\frac{5}{8}$ lin. in breadth, distinctly narrowed behind, finely and sparingly punctured, with a space along the middle free from punctures. Elytra of a rather faint bluish-green colour, $\frac{7}{8}$ lin. in length, the shoulders distinct and the sides parallel, their punctuation rather deep and close, but not coarse. Hind body red, with the apical segments infuscate. Legs with the coxæ and basal portion of femora yellow; the front femora entirely yellow, the four posterior ones and all the tibiæ and tarsi strongly infuscate.

The male has the usual excision on the ventral plate of the 7th segment of the hind body. In the female this plate ends in the middle in a large sharply-pointed tooth, and each outer angle is also produced and acuminate, a broad space being left between the central and each lateral tooth.

Tapajos and Pará, two ♂, two ♀ individuals.

I have also another female from Pará, which has the punctuation of the head coarser, and the sides of the central tooth of the 7th abdominal segment more oblique in direction, so that the tooth is broader at the base; whether this be a distinct species or a mere variety I am unable to decide without an examination of more specimens.

5. *Pæderus protensus*, n. sp. Angustulus, rufus, capite piceo, metasterno abdominisque apice fuscis; femoribus apice nigricantibus, tibiis, tarsis antennarumque articulis 4—7 infuscatis; elytris viridi-cyaneis, parallelis, thorace longioribus, crebre fortiter punctatis. Long. corp. 4 lin.

Antennæ 1¾ lin. in length, slender, yellow, with the middle joints a little infuscate. Palpi yellow; mandibles red. Head pitchy, shining, rather sparingly and irregularly punctured. Thorax shining red, rather longer than broad, a good deal narrowed behind, very sparingly punctured. Scutellum red. Elytra shining bluish-green, ⅞ lin. in length, narrow and parallel, rather coarsely and deeply punctured, the punctures rather dense. Hind body red, with the apical segments pitchy. Mesosternum, coxæ and femora yellow, the apex of the four hind femora blackish; the tibiæ and tarsi infuscate-red, the tarsi slender.

The male has the usual deep excision on the ventral plate of the 7th segment. In the female this plate ends in a broad triangular projection, and each outer angle forms a short broad tooth.

Pará; four specimens collected by Mr. Rogers.

Obs.—This species appears to be closely allied to *P. mutans*, but is rather narrower and more slenderly formed and has the head darker. It is readily distinguished in the male sex by the structure of the ædeagus, which is produced at the extremity into a slender beak-like process. In the female the teeth of the ventral plate of the 7th segment are rather shorter and broader, and so appear less acuminate.

6. *Pæderus amazonicus*, n. sp. Rufus, capite nigricante subopaco; elytris cyaneis, thorace longioribus, crebre punctatus; metasterno, abdominis segmentis ultimis, femorumque apicibus infuscatis. Long. corp. 3½ lin.

Rather slender and elongate. Antennæ long, quite

yellow. Palpi yellow; mandibles red. Labrum pitchy, with a small notch in the middle; on either side of this a little prominent, and outside the prominence slightly emarginate. Head dull blackish, with the neck red; the surface finely coriaceous, sparingly and finely punctured. Thorax narrow, a good deal longer than broad, scarcely narrowed behind, bright red; finely and sparingly punctured, with an impunctate space along the middle. Elytra parallel, with the shoulders distinct, a little longer than the thorax, rather finely, moderately closely punctured, bluish or greenish. Hind body red, with the apical segments infuscate. Legs yellow; the apex of the hind femora distinctly, of the middle ones less distinctly, infuscate.

In the female the ventral plate of the 7th segment of the hind body is produced, so as to form a projecting triangle at the extremity; the apex of this triangle is scarcely pointed, the sides do not project in the form of teeth.

Ega and Tapajos; several individuals.

This species greatly resembles our European *P. fuscipes*, but is remarkably easily distinguished therefrom by the coriaceous surface of the head; it has also the antennæ and palpi differently coloured, these being clear yellow, and the elytra are less closely and distinctly punctured.

7. *Pæderus punctiger*, n. sp. Rufus, capite, pectore, abdominisque apice nigris, antennis fuscis, basi rufo; coxis femorumque basi rufescentibus, horum apice nigricantibus; tibiis tarsisque rufo-fuscis, elytris cyaneis, prothorace longioribus; capite thoraceque crebre punctatis. Long. corp. 4 lin.

Antennæ 1¾ lin. in length, the three basal joints reddish, the rest infuscate; 3rd joint much longer than 2nd, 4—10 each distinctly shorter than its predecessor, 10th much longer than broad. Mandibles red, infuscate at the apex; palpi red, infuscate towards the extremity. Head narrow, only about half as broad as the elytra, much narrowed behind the eyes, so that the constriction at the neck is but little; it is black in colour, with a blue tinge on its upper side; it is rather closely punctured, except on the middle of the hinder half, where the punctures are but sparing. Thorax small, greatly narrower than the elytra; towards the front it is a good deal narrowed; it is red in colour, and distinctly, regularly and rather closely punctured, the punctures being absent on a narrow space along the middle. Elytra much longer than the thorax, ⅞ lin. in length, dark bluish, rather finely but not densely punctured. Hind body dull red, with the two apical segments black. The coxæ are dark red, the femora blackish, with the basal part reddish; the tibiæ and tarsi obscure or infuscate red.

The male has a narrow deep excision, the sides of which are parallel, on the ventral plate of the 7th segment.

Carraranen, April, 1874; a single male brought back by Dr. Trail.

Obs.—The form and sculpture of the head and thorax render this species very easy to distinguish.

SUNIUS.

This genus is one of the most troublesome, in the present state of our knowledge of the *Staphylinidæ*, to any one dealing with a limited fauna, owing to the variety of forms included in the genus itself, while at the same time a number of closely allied forms have been detached as distinct genera. The nineteen species here described as new species of the genus possess all, I believe, in common the following characters: terminal joint of maxillary palpi minute, anterior coxal cavities closed by the junction of the side pieces of the thorax with the large horny prosternum, 4th joint of tarsi consisting of a membrane embracing the under surface and sides of part of the 5th joint. This combination of characters is also found, I believe, in the genera *Acanthoglossa, Stiliderus, Neognathus, Mesumius, Nazeris, Sunides, Stilicopsis, Mecognathus* and *Dibelonetes;* but in the present state of our knowledge it seems to me that these genera can scarcely be maintained as distinct, especially while so many heterogeneous forms are still, as I have said, included in *Sunius*. Of the sixteen species here described the first six, viz., *S. amicus, S. vittatus, S. serpens, S. ventralis, S. strictus* and *S. marginatus*, have quite the facies of our European *S. filiformis*, and are probably structurally very closely allied thereto; while the next four, *S. brevis, S. modestus, S. crassus* and *S. pictus*, must be placed in the section " Spurii " of Erichson, their short, broad form giving them an entirely different facies from the *S. filiformis* group. *S. confinis* and *S. catena* have the thorax more narrow and elongate than in the preceding species, and appear to afford a connecting link between *Stilicopsis* and *Sunius*. The *Sunides boreaphiloides* of Motsch. from Columbia, which is referred in the Munich Catalogue to the genus *Lithocharis*, appears to be an insect allied to *S. confinis*. The next two species, *S. bidens* and *S. bispinus* are possibly allies of the *Dibelonetes biplagiatus*, Sahl.; I say "possibly," because no characters are mentioned by Sahlberg which can be considered to distinguish his genus from *Sunius*, and I am in great doubt as to which one of several allied forms he intended so to name. The two following species, viz., *S. spinifer* and *S. celatus*, are very possibly but sexual forms of one species; they are very remarkable from the great development of the two spines of the labrum. *S. insignis*, the last of the species here described, is very peculiar on account of the abruptly constricted anterior portion of the thorax. From the above remarks it will be inferred that *Sunius* and its allies offer a prominent example of that insuperable difficulty in which those occupied with descriptive zoology find themselves constantly involved; for while it is clear that the only idea that can be formed of a genus is that of a limited aggregate of existing species, and consequently that no genus can be known till all the existing species of it are known, and till whether the characters assigned to it are naturally limited or not is known,—while this, I say, is clear, yet we are obliged to proceed in our actual descriptions on the absurd and "unthinkable" hypothesis that we know the genus before we know the species. I do not think naturalists have yet fully recognized this difficulty, but certainly until they have recognized it and are prepared to deal with it, it will be impossible that zoology can take the place it is entitled to as a most charming and important educational science. The present method of systematic zoology is certainly irreconcileable with a system of synthetic and inductive science, however well adapted it may have been to a period when educated minds were under the confusing domination of metaphysical inquiries.

1. *Sunius amicus*, n. sp. Nigricans, opacus, dense punctatus; antennis, palpis, pedibusque fere albidis, elytrorum apice anguste testaceo; abdomine subparallelo. Long. corp. 2⅓ lin.

Antennæ very pale, not quite so long as head and

thorax, slender and scarcely thickened towards the apex; 2nd joint half as long as 1st; 3rd distinctly longer than 2nd, slightly longer than 4th; 4—7 scarcely differing from one another, 7—10 each very slightly shorter but scarcely stouter than the predecessor, 11th one and a half times the length of 10th. Head scarcely broader than thorax; eyes rather large and prominent, blackish or pitchy, quite dull; punctuation dense, very indistinct, consisting of large umbilicated punctures, separated by very fine interstices, with a fine pale pubescence and outstanding black setæ. Thorax about as broad as the elytra, a good deal narrowed behind, but broad at the base; colour, sculpture and pubescence similar to those of the head. Elytra only slightly longer than the thorax, pitchy, with a narrow band at the apex yellow; densely punctured, with rather fine granular punctures, quite dull. Hind body elongate and narrow, only a very little narrower at the extremity than at the base, dull, densely and finely punctured, with a fine but very distinct ashy pubescence. Legs very pale yellow.

In the male the 6th segment of the hind body has an ill-defined channel along the middle, which reaches neither the base nor the hind margin; the latter is hardly perceptibly emarginate, but is furnished in the middle with a row of very short, fine black setæ; the hinder part of the ventral plate of the next segment bears a rather deep excision.

Tapajos; one male and three female individuals.

Obs. I.—Rather closely allied to our European *S. diversus*; this species is readily distinguished, however, therefrom by its duller surface, longer antennæ, broader thorax and the narrower band of the elytra.

Obs. II.—Besides these individuals, I have from the same locality two specimens (♂ and ♀) which do not appear to me to differ from them in any material respect except that of colour. These two individuals are entirely of a pale-fulvous colour, and appear at first sight, therefore, very different from the dark individuals above described; but I consider them merely a variety of the species.

2. *Sunius vittatus*, n. sp. Elongatus, angustulus, parallelus, dense punctatus, antennis, palpis, pedibusque fere albidis, niger; elytrorum apice vittisque duabus longitudinalibus testaceis. Long. corp. 2 lin.

Extremely closely allied to *S. amicus*, but readily distinguished by the colour of the elytra, each of which has the outside occupied by a large black patch, which does not quite reach the apex; the suture also very narrowly blackish, the space between these black marks, as well as the apex, yellow. It is also rather more slender than *S. amicus*, and the male characters are different.

In the male the ventral plate of the 6th segment of the hind body has a broad longitudinal impression on the middle, its hind margin being quite simple; the hinder part of the ventral plate of the next segment bears a rather deep excision.

Tapajos; three males, two females.

3. **Sunius serpens**, n. sp. Angustus, testaceus, elytris maculâ mediâ abdominequc segmento 6º fuscis; dense punctatus, opacus; abdomine apicem versus latiore. Long. corp. 1¾ lin.

Antennæ not quite as long as head and thorax, pale yellow; 1st joint hardly longer than 3rd; 3rd much longer than 2nd, 4—6 differing little from one another; 7—10 each slightly shorter and scarcely broader than its predecessor, 11th distinctly thicker than 10th. Head broader than thorax, quite as broad as elytra, yellowish in colour, very densely and indistinctly punctured, dull; the punctures very shallow, the interstices very fine; eyes large, placed at the middle of the sides. Thorax longer than broad, a little narrower than the elytra, almost oval, the sides much narrowed behind; colour and sculpture similar to those of the head. Elytra narrow, rather longer than the thorax; rather paler in colour than the other parts, each with a dark oval spot on the middle nearer to the apex than to the base; rather coarsely but indistinctly and not densely punctured, not so dull as the other parts. Hind body elongate, very narrow at the base, broader towards the extremity, yellowish; the 6th segment blackish except at the extremity, quite dull, densely and finely punctured, with a very fine dense pubescence. Legs very pale.

The male has a rather deep excision at the extremity of the ventral plate of the 7th segment of the hind body.

Tapajos; numerous examples.

Obs.—This species greatly resembles the European *S. bimaculatus*, but is rather more slenderly formed, is of a more uniform pale colour, and the sculpture of the upper surface is much finer. A closely allied but distinct species is found at Rio de Janeiro.

4. *Sunius ventralis*, n. sp. Fulvus, antennis, pedibus, palpisque fere albidis, abdomine segmento 6° dorsali basi obscuriore; capite thoracequo opacis, dense minus distincte punctatis; elytris parcius punctatis, nitidulis; abdomine apicem versus dilatato, parcius punctato, evidenter pubescente. Long. corp. 1¾ lin.

About the size of *S. angustatus*, but rather more slender. Antennæ long and slender, as long as head and thorax, scarcely thickened towards the apex, very pale yellow; 3rd joint much longer than 2nd. Labrum very short, with a slight notch in the middle, it is a little angular on each side of the notch. Head considerably broader than the thorax, about as broad as the elytra; eyes large, and placed at the middle of the sides, remote from the antennæ; the surface densely but indistinctly punctured, with the punctures large and ocellated, the interstices extremely fine. Thorax rather small, more than half as broad as the elytra, longer than broad, a good deal narrowed towards the base, and rounded in front; sculpture similar to that of the head. Elytra rather broad, short and convex, a little longer than the thorax, yellowish in colour, paler at the apex than at the base, on the middle of each an obsolete trace of a darker mark; the punctuation sparing at the extremity, closer at the base. Hind body convex, narrow at the base, a good deal broader towards the extremity, shining but for the long, fine pubescence; rather coarsely punctured, the punctures consisting of three rows of transversely placed punctures on each segment; these rows very distinct on the under side, much less distinct on the upper. Legs very pale yellow.

In the male the 6th segment of the hind body has on the under side a large, impunctate, longitudinal impression, and the following segment bears a small notch at the extremity.

Tapajos; a single male.

Obs.—This species is interesting from the peculiar, *Œdichirus*-like punctuation of the hind body.

5. *Sunius strictus*, n. sp. Nigricans, antennis, palpis, pedibus, elytrorumque apice pallide testaceis; elytris thoracis longitudine, rugulose punctatis. Long. corp. 2 lin.

Very similar at first sight to *S. angustatus*. Antennæ yellow, as long as head and thorax; 1st joint rather stout, 2nd more slender and a good deal longer than 3rd; 4—7 slender, 8—10 each a little thicker than its predecessor, 10th more than twice as long as broad, 11th distinctly stouter and a good deal longer than 10th. Mandibles elongate, yellow. Head broader than the thorax, about as broad as the elytra, blackish, with reticulate punctuation, which becomes more rugulose on the vertex. Thorax small, rather longer than broad, much narrower than the elytra; much narrowed towards the base, and more abruptly towards the front, blackish, with a coarse, reticulate sculpture. Elytra broad and short, scarcely longer than the thorax, black, with the apical portion pale yellow, the pale part slightly broader near the suture than at the sides; they are densely, rather coarsely and rugulosely punctured. Hind body blackish, with the hind part of the 6th and following segment paler; it is rather stout and convex, but a good deal narrowed at the base; it is coarsely punctured. The legs are rather short, and almost white.

Rio Purus; a single female, taken by Dr. Trail on the 13th October, 1874.

Obs.—This species much resembles our European *S. angustatus*, but is slightly larger, the antennæ are considerably longer, the elytra are shorter and more rugulosely punctured, and the pale colour does not extend so far forwards along the suture; the hind body is more constricted at the base, and the eyes are a good deal larger.

6. *Sunius marginatus*, n. sp. Niger, fortiter punctatus, antennis elongatis, pedibusque pallide testaceis: elytris thorace longioribus, apice late testaceis. Long. corp. 2 lin.

Antennæ slender, quite as long as head and thorax; the 10th joint scarcely visibly thicker than the preceding ones, and the 11th only very slightly thicker than 10th. Head coarsely punctured, and with reticulate sculpture. Thorax with reticulate sculpture, of which the meshes are large and distinct. Elytra rather longer than the thorax, with a coarse and rather distinct punctuation, black, with the hind margin yellow, the yellow colour extending much farther forwards along the suture than elsewhere. Hind body rather coarsely punctured, and with distinct, rather coarse pubescence and setæ.

Pará; a single female captured by Mr. Rogers.

Obs.—The resemblance of this species to *S. angustatus* is at first extreme; not only is the general form and outline similar, but the colours of the elytra are similarly disposed; nevertheless, the two species are very different when compared. *S. marginatus* has the antennæ very much longer, the eyes larger, the thorax more elongate, and the sculpture coarser and less dense. *S. marginatus* is closely allied to *S. strictus*, but has the sculpture coarser, less dense, and therefore more distinct; the elytra rather longer, and the colours differently disposed, and the intermediate joints of the antennæ rather more elongate.

7. *Sunius brevis*, n. sp. Brevior, latiusculus, parallelus, testaceus, capite thoraceque rufescentibus, elytris maculâ laterali nigrâ; opacus, dense punctatus; capite subquadrato, oculis prominulis; thorace transverso. Long. corp. 1⅓ lin.

Antennæ pale yellow, rather slender, a little shorter than head and thorax; 1st joint elongate, as long as the three following together, 3rd a little longer than 2nd; the joints from 4—10 slightly thickened, the 10th being distinctly broader and shorter than the 4th, but not so broad as long; 11th joint rather stout, and a good deal longer than 10th. Labrum large, oblique on each side, so as to be prominent in the middle, the most projecting part with a slight notch. Mandibles moderately long; the left with one, the right with two teeth. Head broad, quite as broad as the elytra; the eyes large and prominent, placed in the middle of the sides, the vertex appearing emarginate in front of the neck; the hind angles about right angles, and a little rounded; the surface dull, densely punctured with large but indistinct punctures, the interstices of which are extremely fine. Thorax about as broad as the elytra, distinctly transverse; the front rounded, the sides a little narrowed behind; the surface densely punctured, in a similar manner to the head. Elytra yellow, paler than the head and thorax; the side of each with an ill-defined spot, distinctly longer than the thorax, rather closely punctured. Hind body broad and short, very finely punctured. Legs pale yellow, rather short.

In the male the hind margin of the ventral plate of the 6th segment of the hind body is slightly emarginate on

each side of the middle; that of the 7th segment is broadly
emarginate.

Ega; one male and one female.

8. *Sunius modestus*, n. sp. Brevior, latiusculus, parallelus, testaceus, capite piceo-rufo, thorace rufescente; dense punctatus, subopacus; capite subquadrato; thorace transverso. Long. corp. 1⅓ lin.

This species is extremely close to *S. brevis*, but the head and thorax are not quite so broad; the elytra are without lateral mark; the head is darker in colour, and its punctuation is more conspicuous, and joints 3—10 of the antennæ are not quite so elongate. These differences are but slight however.

In the male the hind margin of the 7th segment is broadly emarginate; and the 6th segment, instead of being emarginate on either side as in *S. brevis*, has a single broad, rather deep emargination, extending from side to side, like that of the 7th segment.

Tapajos; one male and one female.

9. *Sunius crassus*, n. sp. Brevior, latiusculus, parallelus, testaceus, capite thoraceque rufescentibus; dense punctatus; capite subquadrato; thorace transverso, elytris hoc paulo longioribus, subnitidis, sat crebre punctatis. Long. corp. 1¼ lin.

Extremely closely allied to *S. brevis*, but not quite so dull; the head and thorax rather more distinctly punctured, the thorax not quite so transverse; the elytra more sparingly punctured and distinctly shining, and without lateral mark, and the antennæ a little shorter. Even more closely allied to *S. modestus*, but with the head rather broader and paler in colour, and the elytra more distinctly less closely punctured.

Tapajos; a single female.

10. *Sunius pictus*, n. sp. Latus, subparallelus, testaceus, capite thoraceque rufescentibus, elytris maculâ laterali suturâque nigricantibus, abdominis segmento 5° infuscato; dense punctatus, fere opacus. Long. corp. 1½ lin.

Antennæ pale yellow, rather slender, shorter than head and thorax; 1st joint almost as long as the three following together, 3rd quite as long as 2nd, each slightly shorter than its predecessor, 7—9 differing little from one another, 10th slightly stouter and shorter than 9th, small, about as long as broad; 11th joint a good deal longer and distinctly stouter than 10th. Head large, subquadrate; the eyes prominent and rather large; the vertex straight, but emarginate at the neck; the hind angles but slightly rounded, reddish, densely punctured, the punctures distinct, and each umbilicate, the interstices extremely fine. Thorax scarcely narrower than the head, and very slightly narrower than the elytra, nearly as long as broad; the front rounded, the sides slightly convergent behind; the colour and sculpture similar to those of the head, but the latter rather finer and less distinct. Elytra rather longer than the thorax, quadrate, rather finely and closely punctured; a large spot on the side of each black, and the suture also blackish, but with this colour not reaching to the scutellum. Hind body broad and short, densely, finely and indistinctly punctured; the basal segments reddish, the 5th infuscate, the 6th pale yellow. Legs pale yellow, rather short.

Ega; a single female.

Obs.—This species is closely allied to *S. brevis*, but is larger, and its thorax is less transverse; I have not been able to examine satisfactorily the mandibles and labrum of the only individual I possess, but from what I can see of them, I have little doubt they are very similar to those of *S. brevis*.

11. *Sunius confinis*, n. sp. Rufescens, antennis, pedibus, elytrisque testaceis; capite thoraceque dense punctatis, fere opacis, hoc elytris multo breviore et angustiore; elytris leviter nitidulis, crebre punctatis. Long. corp. 1½ lin.

Antennæ pale yellow, moderately slender, shorter than head and thorax; 1st joint almost as long as the three following together; 3rd joint about as long as, but a good deal thinner than 2nd, 5—10 scarcely differing in length, each about as long as broad, and just a little broader than its predecessor; 11th distinctly stouter than the 10th. Labrum with two short but distinct denticles in the middle. Mandibles moderately long and slender, their teeth very indistinct. Head rather long and narrow, distinctly broader than thorax, and narrower than elytra; the eyes placed in the middle of the sides, convex and prominent, the hind angles rounded; reddish in colour, densely but not very distinctly punctured, the interstices extremely fine. Thorax rather small, a little longer than broad, much narrower than the elytra; its greatest width in front of the middle, thence much narrowed towards the front, and slightly towards the base; colour and sculpture similar to those of the head. Elytra rather elongate, much longer than the thorax, a little narrowed towards the shoulders, yellow, distinctly and moderately closely and finely punctured. Hind body broad and rather short, tawny in colour, dull, very finely punctured and pubescent. Legs pale yellow, moderately long.

Ega; a single female.

12. *Sunius catena*, n. sp. Rufescens, antennis, pedibus, elytrisque testaceis; capite thoraceque dense punctatis, opacis, hoc sat elongato, capite angustiore; elytris thoracis longitudine, crebre punctatis, leviter nitidulis. Long. corp. 1½ lin,

Antennæ pale yellow, moderately slender; 1st joint as long as the three following together; joints 6—10 each slightly broader than its predecessor, the 10th almost transverse, 11th rather stouter than 10th. Labrum large, with two very approximate denticles in the middle, and with an extremely small one on either side of these. Head large, broader than the thorax, and quite as broad as the elytra, the hind angles quite rounded, the eyes prominent; reddish in colour, densely punctured, with the interstices very fine. Thorax a good deal narrower than the elytra, rather longer than broad; its greatest breadth in front of the middle, thence much narrowed towards the front, and distinctly towards the base; reddish, quite dull, densely punctured. Elytra quite as long as the thorax, quadrate, yellow, distinctly, moderately closely punctured, a little shining. Hind body broad, rounded at the sides, yellowish, densely and finely punctured. Legs pale yellow.

In the male the hind margin of the ventral plate of the 6th segment of the hind body is broadly but shallowly emarginate; the 7th segment is concealed in the example described.

Ega; a single male.

Obs.— This species is readily distinguished from *S. confinis* by its broader form and shorter elytra. I have been unable to keep open the mandibles for proper examination, but I think they are longer and have the teeth more distinct than in *S. confinis*. The insect greatly suggests by its form and colour a *Boreaphilus*, and makes a great approach to *Stilicopsis paradoxus*, Sachse.

13. *Sunius peltatus*, n. sp. Rufo-testaceus, antennis pedibusque testaceis; elytris abdominisque segmento 5° infuscatis, illis apice late testaceis; capite magno, subquadrato, anterius dense punctato, vertice sublævigato; thorace suborbiculato. Long. corp. 1¾ lin.

Antennæ slender, pale yellow, shorter than head and thorax, not thickened towards the extremity; 1st joint elongate, about as long as the three following together; 2nd joint rather slender, 3rd distinctly longer than 2nd, 4—10 each slightly shorter than its predecessor; 11th longer, but scarcely broader, than 10th, oblique at the extremity. Mandibles elongate and slender, the left one with two teeth, of which the upper one is elongate. Labrum much produced towards the front in the middle, the prominent part terminating in two distinct teeth. Head much larger than thorax, and even broader than the elytra; the eyes large and convex, placed in the middle of the sides, the hind angles not much rounded, the vertex emarginate in front of the neck; reddish in colour, closely finely and not very distinctly punctured, with extremely fine interstices, the punctuation becoming obsolete in the middle towards the vertex, so that in front of the emargination it is represented only by some fine granules. Thorax small, much narrower than the elytra, not longer than broad; the greatest width in front of the middle, thence much narrowed towards the front, and distinctly towards the base; reddish in colour, densely punctured, with very fine interstices. Elytra rather broad and short, a little longer than the thorax, infuscate, so as to approach blackish in colour; the extreme base obscurely reddish, the apex broadly, and abruptly, pale yellow; rather closely and distinctly punctured. Hind body broad, very finely punctured; the basal segments reddish, the 5th infuscate, the apical ones yellowish. Legs pale yellow.

Ega; a single female.

Obs.—This species in its general form makes a distinct approach to *Stilicus*.

14. *Sunius palpalis*, n. sp. Antennis, palpis, pedibusque pallide testaceis, femoribus quatuor posterioribus apice

late fuscis; capite thoraceque infuscato-rufis, illo vertice rotundato, hoc apicem versus fortiter angustato; elytris quadratis, summo basi rufescente, apice late testaceo, medio nigricantibus; abdomine nigricante, segmentis duobus basalibus rufescentibus, apice testaceo. Long. corp. 1¾ lin.

Antennæ elongate and slender, almost geniculate, not thickened towards the extremity; 1st joint elongate, as long as the three following together; 2nd joint rather long, 3rd more slender than, but about as long as 2nd, 5—10 each a little shorter than its predecessor, 11th rather long, slightly stouter than, and twice as long as the 10th. Mandibles elongate, the left one with two teeth, of which the upper one is elongate, the right one with three rather large teeth. Labrum large, the middle prominent, and with two rather long and slender teeth, and on each side of these with a small obscure one. Maxillary palpi pale yellow; 2nd and 3rd joints elongate and slender, the 3rd being about four times as long as broad. Head rather large, broader than the thorax and quite as broad as the elytra, the vertex rounded, the neck narrow, the eyes remote from the antennæ, the colour pitchy or infuscate red; the surface rather coarsely punctured, with the usual *Sunius* punctuation. Thorax a good deal narrower than the elytra, rather longer than broad; the greatest width in front of the middle, thence abruptly narrowed to the front, and a good deal towards the base; in colour and sculpture similar to the head. Elytra quadrate, a little longer than the thorax, blackish, with the extremity rather broadly pale yellow; the base reddish, moderately closely, quite distinctly punctured, a little shining. Hind body finely punctured; the two basal segments reddish, the three following ones blackish, the extremity yellow. Legs long and slender, the four hinder femora with a broad band of dark colour at the extremity.

Ega; two female individuals.

Obs.—This species in its general form differs little from **Stilicus**.

15. *Sunius bidens*, n. sp. Elongatus, angustulus, subcylindricus, testaceus, elytris medio puncto fusco notatis; antennis elongatis, tenuibus; labro medio obtuse bidentato. Long. corp. 2½ lin.

Antennæ almost white, very slender, rather longer than

head and thorax; 1st joint elongate and slender, not quite so long as the three following together; 2nd joint rather long; 3rd joint elongate and very slender, longer than 2nd; 4—10 each very slender, each slightly shorter than its predecessor, even the 10th three times as long as broad; 11th slender, but distinctly stouter than the preceding ones, rather longer than 10th. Labrum large, with two short, obtuse denticles in the middle, forming its most prominent part; slightly emarginate on each side of these, and thence falling away on either side as an oblique sinuation to the lateral angle. Mandibles elongate, abruptly curved, stout at the base; the left in the middle with a large stout tooth, which is slightly emarginate on its inner side; the right one with two rather large teeth, of which the upper one is the longer. Head broader than the thorax, quite as broad as the elytra; the eyes large, and very outstanding in front; the sides behind the eyes rounded and narrowed in a curve to the neck, yellowish in colour, rather flat; the surface with *Sunius* punctuation, forming rather large meshes, but immediately in front of the neck the sculpture obsolete; the underside almost impunctate. Thorax a good deal narrower than the elytra, the sides in front of the middle much rounded and narrowed to the neck, only very slightly curved, and narrowed from the widest part to the base; its colour and sculpture similar to the head. Elytra elongate, one and a half times the length of the thorax, yellowish, each with a dark spot on the middle, moderately finely and closely punctured. Hind body elongate and slender, dull, but only sparingly and indistinctly punctured. Legs elongate, very pale.

In the male the hind margin of the ventral plate of the 6th segment is broadly but slightly emarginate; the 7th has a very large and deep semicircular excision.

Ega; a single individual.

16. *Sunius bispinus*, n. sp. Elongatus, subcylindricus, testaceus, elytris abdomineque minus distincte fusco-variegatis; antennis elongatis, tenuibus, labro medio breviter bispinoso. Long. corp. 2¾ lin.

Antennæ almost white, elongate and slender; joints 8—10, though very slender, each distinctly stouter than its predecessor; 11th joint a good deal stouter and a little longer than the 10th. Labrum large, with two short, distinct spines in the middle. Thorax rather broad, but much narrower than the elytra, a good deal longer than broad. Elytra distinctly longer than the thorax, rather coarsely, closely and deeply punctured, marked each with an indistinct fuscous spot on the middle, and with a common one on the suture, towards the extremity. Hind body indistinctly and sparingly punctured, each segment with a longitudinal fuscous mark on each side, close to the margin, and the 6th with the base transversely marked with the same colour.

In the male the hind margin of the ventral plate of the 6th segment of the hind body is broadly but not deeply emarginate; the 7th has a rather broad and moderately deep subtriangular excision, the lower angles of which are quite rounded.

Tapajos; a single male.

Obs.—This species appears at first sight to be very closely allied to *S. bidens*, but when examined is found to be very distinct. *S. bispinus* is much broader, has the teeth of the labrum more elongate, the 10th joint of the antennæ stouter, and the notch of the 7th segment of hind body very different in form.

17. *Sunius spinifer*, n. sp. Elongatus, castaneus, capite thoraceque plus minusve infuscatis, illo elongato, labro spinis duabus elongatis armato; elytris nitidulis, fortiter punctatis, maculâ mediâ fuscâ; antennis tenuibus, perelongatis. Long. corp. $3\frac{1}{2}$ lin.

Antennæ very slender and elongate, nearly $1\frac{1}{4}$ lin. in length, quite filiform; 3rd joint very elongate, about as long as the basal joint; joints 4—10 differing from one another only in those nearest the apex being a little shorter; 11th joint elongate, pointed, slightly stouter and but little longer than 10th. Labrum large, in the middle with two straight, slender, elongate spines, and angulate on each side of these. Mandibles extremely elongate, toothed near the base with two teeth, of which the lower is very small. Head very elongate, the eyes convex, placed rather nearer to the antennæ than to the vertex, the sides behind the eyes narrowed in a gradual curve to the neck; pitchy in colour, with the usual *Sunius* sculpture, which is rather coarse and conspicuous; beneath with moderately coarse impressed punctures. Thorax much narrower than the elytra, longer than broad, transversely convex; the greatest breadth in front of the middle, thence much narrowed towards the front and slightly towards the base; blackish in colour (or reddish if immature), coarsely sculptured in a similar manner to the head, the sides towards the front appearing serrate. Elytra rather longer than the thorax, of a shining castaneous colour, with a dark spot on the middle, rather coarsely and deeply but not densely punctured. Hind body with the segments sericeous and coriaceous at the base, the hind portion of each shining and punctured, the punctures rather obsolete but moderately coarse; beneath entirely with a peculiar silky lustre and very obsoletely punctured. Legs yellow; the apex of the femora slightly infuscate.

In the male the ventral plate of the 6th segment is considerably shorter than the dorsal plate and is a little emarginate, and the ventral plate of the following segment is broadly but not deeply excised. The ædeagus itself is provided at the extremity with a slender elongate appendage, which is quite as long as the body of the organ; this appendage is sinuate in the middle, and furnished at the apex with an abruptly inflexed additional piece, giving its apex the form of a hook.

Ega; three individuals,—two males, one female.

18. *Sunius celatus*, n. sp. Elongatus, castaneus, capite thoraceque nigris, illo elongato, labro spinis duabus elongatis armato; elytris nitidulis, fortiter punctatis, maculâ mediâ fuscâ; antennis tenuibus perelongatis. Long. corp. 3½ lin.

In the male the ædeagus is furnished with an elongate and rather broad strap-like appendage, which is shorter than the body of the organ, a little constricted in the middle and furnished at the extremity with a very small abruptly inflexed additional piece.

Tapajos; a single male.

Obs.—The single individual before me seems to me to offer no certain character to distinguish it from *S. spinifer*, except the difference in the ædeagus; as the form and length of the appendage of this organ is quite the same in the two males from Ega, I have thought it advisable to consider the Tapajos individual as indicating a different species, though further researches on an additional number of specimens will be necessary before this can be considered to be certainly established.

19. *Sunius insignis*, n. sp. Castaneus, nitidulus, an-

tennis pedibusque testaceis, elytris oblique fusco-signatis;
thorace lateribus rotundatis, anterius subito constrictis,
crebre punctato, utrinque longitudinaliter impresso; abdomine latiusculo, minus distincte punctato. Long. corp.
2 lin.

Antennæ pale yellow, of the ordinary *Sunius* structure,
about as long as head and thorax, slender; 3rd joint much
longer than 2nd, 4—10 each shorter than its predecessor,
8—10 each slightly less slender than its predecessor, 10th
longer than broad, 11th rather stouter and a good deal
longer than 10th. Labrum very large, the middle part
much produced, and the most projecting part truncate,
without notch or teeth. Mandibles moderately long, much
curved; the left with two, the right with three teeth in the
middle. Head shining chestnut in colour, of the usual
Sunius form, but more gradually narrowed behind towards
the slender neck; it is only sparingly and indistinctly
punctured. Thorax narrower than the elytra, of a singular, somewhat flask-shaped, form; the sides of the hinder
portion rounded, then abruptly narrowed in front of the
middle to make a slender neck; the middle part is more
elevated than the sides, so that there is the appearance of
a longitudinal depression on each side; it is of a brownish
or chestnut colour and is rather coarsely punctured; it has
a narrow impunctate line along the middle, and the lateral
portions are only indistinctly punctured. Elytra about as
long as thorax, rather shining, of a chestnut colour, each
with a darker oblique mark across the middle, and the
outer angle a little paler; they are rather deeply, somewhat
coarsely and closely punctured. Hind body broad, chestnut-yellow, a little shining; the front portion of the two or
three basal segments rather distinctly punctured, the rest
scarcely visibly punctured. Legs very pale yellow; the
first joint of the hind tarsus about as long as the other four
together.

A single female of this extremely remarkable species
was captured by Dr. Trail on the 5th November, 1874;
but he has not transmitted to me the exact locality.

TÆNODEMA.

This genus consists at most of seven or eight described
species, two only having been known to Erichson; nevertheless I have described here eighteen species, and have
quite a dozen others in my collection. The genus appears

to me one of the most characteristic of the South American *Staphylinidæ*; it has not yet occurred in Mexico or away from the tropical parts of the South American continent, and I know of no similar insects in the Old World. The species here described show a considerable variation of size, form and appearance, though some of them are excessively closely allied *inter se*. They are apparently very rare, and I have had great difficulty in making up my mind as to whether certain of the forms should be treated as species or varieties; it is possible that I may in some cases have come to wrong conclusions on this point, although I have made a very careful examination in each case before coming to a decision.

1. *Tænodema plana*, n. sp. Nigra, capite, thorace, elytrisque æneis, abdomine segmentis 4^0, 5^0, 7^0que ex parte rufis: thorace parce, irregulariter, dorso sub-biseriatim punctato. Long. corp. 11 lin.; lat. (elytrorum) 2 lin.

Antennæ black, rather slender, nearly as thick at the extremity as at the base, about as long as head and thorax (3 lines in length); 3rd joint nearly twice as long as 2nd, 11th slightly longer than 10th. Head distinctly narrower than the thorax, with a large shallow impression on each side in front; its surface with the middle and vertical portions nearly impunctate, the impressions coarsely punctured; some coarse scattered punctures behind the labrum, and a few very coarse rugose punctures on the inner side of the hind part of the eye; besides the coarse punctures the surface is sparingly sprinkled with some minute punctures. Thorax about as long as broad, slightly rounded at the sides, dark brassy, shining, sparingly and rather finely punctured; the punctures irregularly distributed, consisting of two longitudinal patches along the middle separated by a broad irregular space, and outside these with some punctures towards the sides, the front part (behind the eyes) slightly raised and smooth; besides the larger punctures with a few minute punctures, which are most visible about the sides. Scutellum narrow and elongate, with a few coarse punctures. Elytra about as long as, and scarcely broader than, the thorax, dark brassy, with some coarse punctures scattered over them; the hinder external angles much rounded, and, as well as the projecting humeral angles, free from punctures. Hind body elongate, black; on the upper side the 4th segment has a large red blotch on each side; the hinder half of the 5th and of the 7th segments red, the basal portions of the segments with coarse but not dense punctures; on the underside black, with the hind part of the 7th and small patches at the angles of the 4th and 5th segments red, very coarsely punctured, each of the terminal segments with a small smooth space towards the extremity. Legs quite black.

Ega; a single female; also a variety from the Upper Amazons, brighter in colour, punctuation slightly coarser, and thorax obsoletely bi-impressed; this variety also represented by only a single female.

The *Staphylinus æneus* of Olivier, from Surinam, is probably a closely allied species.

2. *Tænodema lævis*, n. sp. Elongata, nitida, nigra, capite thoraceque superne viridi-æneis, elytris violaceo-cæruleis; abdominis apice rufo, antennarum basi palpisque testaceis. Long. corp. 8 lin.; lat. elytrorum $1\frac{1}{4}$ lin.

Mas: abdomine segmento 7° ventrali apice medio triangulariter exciso.

Antennæ rather slender, $1\frac{3}{4}$ lin. in length, black, the two basal joints yellow but infuscate inwardly; 3rd joint twice as long as 2nd, the apical joints slender. Palpi obscure yellow; mandibles pitchy. Head above shining brassy, the front part coarsely punctured; a space in front of the vertex nearly impunctate, the vertex on either side coarsely punctured. Thorax a little longer than broad, $1\frac{1}{4}$ lin. in length, 1 lin. in breadth, nearly parallel-sided, similar in colour to the head, with a longitudinal patch of punctures on each side the middle, with some other coarse punctures at the side about one-third of the length from the front, and three or four others behind these, and also with some fine punctures scattered sparingly over the surface. Scutellum same colour as head and thorax, with a few rather coarse punctures. Elytra scarcely longer than the thorax, violet, but blue at the base, smooth and shining, very sparingly punctured, the punctures becoming extremely fine and sparing towards the extremity. Hind body shining black; the 7th segment (except its base) and the 8th bright-orange colour, each segment at its base with a few coarse but obsolete punctures. Legs and under surface (except extremity) black; under face of hind body coarsely but not densely punctured; front tarsi infuscate-yellow.

St. Paulo; two individuals, ♂ and ♀.

Apart from its beautiful colour, this species, though clothed with long black setæ, is remarkable for the very scanty pubescence of the upper surface.

3. *Tænodema recta*, n. sp. Nigra, nitida, capite, thorace, elytrisque viridi-æneis, his apicem versus cyanescentibus; palpis antennarumque articulis basalibus testaceis, his intus infuscatis, tarsis anterioribus fusco-testaceis; capite thoraceque fortiter inæqualiter, elytris parcius subtiliter, punctatis. Long. corp. 7 lin.; lat. elytrorum 1⅛ lin. (vix).

Mas latet.

Elongate and narrow; antennæ rather slender, 1½ lin. in length, not thickened towards the extremity, black, with the two basal joints yellow, but infuscate inwardly; 3rd joint about twice as long as 2nd. Palpi yellow; mandibles pitchy black. Head on the upper surface pale brassy, coarsely punctured, only a narrow transverse space between the eyes being impunctate; the punctures are umbilicate, each bears a fine pale hair. Thorax nearly as long as broad (about 1 lin. broad, and $\frac{13}{8}$ long); it is a little rounded at the sides, and slightly narrowed behind, similar in colour to the head; the surface is rather closely but unequally covered with punctures, the punctures wanting on a space on each side near the front angles, also on a very small longitudinal space in front of the base in the middle, and more sparing about the margins than they are on the disc; the punctures are of unequal sizes, the larger ones being umbilicate. Scutellum brassy, punctured. Elytra quite 1⅛ lin. in length, shining metallic, the basal part similar in colour to the head and thorax, the apical part blue; they are sparingly punctured, the punctures are all fine, those at the base being coarser than the extremely fine ones at the extremity. Hind body long and slender, entirely black; on the upper side the basal segments are rather coarsely but not closely nor deeply punctured on their anterior portions; the 6th and 7th segments more finely and sparingly punctured, the latter towards the extremity with the setæ shorter and coarser; on the under surface the punctuation of the basal segments is deeper and coarser, but on the 7th segment it is finer than on the upper surface, as the setigerous punctures are not so conspicuous; on both faces it is furnished with conspicuous black setæ of various lengths, but is otherwise quite destitute of pubescence; the upper surface of the front tarsi is infuscate-yellow.

Ega; a single female individual.

4. *Tænodema lenta*, n. sp. Nigra, nitida, capite, thorace, elytrisque viridi-æneis; palpis, antennarum basi, tarsisque anterioribus testaceis; capite thoraceque fortiter, elytris parce subtiliter, punctatis. Long. corp. 7½ lin.; lat. elytrorum 1 lin.

Mas: abdomine segmento 7° ventrali apice late, paulum profunde emarginato.

Elongate and narrow; antennæ slender, about 1¼ lin. in length; the two basal joints yellow, very slightly infuscate inwardly; the apical joint pitchy, 3rd one and a half times as long as 2nd. Palpi yellow; mandibles pitchy at the base, red at the apex. Head small, on the upper surface shining brassy, densely and coarsely punctured, a transverse space in front of the vertex impunctate; the punctures are umbilicate, and bear each a fine pale hair. Thorax ⅞ lin. long, and not quite 1 lin. broad, distinctly rounded at the sides, similar in colour to the head, very shining, coarsely punctured, the punctures irregularly distributed, wanting on a space near the front angles; two elongate patches of punctures along the middle separated by an indistinct space, bearing finer punctures, which are absent in front of the base; the marginal portions coarsely but not densely punctured, the larger punctures umbilicate. Scutellum concolorous, slightly punctured. Elytra over 1 line in length, similar in colour to the thorax, very shining, greenish brassy, very sparingly and finely punctured, the apical punctures finer than the basal ones. Hind body elongate, and very narrow, black, the basal segments coarsely but not deeply nor densely punctured, their hind margins impunctate; 6th and 7th segments very sparingly punctured, the hind margin of the latter obscure reddish; the under surface coarsely punctured, the punctuation of the 7th segment rather coarse but obsolete, the hind margin reddish; the terminal segment obscurely reddish beneath, pitchy above; front tarsi yellow.

The male has the hind margin of the 7th ventral segment broadly but not deeply emarginate at the extremity.

Ega; a single male.

Very closely allied to *T. recta*; rather more slender, 3rd joint of antennæ shorter, the front parts more shining.

5. *Tænodema dubia*, n. sp. Nigra, nitida, capite, thorace, elytrisque viridi-æneis; palpis, antennarum basi tarsisque anterioribus testaceis; capite thoraceque fortiter, elytris parce subtiliter, punctatis. Long. corp. 6½ lin.

Mas: abdomine segmento 7° ventrali apice late, paulum profunde emarginato.

Elongate and narrow; antennæ slender, 1¼ lin. in length, black, with the two basal joints yellow and infuscate inwardly; 3rd joint one and a half times as long as 2nd. Head small, coarsely and densely punctured, except on a space in front of the vertex. Thorax ¾ lin. in length, and hardly broader than long, very little rounded at the sides, and slightly narrowed behind. Scutellum with a few fine punctures. Elytra 1 lin. in length. Hind body nearly concolorous at the extremity.

Ega; a single male.

This may possibly prove to be only an individual variation of *Tænodema lenta*; it differs only from it in being rather smaller, and having the front parts rather narrower, the thorax being a little less rounded at the sides, rather more coarsely punctured along the middle, the punctuation of the hind body rather less, and the legs a little shorter.

6. *Tænodema quadrata*, n. sp. Nigra, nitida, capite, thorace, elytrisque viridi-æneis, antennarum basi, palpis, tarsisque anterioribus testaceis; antennis apice piceis; capite thoraceque fortiter punctatis, elytris subtiliter punctatis, punctis suturalibus magis numerosis. Long. corp. 6 lin.

Antennæ about 1¼ lin. in length, the two basal joints yellow, the intermediate ones nearly black, the apical joint pitchy yellow; 3rd joint about one and a half times as long as 2nd. Head small, rather closely punctured, a space in front of the vertex impunctate, the punctures on the front part forming a patch in the middle and a patch on each side, but the three patches only a little separate; vertex rather finely punctured. Thorax not quite so long as broad, nearly ⅞ lin. in length, a little rounded at the sides, not narrowed behind, shining, coarsely punctured, the punctures more sparing towards the margins, and wanting on a space near the front angle. Scutellum finely but rather closely punctured. Elytra about 1 line in length and rather broader than long, brassy-green, shading into bluish at the sides and extremity, very shining, finely and sparingly punctured, the punctures almost wanting towards the extremity, and closer together along the suture, especially at the base. Hind body quite black. Legs rather short.

Ega; a single female.

This species is again very closely allied to *T. lenta*, but is shorter but not narrower; the elytra are distinctly more quadrate, and the legs a little shorter.

7. *Tænodema tarsalis*, n. sp. Nigra, nitida, capite, thorace, elytrisque viridi-æneis; antennarum basi, palpis, tarsisque anterioribus testaceis; capite thoraceque fortiter, elytris parce subtiliter, punctatis, punctis suturam versus magis numerosis. Long. corp. 5½ lin.

Antennæ slender, just over a line in length, black, with the two basal joints yellow, the apical joint nearly black; the 3rd joint one and a half times as long as the 2nd. Head small, above shining green, the front part coarsely punctured, not so closely along the middle as in the lateral depressions; a space in front of the vertex impunctate, the vertex punctured, but the punctures almost absent from its middle. Thorax about as long as broad, a little rounded at the sides, scarcely at all narrowed behind, shining, brassy-green, coarsely punctured, the punctures sparing towards the margins; a space near the front angles quite impunctate. Elytra about ⅞ lin. in length, scarcely so broad as long, greenish-brassy, more bluish at the sides and extremity, finely and sparingly punctured. Hind body entirely black; legs rather short.

Ega; a single female.

I am doubtful whether this will prove a distinct species from *T. quadrata* or not; it is a little smaller than *T. quadrata*, the head, thorax, and especially the elytra, are a little narrower; the antennæ are a little shorter and the front tarsi rather less dilated.

8. *Tænodema bella*, n. sp. Nigra, nitida, capite thoraceque viridi-cyaneis, elytris cyaneis, antennarum basi palpisque testaceis; prothorace disco plano, dense fortiter punctato; elytris parcius basi sat fortiter punc-

tatis, versus angulos externos posteriores fere lævigatis. Long. corp. 5½ lin.

Antennæ slender, with joints 1 and 2 yellow, 3—7 black (the rest wanting); basal joint rather short, scarcely longer than the 3rd joint, this being hardly one and a half times the length of the 2nd joint. Palpi yellow; mandibles pitchy, reddish at the extremity. Head small, shining, bluish-green, densely and very coarsely punctured, bearing a few fine pale hairs; transverse space between the eyes smooth and shining. Thorax about ¾ lin in length, and about the same in breadth, slightly rounded at the sides, and slightly narrowed behind; the disc flattened and coarsely and densely punctured, the margins with only a few scattered punctures, this free space being largest on the front near the angles; the colour is shining bluish-green, like that of the head. Elytra about as long as the thorax, rather broader than long, shining blue, rather finely and sparingly punctured; the punctures rather coarser and closer at the suture behind the scutellum than elsewhere, becoming finer towards the extremity and sides, so as to be wanting at the hind angles. Hind body black and shining, sparingly punctured, the hind parts of the segments impunctate, the 6th and 7th only very finely punctured, the outstanding black setæ well marked. Legs short and stout, black, with the front tarsi pitchy above, pale beneath; under face of hind body more coarsely punctured than the upper.

Ega; a single female.

This species, though it greatly resembles, at first sight, *T. tarsalis* and the allied species, may be readily distinguished by the dense and regular sculpture of the thorax, the punctures being all large and crowded together on the middle, instead of being of unequal sizes and unevenly distributed.

9. *Tænodema cinerea*, n. sp. Elongata, nigra, capite, thorace, elytrisque minus distincte metallescentibus, antennarum basi postice obscure testaceo; pedibus piceis; cinereo-pubescens, prothorace spatio utrinque, elytris suturâ apiceque, abdomine segmentorum marginibus posterioribus lævibus. Long. corp. 7¾ lin.

Mas: abdomine segmento 7º ventrali apice paululum emarginato.

Antennæ 2⅛ lin. in length, a little thickened towards

the extremity, pitchy black, with the two basal joints yellow behind; 3rd joint considerably shorter than 1st, and about twice as long as 2nd. Palpi pitchy, with the upper border of the last joint yellowish. Head coarsely punctured and clothed with an ashy pubescence, with a small space in front of the vertex smooth and shining. Thorax longer than broad, 1¼ lin. in length, 1 lin. broad, straight at the sides and scarcely at all narrowed behind; very densely and coarsely punctured, clothed with an ashy pubescence, the punctures leaving the hind margin, and a well-defined space on each side at the front, free and smooth; the base in the middle has also a small longitudinal smooth space. Scutellum elongate and narrow, almost impunctate. Elytra slightly longer than thorax, about 1¼ lin. in length, coarsely punctured and clothed with an ashy pubescence, but with the hind margin, the suture, except at the base, and the humeral angles, smooth and free from pubescence. Hind body elongate, clothed with an ashy pubescence; the 5th, 6th and 7th segments nearly entirely covered, the 2nd, 3rd and 4th with the hind margins broadly free, the 7th and 8th at their extremities ferruginous; under side covered also with a pale pubescence, which is more scanty than on the upper side. Legs elongate, pitchy at the base; tibiæ black; front tarsi pitchy above, the hairs of their under surface white. The male has the hind margin of the ventral plate of the 7th segment broadly but slightly emarginate; in both sexes this plate is a little emarginate on each side, so that its lateral margin projects as a short tooth.

Ega; three specimens, 2 ♂, 1 ♀.

10. *Tænodema vicina*, n. sp. Elongata, nigra, capite, thorace, elytrisque cyanescentibus: palpis, pedibus, antennisque piceis, pedibus basi rufescentibus, antennis basi postice testaceo; cinereo-pubescens, prothorace spatio utrinque, elytris suturâ apiceque, abdomine segmentorum marginibus posterioribus lævibus. Long. corp. 7½ lin.

This species is excessively closely allied to *T. cinerea*; it has the antennæ distinctly shorter, viz., 1⅞ lin. in length, the front parts are more distinctly blue, the legs paler at the base and a little shorter, the punctuation of the thorax approaching more nearly to the hind margin.

Amazons; a single female without special locality.

11. *Tænodema similis*, n. sp. Elongata, nigra, antennis pedibusque basibus rufis; capite, thorace, elytrisque cyanescentibus; cinereo-pubescens, prothorace spatio utrinque, elytris suturâ apiceque et abdomine segmentorum marginibus posterioribus lævibus. Long. corp. 7 lin.

Antennæ about 2 lin. in length, scarcely thickened towards the extremity, pitchy, with the two basal joints yellow behind; 3rd joint much shorter than 1st, about one and a half times as long as 2nd. Palpi pitchy yellow, with the upper margin of the last joint yellow; mandibles pitchy, reddish at the apex. Head coarsely and densely punctured and clothed with a pale pubescence, a transverse space in front of the vertex smooth and shining. Thorax 1 lin. long and about $\frac{7}{8}$ broad, very densely punctured, a small space on each side in front and the hind margin free (in a fresh specimen probably densely clothed with ashy pubescence). Elytra fully 1^1 lin. in length, densely punctured; the suture (except at the base) has a very narrow space free from punctures, and the hind margin is also free, but the punctures almost completely cover the extreme base and front angles (in fresh specimens the punctured parts probably covered with a dense pale-ashy pubescence). Hind body black; the segments punctured and pubescent, except on their apical portions; 8th segment and hind margin of 7th reddish. Legs pitchy, with the femora red.

In the male the hind margin of the ventral plate of the 7th segment is a little emarginate at the extremity.

Ega; a single male individual.

Very closely allied to *T. cinerea*, but narrower, with the thighs clear red, the front parts on the upper surface quite blue, and the thorax and elytra more evenly covered with punctures. The male has the ædeagus considerably longer than in *cinerea* and rounded at the extremity, whereas this is acuminate in *cinerea*. The single individual, which is all I possess for comparison, has the upper surface very rubbed, but I have no doubt, from the few hairs that remain, that in the fresh state the pubescence is very similar to that of *T. cinerea*.

12. *Tænodema rudis*, n. sp. Nigra, capite, thorace, elytrisque minus distincte metallescentibus; pedibus piceo-rufis, basi dilutioribus; antennarum basi testaceo, palpis piceo-testaceis, abdominis apice ferrugineo; cinereo-pubes-

cens, prothorace spatio utrinque, elytris apice suturâque apicem versus, abdomineque segmentorum marginibus posterioribus lævibus. Long. corp. 6½ lin.

Antennæ 1⅔ lin. long, distinctly thickened towards the extremity, pitchy, with the two basal joints yellow but infuscate in front; 3rd joint one and a half times as long as 2nd. Palpi pitchy yellow. Head coarsely punctured and with a pale pubescence, a small space in front of the vertex smooth and shining. Thorax 1 line (scarcely) in length, ⅞ lin. broad, coarsely and very densely punctured, with a space in front on each side smooth and shining; the hind margin also free from the dense punctuation, the surface bearing an ashy pubescence. Elytra 1¼ lin. long, coarsely punctured, bearing a pale pubescence; the hind margin and the suture except at the base free. Hind body black, the segments punctured and pubescent except at their hind margins, the apex reddish. Legs, including the coxæ, reddish; the tibiæ infuscate.

Ega; a single female individual.

This species is again very closely allied to *T. cinerea*, but besides being considerably smaller, it has the legs shorter and paler, and the punctuation of the thorax and elytra a little coarser. The individual described is much rubbed, but I do not think the pubescence would differ materially in fresh individuals from that of *T. cinerea*.

13. *Tænodema filum*, n. sp. Elongata, perangusta, nigra, antennarum basi, palpisque rufis; capite thoraceque viridi-cyaneis, dense fortiter punctatis, illo spatio frontali lævi; elytris cyanescentibus, elongatis, fortiter minus crebre punctatis, punctis ad apicem fere nullis. Long. corp. 6 lin.

Antennæ long and slender, 1¼ lin. in length, pitchy black, the two basal joints reddish, the following ones pitchy red; 3rd joint not one and a half times the length of the 2nd, 10th much longer than broad. Mandibles reddish; palpi yellow. Head small, of a greenish metallic colour, very densely, very coarsely and regularly punctured, with fine, sparing, pale hairs, a space between the eyes smooth and shining; the surface is a little convex, the antennal depressions absent. Thorax ⅔ lin. in length, and scarcely so broad, slightly rounded at the sides, evenly convex, coarsely and densely and deeply punctured; the punctures more sparing towards the sides, and especially so at the base, which is shining, but there is no impunctate space on the front; the colour is similar to that of the head, but a little bluer. Elytra elongate, 1¼ lin. in length, slightly broader than the thorax, dark blue, shining, rather coarsely but sparingly punctured, the punctures disappearing at the outer hind angle. Hind body elongate and very narrow, shining black; the basal segments rather coarsely but not closely punctured, their hind margins impunctate; apical segments finely and sparingly punctured, hind margin of 7th obscurely reddish. Legs black, short; front tarsi dusky red above; under face of hind body more coarsely punctured than the upper.

St. Paulo; a single individual, which is apparently a female.

14. *Tænodema producta*, n. sp. Elongata, perangusta, nigra, antennis piceis, basi palpisque rufis; capite, thorace, elytrisque cyanescentibus, illis dense fortiter punctatis; elytris fortiter, minus crebre punctatis, punctis ad apicem fere nullis. Long. corp. 6 lin.

Very closely allied to *T. filum*, but without any smooth impunctate frontal space; it is even a little more slender, the thorax a trifle smaller, and the elytra a trifle shorter, but in other respects resembles almost exactly *T. filum*.

Amazons; a single specimen, without special locality. It is apparently a female.

15. *Tænodema laticornis*, n. sp. Parallela, nigra, antennarum basi, palpis, abdominisque apice rufis; capite thoraceque cyaneo-viridibus, dense fortiter æqualiter punctatis; elytris cyaneis, fortiter crebre punctatis. Long. corp. 5½ lin.

Antennæ ⅞ lin. in length, the two basal joints red, the rest nearly black; 1st joint scarcely longer than 3rd; joints 4—10 differing little from one another in length, but each a little broader than its predecessor,—each of these joints is much broader at the extremity than at the base, they are broad and flat, the 10th broader than long; the 11th joint small, a good deal narrower than the 10th. Palpi yellow, mandibles small, each with a small, simple tooth in the middle, pitchy in colour. Head with the upper surface convex, the eyes small (in comparison with other species of the genus); the colour greenish-blue, the surface very densely punctured, and with fine grey hairs; a very small, ill-defined space behind the middle, smooth and shining. Thorax ⅔ lin. in length, and hardly so broad, not in the least curved at the sides, but a little narrowed behind; the whole surface covered with dense coarse punctures, a very narrow, longitudinal, smooth line along the middle; the colour is similar to that of the head, and it bears a scanty, pale pubescence. Scutellum short and broad, rounded at the extremity, punctured. Elytra about as long as the thorax, dark blue in colour, the punctuation rather coarse and moderately close, finer and more sparing towards the extremity. Hind body shining black, with the 8th segment and the hind margin of the 7th red; the connecting membrane of segments 2—5 white, but that between the 4th and 5th segments broadly infuscate in the middle; the segments are rather coarsely punctured. The legs are short and stout; the front and middle tarsi reddish, the front femora and tibiæ pitchy; the trochanters reddish, elsewhere nearly black.

Tapajos; a single specimen, apparently a female.

16. *Tænodema serpens*, n. sp. Nigra, capite, thorace, elytrisque plumbeis, fere opacis, dense fortiterque punctatis; antennarum basi, palpis, abdominisque apice rufis, pedibus anterioribus piceis. Long. corp. 4¼ lin.

Mas: abdomine segmento 7° ventrali apice medio excisione semicirculari magnâ, basi rigide nigro-setoso.

Antennæ moderately long, distinctly thickened towards the extremity, blackish; 1st and 3rd joints short, of about equal lengths, 4—10 each a little broader than its predecessor, 10th about as long as broad. Palpi yellow; mandibles red. Head with the surface convex, faintly metallic, very densely punctured, bearing a pale, upright pubescence; a very narrow space above the middle, smooth and shining. Thorax rather longer than broad, almost straight at the sides and slightly narrowed behind, transversely convex; the whole surface very densely and coarsely punctured, bearing a pale, upright pubescence. Scutellum small, coarsely punctured. Elytra slightly longer than the thorax, closely and coarsely punctured, a little shining, with a pale, upright pubescence. Hind body black, reddish at the extremity, sparingly clothed with a pale pubescence, most distinct on the 7th segment; the four basal segments coarsely and closely punctured, the two following ones much more indistinctly. Front legs pitchy, with the tarsi paler; middle legs darker than the front ones, but not black; hind legs black, with the trochanters pitchy: under face of hind body scarcely so closely and coarsely punctured as the upper.

The male has a broad and deep excision at the hind part of the ventral plate of the 7th segment of the hind body, the front portion being filled with coarse, black, straight setæ.

Ega; two individuals, ♂, ♀.

Allied to *T. laticornis*, but more slender, the antennæ not so broad; the front parts duller in colour, and the punctuation of the upper surface denser.

17. *Tænodema tecta*, n. sp. Nigra, subdepressa, opaca, cinereo-pubescens, antennarum basi palpisque testaceis, pedibus piceis; dense punctata. Long. corp. 4 lin.

Mas: abdomine segmento 7° ventrali apice late, paulum profunde exciso, excisione rigide nigro-setosâ.

Antennæ ⅞ lin. long, broad and flat, pubescent; two basal joints yellowish behind, infuscate in front, 4—10 each broader than its predecessor, the 10th much broader than long. Palpi yellow, the apical joint infuscate at the base; mandibles red. Head broad and short, its surface closely and rather coarsely punctured, with a conspicuous, upright, whitish pubescence. Thorax quadrate, about ¾ lin. long; the whole surface densely covered with moderately coarse punctures, and bearing a whitish pubescence. Elytra quadrate, flat, not longer than the thorax, rather densely punctured, sparingly clothed with white pubescence, which is replaced towards the extremity by a dark pubescence, the apical portion slightly shining. Hind body rather broad, black, coarsely and closely punctured, bearing a white pubescence, which is scanty on the basal segments, but dense and very conspicuous on the apical ones; the extreme apex obscure red, but the colour concealed by the pubescence. Under surface rather shining, the punctuation on the hind body rather coarse but not dense, the pubescence scanty. Legs pitchy, front tarsi paler, and only moderately dilated, hind tarsi elongate, nearly as long as the tibiæ.

In the male the ventral part of the 7th segment of the hind body has a broad but not deep notch at the extremity, the border of the notch being densely set with coarse, rigid, black setæ.

Ega: one specimen.

18. *Tænodema lurida*, n. sp. Nigra, supra opaca, dense punctata, cinereo-pubescens, antennarum basi palpisque rufis; pedibus piceis, anterioribus rufescentibus. Long. corp. 4 lin.

Antennæ thickened towards the extremity; the two basal joints yellow; the 3rd pitchy, the following ones nearly black, 10th strongly transverse. Palpi yellow, the basal portion of the last joint infuscate; mandibles red. Head very densely and coarsely punctured, bearing a pale pubescence; a narrow transverse space in front of the vertex, smooth and shining. Thorax transversely convex, about as long as broad, slightly rounded at the sides and slightly narrowed behind, coarsely and very densely punctured. Elytra slightly longer than the thorax, densely punctured; the suture and the basal portion bearing a pale pubescence, as also the inflexed side, but the larger portion of the hind part of each elytron with a fine dark pubescence. Hind body black, with the extremity pitchy red; the segments coarsely and closely punctured, the hind margins smooth and shining in the middle; the 6th and 7th segments more finely punctured than the basal ones; the basal segments are clothed with some fine white hairs, and these become more dense and distinct on the 6th and 7th segments, but the hind margin of this latter is quite smooth and shining. Anterior legs reddish, middle ones pitchy red, with the tibiæ darker; hind ones nearly black, with the trochanters red; under surface shining, hind body coarsely and evenly not densely punctured.

St. Paulo; a single individual, which I believe to be a female; the ventral plate of the 7th segment of the hind body is slightly elongate.

This species is closely allied to *T. tecta*; it is a little more slender and less depressed, the antennæ are slightly longer, the pale pubescence of the upper surface is less conspicuous, and is absent from a larger portion of the elytra.

Pinophilus.

This genus, as at present understood, consists of about fifty species, inhabiting the warmer parts of the Old and New Worlds in about equal proportions: it is probable, however, that the species are really more numerous in the New World than in the Old; for while in Europe the genus is only represented by an eastern species that has extended its habitat to the island of Sicily, in America, north of Mexico, several species occur, some of them, indeed, being apparently not uncommon there.

Mr. Bates has discovered a remarkably fine series of the genus, no less than twenty-four species being here described; of these *Pinophilus dux* is to be ranked amongst the largest of the *Staphylinidæ*, while other species are insignificant in size. One of the species, *P. mimus*, is remarkable from its great general resemblance to *Lathrobium opalescens*, the two species being, I believe, found living together.

As is also the case with some of the Amazonian species of *Lathrobium* and other genera, certain of the species here described bear the most complete resemblance to one another in their general characters, but are distinguished by well-marked external and internal sexual characters.

1. *Pinophilus dux*, n. sp. Robustus, niger, capite thoraceque nitidis, vage punctatis; elytris subnitidis, fortiter minus crebre punctatis, abdomine crebre sat fortiter punctato; antennis tarsisque obscure rufis. Long. corp. 14½ lin.

The largest and most powerful species of the genus, being considerably broader and slightly longer than *P. tenebrosus*. Antennæ rather short and stout, obscure reddish, the basal joints pitchy red. Head short and broad, shining black, sparingly and irregularly punctured; the punctures coarse, almost ocellated, and bearing fine setæ. Thorax just about as long as broad, a little narrowed behind, with all the angles rounded, shining black, coarsely and sparingly punctured, with punctures similar to those of the head, and irregularly distributed, a longitudinal irregular space along the middle being free from punctures; outside this an irregular double row of punctures, then a narrow very irregular space almost free from punctures, then a broad space at the sides with scattered punctures. Elytra slightly narrower than the thorax, and scarcely longer; rather coarsely and sparingly punctured, and very finely and sparingly pubescent. Hind body elongate and parallel; the segments rather coarsely and closely punctured and finely pubescent, the apical ones with the punctures coarser and more elongate than the others, but not rugose; the apex rather deeply emarginate.

Ega; a single female specimen.

This species appears to be closely allied to *P. torosus*, Er., a species which I know only by description. *P. dux* would seem to be larger than *torosus*, and it has not the apical segments of the hind body rugose; other details of Erichson's description of *P. torosus* do not seem very applicable to *P. dux*, so that if a comparison of the two species were made it is possible they might be found to differ in other minor points.

2. *Pinophilus ater*, n. sp. Niger, antennis, palpis, tarsis, abdominisque apice obscure rufis ; capite thoraceque pernitidis, parce punctatis; thorace oblongo, dorso biseriatim punctato, præter punctos majores punctis minimis adsperso; elytris thorace vix longioribus, subnitidis, fortiter sat crebre punctatis ; abdomine fortiter sat crebre punctato. Long. corp. 10 lin.

Elongate and parallel; antennæ nearly twice as long as the head, dull red ; the basal joints stout, the 3rd joint distinctly longer than 2nd. Mandibles with a long tooth. Head shining black, with a few large punctures about the middle, and also some very minute punctures; the hind angles rather closely punctured. Thorax distinctly longer than broad, quite as broad as the elytra, quite straight at the sides, with the hind angles much rounded ; very shining black, sparingly punctured, near the middle two irregular rows of punctures, and with some other scattered punctures about the sides and front, and the surface besides is sprinkled with very minute punctures. Elytra scarcely longer than the thorax, rather coarsely but not densely punctured. Hind body rather coarsely and closely punctured, its extremity reddish, truncate, and with a small spine on each side the truncation ; the apical segments rather more sparingly punctured than the preceding ones. The legs are pitchy, with the tarsi reddish ; the metasternum has a fine channel along the middle.

Tapajos; a single female specimen, from which the pubescence has been removed. This species is about the size of our *Ocypus ater*. It is larger and more parallel than the North American *P. picipes*, and differs therefrom by the large tooth of the mandibles.

3. *Pinophilus rectus*, n. sp. Elongatus, parallelus, niger, antennis, palpis, pedibusque rufis ; capite thoraceque nitidis, hoc oblongo, crebre irregulariter punctato, lineâ

mediâ angustâ lævi; elytris crebre fortiter punctatis; abdominis apice breviter bispinoso. Long. corp. 8 lin.

Mas: femoribus posterioribus subtus medio dente acuto. Fem.: femoribus muticis.

Antennæ nearly as long as head and thorax, red. Mandibles with an elongate tooth. Head slightly narrower than thorax, shining, the upper surface coarsely but sparingly punctured, except at the hind angles, where the punctuation is dense and rugose; it has also a few very fine punctures. Thorax as broad as the elytra, elongate and parallel, longer than broad, quite straight at the sides, moderately coarsely punctured, shining, the punctures irregularly distributed, leaving a narrow space along the middle smooth; outside this a space where the punctures are crowded together, then again an ill-defined space, most distinct towards the front, free from punctures, the sides again more closely punctured especially on the hinder part. Elytra oblong, quite as long as the thorax, rather coarsely and closely punctured. Hind body closely and rather coarsely punctured, generally reddish at the extremity; the apex of the upper terminal plate is not truncate, but a little rounded or produced in the middle, with a short spine on each side. Legs clear red.

In the male the trochanters of the hind legs are acuminate at the extremity, and the thighs are armed in the middle beneath with a sharp tooth; the 6th ventral segment is depressed in the middle towards the extremity, the hind margin narrowly emarginate; the 7th segment has a rather broad and deep emargination at the extremity, in front of which it is flattened or depressed.

Santarem and Ega. The specimen from Ega has the extremity of the hind body concolorous, whereas it is reddish in the specimens from Tapajos.

4. *Pinophilus æqualis*, n. sp. Nigricans, antennis, palpis, pedibusque ferrugineis, capite thoracequc nitidis, illo vertice dense punctato; thorace oblongo, crebre minus fortiter punctato, lineâ mediâ angustâ lævi; elytris opacis, dense subtiliter rugoso-punctatis; abdomine crebrius minus fortiter punctato, apice unispinoso. Long. corp. 7½ lin.

Mas: abdomine segmento 7° ventrali medio profunde emarginato.

Antennæ obscure red, slender, as long as head and

thorax; 3rd joint distinctly longer than the elongate 2nd joint. Head distinctly narrower than the thorax, with two coarse punctures in the middle immediately behind the labrum; on each side the middle near the front with a patch of about a dozen coarse punctures, the front part shining and with some scattered minute punctures; the vertex and inner margin of the eyes densely and coarsely punctured and not shining. Thorax slightly longer than broad, straight at the sides, and only very slightly narrowed towards the base; rather shining, closely but not coarsely punctured, with a very narrow line along the middle smooth; also with a small smooth space near the front angles, and a second just behind it. Elytra about as long as the thorax, and scarcely broader, with a very dense moderately fine punctuation, quite dull. Hind body quite dull, rather densely pubescent, closely and rather finely and indistinctly punctured; its extremity obscure red, terminated in each sex by a short point or spine. Legs obscure reddish.

In the male the ventral plate of the 7th segment has in the middle a rather deep notch, which is broad at the opening, narrow and rounded at the summit.

My specimens of this species bear no other locality than " Amazons."

5. *Pinophilus mimus*, n. sp. Parallelus, piceus, capite thoraceque nitidis, illo vertice parce punctato; thorace quadrato, subtiliter crebre punctato, lineâ mediâ angustâ, minus discretâ, lævi; elytris crebre sat fortiter punctatis, vix nitidis; abdomine opaco, fusco-pubescente, minus dense punctato, apice rufo; antennis, palpis, pedibusque rufis; abdomine segmento 5° ventrali apicem versus lineâ transversâ impressâ. Long. corp. $5\frac{1}{2}$—6 lin.

Mas: segmento 7° ventrali apice medio emarginato, emarginatione anterius angustâ.

Antennæ red, very slender, elongate, quite as long as head and thorax; 3rd joint considerably longer than 2nd. Head small, shining, with a purplish reflection, the front part very sparingly punctured; the vertex sparingly but distinctly punctured, the punctures not coarse. Thorax quadrate, similar in colour to the head, evenly, rather sparingly and finely punctured, with an indistinct line along the middle, and a small space near the front angles, impunctate; the surface is very shining, the pubescence being very fine and scanty. Elytra slightly longer than the thorax, evenly, rather closely, and moderately finely punctured. Hind body quite dull, with a long and distinct pubescence; its punctuation moderately coarse, but indistinct and not dense, the extremity red, the hind margin simply rounded; the 5th segment on the underside is impressed near the extremity with a not very distinct transverse line. Legs red.

The male has a rather deep notch, the anterior part of which is very narrow in the middle of the hind margin of the 7th ventral segment.

Tapajos; several specimens.

6. *Pinophilus modestus*, n. sp. Parallelus, piceus, capite thoraceque pernitidis, illo parce punctato; thorace oblongo, sat crebre subtiliter punctato, lineâ mediâ, minus discretâ, lævi; elytris sat crebre fortiter punctatis, subnitidis; antennis pedibusque rufis. Long. corp. 7½ lin.

Antennæ red, slender and elongate, quite as long as head and thorax; 3rd joint one and a half times the length of 2nd. Head almost as broad as thorax, very shining, the upper surface with a few coarse punctures, most numerous about the hind angles, but nearly wanting in the middle of the vertex. Thorax oblong, longer than broad, quite straight at the sides, shining, rather finely and not densely punctured, the punctures leaving an irregular space along the middle, and a round space near the front angles, impunctate. Elytra slightly longer than thorax, rather coarsely and not densely punctured. Hind body not densely punctured, reddish at the extremity.

Ega; a single female.

This species, though closely allied to *P. mimus*, is abundantly distinct by its larger size, more elongate thorax, and the more sparingly punctured front parts of the upper surface; the extremity of the hind body is simple, as in *P. mimus*, but I have not ascertained whether the 5th segment bears a transverse line or not.

7. *Pinophilus tenuis*, n. sp. Angustulus, parallelus; capite thoraceque pernitidis, fere impunctatis; rufescens, capite, elytris, abdomineque obscurioribus, hoc longius pubescente, minus distincte punctato. Long. corp. 4 lin.; lat. prothoracis vix ultra ½ lin.

Mas: abdomine segmento 7° ventrali apice obsolete emarginato.

Antennæ red, scarcely so long as head and thorax, rather stout, not thinner at the extremity; 3rd joint a little longer than 2nd. Head small, shining, dark reddish, with a very few punctures. Thorax longer than broad, about as broad as the elytra, quite straight at the sides, shining red, with a few fine punctures, viz., four distant ones in a row on each side of the middle, and five or six on each side near the margin. Elytra a little longer than the thorax, infuscate, rather finely and sparingly punctured, but slightly shining. Hind body dull, infuscate, reddish at the extremity, the punctuation indistinct, moderately close. Legs yellowish-red; extremity of hind body simply rounded.

The male has only a very shallow notch at the extremity of the ventral plate of the 7th segment of the hind body.

8. *Pinophilus distans*, n. sp. Niger, fere opacus, antennis, palpis, pedibusque testaceis, abdomine apice rufo; vertice dense punctato; thorace oblongo, elytrisque dense minus fortiter punctatis. Long. corp. 6¼ lin.

Mas: abdomine segmento 7º ventrali fere simplice, apice obsolete emarginato.

Antennæ yellow, slender, moderately long; 3rd joint a little longer than 2nd. Palpi pale yellow; mandibles pitchy. Head with the vertex densely punctured; in front of this with a transverse, quite shining space, the anterior part with coarse punctures on each side, and two moderately large punctures on the front edge of the clypeus in the middle. Thorax one-fourth longer than broad, quite straight at the sides, densely, moderately coarsely and evenly punctured, the punctures covering the whole surface and leaving no smooth spaces. Elytra scarcely longer than the thorax; their punctuation very similar to that of the thorax, their hind margin obscurely red. Hind body moderately coarsely and not closely punctured, the extremity broadly red. The legs are pale yellow; the anterior tarsi extremely broad.

The male has the ventral plate of the 7th segment almost simple; it is not produced, and only very obsoletely emarginate at the extremity; the ventral plate of the 8th segment is visible and rather broad, and remains part of the sheath, it not being modified to form part of the intromittent organ; the form of the latter is very peculiar.

Amazons; a single male individual, without more special

locality; the species appears allied to *P. javanus*, Er., and its Indian allies, rather than to the other Austro-Columbian species; this makes me think it possible there may be an error in the indication of its locality.

9. *Pinophilus incultus*, n. sp. Niger, omnino opacus, antennis, palpis, pedibusque pallidis; thorace quadrato, obsoletissime punctato; elytris creberrime sat fortiter ruguloso-punctatis. Long. corp. 4½ lin.

Mas: abdomine segmento 7° ventrali apice medio emarginato.

Antennæ yellow, very slender, 3rd joint about as long as 2nd. Head short, black, mandibles and front of clypeus pitchy red, dull, the front part with large distant punctures, the vertex on each side coarsely but obsoletely punctured. Thorax not quite so long as broad, slightly narrowed behind, very dull, sparingly and extremely obsoletely punctured; in the middle in front of the base with a short, very obscure, longitudinal elevation. Elytra a little longer than the thorax, quite dull, densely punctured; the punctures moderately coarse, the interstices rugulose. Hind body narrow, quite dull, indistinctly punctured, the punctuation obscured by a coarse pubescence. Legs pale yellow, with the coxæ dusky reddish, the front tarsi very broad.

In the male there is a small but distinct notch at the extremity of the ventral plate of the 7th segment of the hind body; the ventral plate of the 8th segment is enfolded by the dorsal plate; it is narrowed towards the extremity, and furnished there with two small hooks placed at right angles to the rest of the plate; in the female this plate is broader and visible, and has a triangular notch at the extremity.

Tapajos; several specimens; one immature specimen is quite pale brown; I have also a male specimen from Ega which differs only in being broader and a little more depressed.

10. *Pinophilus proximus*, n. sp. Niger, opacus, antennis, palpis, pedibusque pallidis; thorace quadrato, parce obsolete punctato; elytris dense fortiter ruguloso-punctatis, subopacis. Long. corp. 4½ lin.

This species differs from *P. incultus* only by some details of punctuation; the thorax is rather more distinctly punctured, the elytra more coarsely and less densely punctured and not so dull. The male has the 7th segment formed as in *P. incultus*, but the notch is narrower in *P. proximus*.

Tapajos; two specimens, one of which is brown, being immature.

11. *Pinophilus angustus*, n. sp. Niger, antennis, palpis, pedibusque pallidis; thorace quadrato, æquali, opaco, parcius obsolete punctato; elytris opacis, fortiter crebre punctatis. Long. corp. 4½ lin.

Narrow; antennæ yellow, slender, moderately long, 3rd joint a little longer than 2nd. Palpi yellow; mandibles, front edge of clypeus and antennal tubercles pitchy. Head dull, front part with sparing coarse punctures, punctuation of the vertex obsolete. Thorax quite as long as broad, quite straight at the sides, not narrowed behind, quite as broad as the elytra; its surface dull, sprinkled with distinct, though very slightly impressed, moderately fine punctures. Elytra slightly longer than the thorax, dull, rather coarsely and moderately closely punctured, the interstices larger than the punctures. Hind body rather slender, its punctuation indistinct, rather coarse, but shallow and not dense, concealed by a coarse pubescence. The legs are pale yellow, with the coxæ reddish; the anterior tarsi very broad.

Tapajos; two specimens, both females; the extremity of the hind body is formed as in *P. incultus* ♀, to which species, as well as to *P. proximus*, the present species is closely allied, but it may readily be distinguished by its more slender form and more sparingly punctured elytra.

12. *Pinophilus oblatus*, n. sp. Elongatus, niger, opacus, antennis gracillimis, palpis pedibusque pallidis; thorace quadrato, obsoletissime punctato; elytris peropacis, dense ruguloso-punctatis; abdomine apicem versus minus discrete rufescente. Long. corp. 6 lin.

Mas: abdomine segmento 7° ventrali dimidio apicali lævi, apice truncato, haud emarginato.

Antennæ very slender, very elongate, the incrassation at the extremity of each joint less than in the other species; 3rd joint slightly longer than 2nd; 8th joint about as long as 3rd (the three apical joints are wanting in the individual described). Head small, dull, with the exception of the front part of the clypeus, which is shining and not so black as the posterior parts; the front part with some coarse punctures, the middle part without punctuation, the vertex with obsolete but rugose and dense punctuation. Thorax quadrate, scarcely longer than broad, straight at the sides; its surface quite dull, and with a very obsolete punctuation. Elytra a little longer than the thorax, quite dull, closely and moderately coarsely, rugosely punctured. Hind body moderately coarsely and closely punctured, the extremity obscurely reddish. Legs yellow, with the tibiæ slightly infuscate.

Ega; a single male specimen. The apical portion of the 7th ventral segment is smooth, being quite free from punctuation or pubescence, and shows no trace of emargination; the ventral plate of the 8th segment is enfolded and concealed by the dorsal plate; it is elongate and slender (when dissected out), its basal portion only slightly broader than the apical half, and it is furnished at the apex with a small curved appendage.

13. *Pinophilus extremus*, n. sp. Niger, opacus, antennis, palpis, pedibusque pallidis; capite anterius rufescente; thorace oblongo, basin versus angustato, crebre sat fortiter sed minus profunde punctato; elytris omnino opacis, profunde sat fortiter et crebre ruguloso-punctatis; abdomine apicem versus rufescente. Long. corp. 6 lin.

Mas: abdomine segmento 7° ventrali apicem versus parcius punctato, apice ipso rotundato.

Antennæ slender, moderately long; 3rd joint distinctly longer than 2nd. Head with the front part red, covered with a very dense and fine punctuation, which becomes more sparing towards the front, so that it is there shining; also with some coarse punctures on each side in front of the middle, and with the vertex coarsely but not deeply punctured. Thorax about as long as broad, distinctly narrowed behind, dull, coriaceous, and with rather coarse but only little impressed punctures, which are rather more distinct about the middle than elsewhere; there is no definite longitudinal space free from these punctures, but just in front of the base is a small, shining, longitudinal elevation. The elytra are slightly longer than the thorax, quite dull black, obscurely reddish at the extremity, covered with a rather deep, but only moderately coarse and close punctuation. Hind body reddish towards the extremity.

In the male the 7th ventral plate is broad and rounded at the extremity, towards which part the punctures become more sparing, so that at the apex, in the middle, they are quite wanting; the ventral plate of the 8th segment forms a deep, elongate trough, terminated at its extremity by a curved, elongate spine.

Tapajos; only one individual.

14. *Pinophilus sulcatus,* n. sp. Niger, parallelus; thorace bisulcato, grosse punctato, interstitiis nitidis; elytris dense fortiter punctatis, fere opacis; antennis gracillimis, elongatis, rufescentibus, basi cum pedibus palpisque pallide testaceis. Long. corp. 8 lin.

Mas: abdomine segmento 7º ventrali sat producto, apice lato, sat profunde latius emarginato.

Antennæ rather longer than head and thorax, extremely slender; 3rd joint distinctly longer than 2nd, basal joint pale yellow, the rest rather dark. Head shining in front, opaque behind, the back part coriaceous, and the hinder angles with obsolete, very coarse punctuation; the front part with sparse, fine punctures, with two coarse ones in the middle, immediately behind the labrum, and with four or five other large punctures obliquely placed on each side, near the front. Thorax about one-fourth longer than broad, a little narrowed behind, with extremely coarse, shallow punctures; the interstices very shining, the punctures opaque, being coriaceous; along the middle with two broad, longitudinal impressions, rather deep behind, indistinct towards the front; in these impressions the punctures are confluent; small spaces near the front and hind angles are free from punctures. Elytra hardly longer than the thorax, coarsely and closely rugosely punctured. Hind body dull, rather closely, moderately finely punctured. Legs pale yellow, coxæ pitchy.

In the ♂ the under part of the 7th segment of the hind body is produced into a short, broad plate, the extremity of which is occupied by a broad and rather deep notch; the ventral plate of the 8th segment forms a trough, furnished at the extremity with a very short projection.

Ega; three individuals, one ♂, two ♀.

15. *Pinophilus duplex,* n. sp. Niger, parallelus, thorace bisulcato, grosse punctato, interstitiis nitidis; elytris dense fortiter punctatis, fere opacis; antennis gracillimis,

elongatis, rufescentibus, basi cum pedibus palpisque pallide testaceis. Long. corp. 8 lin.

Mas: abdomine segmento 7° ventrali longius producto, apice angusto, parum emarginato.

This species resembles extremely the *P. bisulcatus*, but differs by its male characters. In other respects it may be said to resemble that species exactly; the single specimen I have seen exhibits some slight differences in sculpture, the punctuation of the upper surface being a little coarser; this is more notable on the thorax than elsewhere, so that the coarse punctures are more confluent, and the shining interstices more reduced in extent.

I have no more special locality for this species than " Amazons," but I suspect it to be from the upper portion of the river.

16. *Pinophilus laxus*, n. sp. Niger, nitidus, antennis, palpis, pedibusque testaceis, geniculis nigris; thorace basin versus angustato, crebre fortiter punctato; elytris fortiter punctatis; antennis elongatis, tenuibus, articulo ultimo præcedente paulo longiore. Long. corp. 7 lin.

Mas: antennis articulo ultimo magis elongato; abdomine segmento 7° ventrali elongato, apice angusto, obsolete emarginato.

Antennæ slender and elongate, yellow; 3rd joint one and a half times as long as 2nd. Mandibles pitchy, with a long tooth; palpi pale yellow. Head black, shining, except at the hind angles, with a few punctures immediately behind the labrum; the middle part with a patch of coarse distant punctures, and the vertex with sparing coarse punctures. Thorax about as long as broad; the sides a little rounded, and distinctly narrowed behind; transversely convex, black and shining, rather coarsely and closely punctured; the punctures not so close in the middle at the base, and leaving also the hind angles, and a small space near the front angles, free. Elytra ample, fully one-third longer than the thorax, rather coarsely but not closely punctured, shining black; the pubescence fine, sparing and very easily removed. Hind body rather slender, sparingly punctured, but only slightly shining. Legs rather long and stout, pale yellow, with the knees infuscate.

The male has the 7th ventral segment produced, narrow

and a little emarginate, shining and impunctate towards the extremity.

Ega, a single specimen; also without special locality two other specimens, male and female, which I suspect represent a closely allied species, but may perhaps be a variety of this species; they have the thorax more sparingly punctured, and the male has the antennæ with the joints a little differently shaped, but this may possibly depend on a slight shrivelling of these delicate structures. *P. palmatus*, Er., is evidently a closely allied species.

17. *Pinophilus aberrans*, n. sp. Niger, nitidus, antennis, palpis, pedibusque testaceis; thorace elytrisque crebre fortiter punctatis; antennis articulo ultimo præcedenti longiore. Long. corp. 6—6½ lin.

Mas: antennarum articulo ultimo valde elongato; abdomine segmento 7° ventrali producto, apice angusto, obsolete emarginato.

Antennæ yellow, rather slender; 3rd joint distinctly longer than 2nd. Palpi pale yellow; mandibles pitchy. Head shining black, with two or three coarse punctures immediately behind the labrum, then a small impunctate space, then across the middle a space with coarse punctures, behind this another impunctate space; the vertex on each side punctured, the hind angles dull. Thorax about as long as broad; the sides nearly straight, and not narrowed till near the hind angles, when they become much narrowed, so that the base is a good deal narrower than the front; transversely convex, rather coarsely and moderately closely punctured, the punctures wanting on a small space near the front angles, and at the hind angles, and sometimes more sparing along the middle. Elytra fully one-third longer than thorax, rather coarsely but not closely punctured. Hind body rather slender, sparingly punctured. Legs pale yellow.

In the male the terminal joint of the antennæ is nearly as long as the three preceding joints together; the ventral plate of the 7th segment is produced and narrowed at the extremity, where it is scarcely emarginate; the apical portion is smooth and impunctate. In the female the last joint of the antennæ is only about one and a half times the length of the 10th joint.

Ega; one ♂, four ♀ individuals.

This species presents a very great resemblance to *P. laxus*, yet the structure of the male intromittent organ is very different.

18. *Pinophilus bicolor*, n. sp. Nitidus, rufus, capite abdomineque nigris, antennis, palpis, pedibusque testaceis; thorace basin versus angustato, cum elytris fortiter parcius punctato. Long. corp. 4½—5 lin.

Mas: abdomine segmento 7° ventrali producto, apice lævi, truncato.

Antennæ slender, moderately long, yellow; 3rd joint about as long as 2nd. Palpi pale yellow; mandibles reddish. Head rather small, black and shining, the vertex coarsely punctured, in front of this a very shining space, the front part with a few coarse punctures, and with some fine ones scattered. Thorax shining red, about as long as broad, slightly rounded at the sides and distinctly narrowed behind, with coarse scattered punctures, wanting on a small space near the front angles and about the hind angles, and not so dense along the middle. Elytra shining red, elongate, deeply and rather coarsely but sparingly punctured. Hind body black, rather coarsely but not closely punctured. Legs short, pale yellow, the coxæ red.

In the male the ventral plate of the 7th segment of hind body is produced; the extremity is truncate, not in the least emarginate, the apical part shining and impunctate; the ventral plate of the 8th segment forms part of the intromittent organ, and is bihamate at the extremity.

Ega; four individuals, two ♂, two ♀.

19. *Pinophilus Batesi*, n. sp. Nitidus, rufus, capite abdomineque apice nigris, thorace piceo-rufo, antennis, palpis, pedibusque pallidis; thorace elytrisque fortiter sat crebre punctatis. Long. corp. 5½ lin.

Mas: abdomine segmento 7° ventrali producto, apice angusto, simplice.

Antennæ slender, moderately long; 3rd joint considerably longer than 2nd; the basal joint pale yellow, the rest rather darker. Palpi pale yellow; mandibles and labrum red. Head shining black, vertex coarsely punctured, clypeus with two large punctures on its front edge in the middle; in front of the eyes a transverse curved patch of ten or a dozen coarse punctures, behind these a very smooth and shining space. Thorax as long as broad, slightly rounded at the sides and a little narrowed behind, red suffused with black, the base being more distinctly red than the other parts; coarsely irregularly and not densely punctured, the punctures leaving a space near the front angles free; in the middle in front of the base the surface is indistinctly elevated, this part also being free from punctures. Elytra one and a third times the length of the thorax, shining red, rather coarsely and sparingly punctured. Hind body red, the two apical segments suffused with black, rather sparingly punctured. Legs pale yellow, with the coxæ red.

In the male the ventral plate of the 7th segment is much produced and narrow at the extremity, the apical portion quite impunctate and shining; the ventral plate of the 8th segment forms part of the intromittent organ; it is elongate and narrow, and furnished at the extremity with two very long processes, which are quite half the length of the rest of the plate.

Tapajos; a single male.

This species bears a very great resemblance to *P. bicolor*, but may be distinguished by the colour of the thorax and hind body; it is remarkable that, though from the external characters the two insects might almost be supposed conspecific, yet the structure of the intromittent organ of the male is extremely different.

20. *Pinophilus debilis*, n. sp. Subdepressus, piceus, capite thoraceque dilutioribus; nitidus, antennis, palpis, pedibusque pallide flavis; thoraceque quadrato, crebre fortiter punctato; elytris crebre profunde punctatis. Long. corp. 3¼—3½ lin.

Mas: abdomine segmento 7° ventrali medio producto, apice angusto, rotundato-acuminato.

Antennæ yellow; 3rd joint shorter than 2nd. Palpi pale yellow. Head pitchy red, shining, with two punctures in the middle behind the labrum; behind these four or five on each side in an irregular longitudinal row, and with some others close to the inner margin of the eye and the vertex, the smooth parts with a few distant extremely fine punctures. Thorax about as long as broad, straight at the sides and not narrowed behind, shining, coarsely punctured, with a space near the front angles free; with no distinct impunctate longitudinal line along the middle, but with the part in front of the base in the middle slightly elevated. Elytra slightly longer than the thorax, deeply, rather coarsely, not densely punctured, the interstices shining. Hind body parallel, the margins of the segments slightly paler, clothed with a fine, rather dense pubescence, rather closely and finely punctured. Legs pale yellow; anterior tarsi very broad.

In the male the ventral plate of the 7th segment is produced, its extremity narrow, the apical portion is pale, smooth and shining; the ventral plate of the 8th segment forms part of the intromittent organ, and is very deeply grooved, its extremity without appendages.

Tapajos; several specimens.

21. *Pinophilus minor*, n. sp. Angustulus, rufescens, nitidulus, antennis, palpis, pedibusque pallidis; thorace oblongo, basin versus leviter angustato, fortiter sat crebre punctato; elytris fortiter profundeque punctatis. Long. corp. 3 lin.

Mas: abdomine segmento 7° ventrali producto, apice obsolete emarginato.

Fem.: abdomine segmento 7° ventrali apice medio incisurâ parvâ.

Antennæ yellow, short; 3rd joint rather shorter than 2nd. Palpi pale yellow; last joint elongate and slender. Head shining, reddish, paler towards the front, with two punctures in the middle in front; behind these with four or five on each side, with two or three along the inner margin of the eyes, and a few at the extreme vertex, elsewhere quite smooth and shining. Thorax scarcely longer than broad, straight at the sides but a little narrowed behind, obscurely reddish, very shining, coarsely but not densely punctured, the punctures wanting towards the front angles; a shining slightly elevated longitudinal space in front of the base in the middle. Elytra a little longer than the thorax, reddish, rather darker at the base, deeply and coarsely punctured, the interstices shining. Hind body slender, the basal segments opaque in front and shining at the hind edge; the apical segments sparingly punctured and rather shining. Legs almost white.

In the male the ventral plate of the 7th segment is produced, the apex of the produced part moderately broad and slightly emarginate; the plate of the 8th segment forms part of the intromittent organ, its extremity forms a sharp spine slightly curved upwards.

In the female there is a small but very sharply defined
notch in the middle of the extremity of the ventral plate of
the 7th segment; the plate of the 8th segment is rather
broad, its extremity emarginate so as to be acuminate on
each side.

Tapajos; four specimens, one ♂, three ♀.

Closely allied to *P. debilis*, but narrower and paler, with
the thorax more coarsely punctured and the sexual characters different in both sexes.

22. *Pinophilus affinis*, n. sp. Nigricans, nitidus, antennis, palpis, pedibusque flavis; thorace quadrato, basin
versus leviter angustato, fortiter punctato; elytris crebre
profunde punctatis. Long. corp. 3¼ lin.

Mas latet.

Antennæ yellow, short; 3rd joint rather shorter than
2nd. Head black, with the mandibles and edge of the
clypeus reddish, punctured as in *P. debilis*, but with the
fine sparing punctures less distinct. Thorax about as
long as broad, slightly narrowed behind, pitchy black,
shining, coarsely punctured, the punctures wanting on a
small space near the front angles; in the middle, in front
of the base, slightly elevated, the punctures leaving this
part free. Elytra about one-third longer than the thorax,
shining, blackish, with the hind margin pitchy red, deeply
and rather coarsely and closely punctured. Hind body
blackish, with the margins of the segments pitchy, rather
closely and finely punctured. Legs pale yellow.

The female has the extremity of the ventral plate of the
7th segment simple, that of the 8th formed as in *P. minor*.

St. Paulo; a single specimen.

Though very closely allied to *P. debilis*, I have no doubt
this is a distinct species; the thorax is more coarsely punctured, and slightly narrowed behind; the last joint of the
maxillary palpi a little shorter, the legs a little longer;
the male would probably offer good distinctive characters.

23. *Pinophilus egens*, n. sp. Rufo-castaneus, nitidulus,
abdomine segmentis duobus ultimis dorso infuscatis, antennis geniculatis, palpis pedibusque albidis; thorace
biseriatim punctato; elytris fortiter punctatis. Long.
corp. 3½ lin.

Mas: tibiis anterioribus mucronatis, abdomine segmento
7° ventrali subproducto, medio obtuso.

Narrow, subcylindric. Antennæ short, geniculate; the basal joint as long as the three or four following joints together, 2nd twice as long as 3rd, 7—10 similar to one another, bead-like. Palpi yellow, long, the apical joint slender and elongate. Head shining, rather paler than the other parts; pale yellow, with a single obscure puncture on the front edge of the clypeus, with a distinct impression on each side, inside the antennal tubercle, in which are two punctures; behind these with two other punctures, and two or three punctures on each side at the extreme back part; the temporal angles less developed than in the other species of the genus. Thorax transversely convex, slightly curved at the sides, a little narrowed behind, a good deal longer than broad; on each side with a row of four or five punctures along the middle, and outside these three or four other punctures on each side. Elytra longer than the thorax, reddish, shining, coarsely, moderately closely punctured. Hind body reddish, with the 6th and 7th segments infuscate in the middle, and their hind margins very pale; the segments are rather closely punctured, but shining. The legs are white; the front tarsi very broad, the hinder ones slender and elongate.

In the male the front tibiæ are furnished at the extremity (or perhaps the tarsi at the base) with a stout process, which projects inwards, and is broader at the apex than the base; the ventral plate of the 7th segment is slightly produced in the middle, that of the 8th segment is internal; the sides of the dorsal plate greatly overlapping one the other at the base.

Tapajos; two specimens.

This is a very curious species, and will probably form a distinct genus.

24. *Pinophilus abax*, n. sp. Capite thoracequc nigerrimis, pernitidis; elytris rufis, grosse punctatis; abdomine opaco, nigro, nigro-pubescente, segmentis basalibus lateribus cinereo-pubescentibus, segmento 7° pallido, medio nigro-maculato; antennis, palpis, pedibusque albidis. Long. corp. 5¼ lin.

Mas: abdomine segmento 7° ventrali valde producto.

Antennæ geniculate, white; 1st joint as long as the three or four following joints together; 2nd joint a little longer than the 3rd, which is almost equal to the 4th; joints 4—11 slender, each a little shorter than its predecessor, each distinctly longer than broad. Palpi elongate, white, the terminal joint slender and elongate. Head much smaller than the thorax, very shining black; the mandibles red; the antennal tubercles reddish; a single fine puncture on the front margin of the clypeus, a depression inside the tubercles, and behind this two punctures on each side, and a few punctures on each side of the vertex. Thorax very shining, black, about as long as broad, transversely convex, not narrowed behind, with an irregular row of four or five coarse punctures on each side the middle; with four or five others outside these, some fine punctures on the front margin, and with a few fine obsolete punctures scattered over the surface. Scutellum black, not punctured. Elytra slightly longer than the thorax, shining red, deeply and very coarsely punctured, the punctures not dense. Hind body short, opaque, the segments obscurely punctured, the three or four basal ones with ashy pubescence on each side; the 7th segment white, with a black patch on the middle. Legs rather stout, white. The sternum, with the coxæ and under face of hind body, reddish.

In the male the ventral plate of the 7th segment has the middle part greatly produced, so as to form a large tongue-like process; it is separated by a deep notch from the lateral portion, and is finely punctured in the middle; that of the 8th segment forms part of the intromittent organ; it is polished and quite smooth, from beyond the middle gradually narrowed to the extremity, where it forms a sharp spine.

Amazons; a single male, without special locality.

This very remarkable species is allied to the *P. egens*, though extremely different from it in appearance. I should suppose the Columbian *P. crassicollis*, Er., may be an allied species.

ŒDODACTYLUS.

The two species here described are only referred wit' doubt to the genus *Œdodactylus*, hitherto represented only by *Œ. fuscobrunneus*, from Chili. They are remarkable by the elongate anterior coxæ, and the great development of the side pieces of the thorax, and may be considered the Austro-Columbian representatives of the Arctogæal and Australian *Procirrus*, to which genus the Chilian species approaches in facies more nearly than do the two Amazonian insects. These, moreover, are discordant in appearance inter se; the sculpture and some of the details of *Œ. anceps* recalling *Sunius*, and I think it quite possible that a real affinity in that direction will be detected. The genus *Procirrus* is one of the most remarkable of the *Staphylinidæ*, from the fact that the very elongate front coxæ are entirely exserted, there being only a small circular opening at the extreme front angle of the thorax for their insertion. This peculiarity appears to result from an unusual and extreme development of the ento-thorax, and from certain parts thereof, that are usually membranous, becoming horny. Certain other members of the *Pinophilini* (*Pinophilus latipes*, e. g.) offer us the existing intermediate stages of this transformation; and it appears to me probable that the gradations of metamorphosis of this part will offer the most important clue to the classification of the members of the group.

1. *Œdodactylus errans*, n. sp. Rufo-brunneus, antennis pedibusque testaceis; dense punctatus, thorace elongato, subnitido; antennis brevioribus. Long. corp. 2¾ lin.

Mas: abdomine segmento 7° ventrali ante apicem lævi, margine posteriore late emarginato; segmento octo processubus duobus deorsum curvatis.

Antennæ quite yellow, short, moderately stout; 3rd joint considerably shorter than 2nd, the apical joints scarcely longer than broad, the 11th truncate. Maxillary palpi yellow, the last joint large, securiform. Head small, narrower than the thorax, closely punctured, a little shining, especially along the middle. Thorax rather long and narrow, narrower than the elytra, narrowed towards the base, closely and moderately coarsely punctured, a little shining; a narrow longitudinal space in front of the base in the middle obscurely continued forwards, shining. Scutellum very small. Elytra at the sides about as long as the thorax; a little emarginate at the extremity, so that along the suture they are shorter than the thorax; densely and deeply, rather coarsely punctured, the very narrow interstices shining. Hind body quite cylindric, each segment narrower at base than at the extremity; rather closely punctured, with elongate pubescence; the apical segments less closely punctured towards their hind margins, so as to be a little shining.

Tapajos; a single individual.

2. *Œdodactylus anceps*, n. sp. Brunneus, opacus, dense punctatus; thorace minus elongato, basi rotundato, antennis pedibusque testaceis. Long. corp. 2¾ lin.

Antennæ yellow, moderately long, slender; 2nd joint longer than the very slender 3rd joint; joints 4—11 each very slender at the base, and longer than broad. Maxillary palpi yellow, last joint less produced inwardly than in *Œ. errans*. Head small, narrower than the thorax, closely and rather coarsely but not deeply punctured, the interstices slightly shining. Thorax about as long as broad, with the base rounded, so that the hind angles have disappeared; it is slightly lobed in the middle in front, the front angles nearly right angles; it is densely punctured, the punctures on the middle shallow, subocellate, the sculpture at the sides and base granular. Elytra a little longer than the thorax, dull, densely punctured with an asperate punctuation. Hind body slender and elongate, pointed at the extremity, densely punctured. Legs rather long, yellow, front tarsi elongate and (for the group *Pinophilini*) narrow.

Tapajos; a single individual. It is, I believe, a male, but the extremity of the hind body is retracted, and I have damaged the specimen in trying to withdraw it.

Œdichirus.

This genus has not been registered as found in the New World, but my collection contains six or eight species from South America. According to a note of Mr. Bates the *Œ. optatus* here described is found on trees.

I believe the *Elytrobæus geniculatus*, Sahlberg, from description (Act. Soc. Fenn. 2, p. 802), to be an insect of this genus, the characters mentioned as separating the genus from *Œdichirus* appearing to me very indefinite. Amongst these, Sahlberg lays stress on the last joint of the antennæ terminating in a spine. On examining the European *Œ. pæderinus*, I find that in some individuals the antennæ end in a short spine or seta; and of the three specimens I possess of *Œ. optatus*, the two females have the extremity of the antennæ truncate, while in the male these organs are terminated by a slender but rather long spine or seta. I have not, however, made any allusion to this in my description, for I am not at all clear that this character is more than illusory. It appears to me probable that the apex of the antennæ is formed by a projecting membrane which bears the spine, and that the membrane can be retracted, in which case the spine disappears, and the extremity of the antenna is apparently truncate.

1. *Œdichirus optatus*, n. sp. Piceus, antennis, palpis, pedibusque testaceis; elytris thorace fere duplo brevioribus, basi valde angustatis; abdomine magno. Long. corp. $4\frac{1}{4}$ lin.

Mas: abdomine segmento 7° ventrali medio profunde angustius exciso.

Antennæ yellow; $1\frac{1}{4}$ lin. in length, 2nd and 3rd joints about equal. Palpi and mandibles yellow. Head broad and short, the front half bearing ten or twelve coarse punctures; the clypeus in front smooth, the vertex with a few coarse punctures on each side; the hind part bounded on either side by a slightly raised carina. Thorax fully $\frac{3}{4}$ lin. in length, and about $\frac{2}{3}$ in its greatest width, very much narrowed behind, of a shining pitchy colour like the head; the sides in front greatly deflexed, the surface bearing very coarse punctures, consisting of two irregular rows of about a dozen punctures each along the middle, these rows separated on the basal portion by a broad smooth space, but on the anterior portion between them about five coarse punctures on each side, so as to occupy there the middle space; near each side is a shorter row of punctures divergent towards the front; between this and the middle three or four punctures in front of the middle, and also with punctures along the base and margins. Scutellum moderately large, smooth. Elytra, from the apex of scutellum to extremity of suture, $\frac{3}{4}$ lin. in length, curved at the sides and much narrowed at the base, rather paler-pitchy in colour than the thorax; their hind angles projecting a good deal behind, the surface bearing coarse ill-defined punctures and a few long hairs. Hind body ample, the segments coarsely punctured, the 6th and 7th segments only obsoletely and sparingly punctured; the surface is but little shining, the basal portion of each segment being coriaceous; the punctures are quite irregularly disposed. Ths legs are entirely pale yellow.

The male has a narrow but very deep notch on the ventral plate of the 7th segment of the hind body; the middle of the plate on each side of this notch has a large ill-defined depression.

Tapajos; one ♂, two ♀ specimens.

PALAMINUS.

The species of this genus hitherto described are only about twelve, and of this number only five are from South America, the others inhabiting North America, Ceylon, and the Cape Verde Islands. Nevertheless the species in South America are excessively numerous, my collection containing about sixty species from that continent. The genus is one of those where the examination and description of the species in a thoroughly satisfactory manner is surrounded with great difficulties. These insects are small and delicate, and the different species bear the greatest resemblance inter se, and apparently in some cases are found together in a gregarious manner. In order to distinguish the species a very careful examination of the sexual characters should be made, and this ought to include an examination of the appendages of the male intromittent organ; these are very different in the different species, and in certain cases are extremely remarkable; bilateral asymetry is very common. The external abdominal sexual characters are in some species striking, while in others they are scarcely present. I cannot mention any general external characters by which the sex of an individual may be distinguished with certainty, but, as a rule, it appears that an excision or notch at the hind margin of the ventral plate of the 7th segment is characteristic of the female, while a conspicuous prolongation of that part generally indicates the individual possessing it to be a male. Erichson (who in his "Genera and Species" has only described four species of the genus) appears to have probably fallen into error in his identification of the sex of the individuals described by him; in the case of one of the species, *P. variabilis*, he enumerates five or six varieties, all of which will, I have no doubt, prove to be distinct species. Twenty species found by Mr. Bates are here described, as well as one other captured by Dr. Trail, and, after comparing them carefully with a large series of species found in the neighbourhood of Rio de Janeiro, I find not a single one is common to the two localities.

1. *Palaminus simplex*, n. sp. Testaceo-ferrugineus, antennis, pedibus, palpisque pallide testaceis; prothorace elongato, basin versus angustato, inaequaliter punctato. Long. corp. 2½—2¾ lin.

Mas et femina: abdomine segmento 7° ventrali simplice.

An elongate and slender species. Antennæ almost white, longer than head and thorax; 3rd joint elongate, longer than the 2nd joint; 11th joint simple, scarcely so long as, and slightly broader than, the 10th. Palpi pale yellow, with the last joint very large. Head yellow, the vertex slightly emarginate, the eyes reaching very near to the back of the head, with a fine margin behind them; the surface is rather coarsely punctured, the punctures, however, not sharply defined, and become more obsolete and sparing towards the vertex. Thorax fully as long as broad, the sides not curved but much narrowed behind; the surface rather coarsely and irregularly punctured, the punctures towards the sides more indistinct than about the middle, the smooth spaces not conspicuous. Elytra slender but considerably broader than, and about one and a half times the length of, the thorax; deeply emarginate behind, so that the outer angle reaches considerably farther back than the suture; ferruginous yellow, slightly darker about the suture; rather deeply and moderately coarsely punctured, the punctuation at base rather close, sparing at the extremity. Hind body slender, darker and less shining than the other parts, ferruginous; the four basal segments coarsely sculptured, the two apical ones smooth. Legs slender and elongate, almost white, basal joint of hind tarsus forming rather more than half the length of the tarsus.

The 7th abdominal segment is nearly similar in the two sexes, and shows no peculiar structure; the terminal joint of the maxillary palpi is, however, considerably larger in the male than in the female, and the dorsal plate of the 7th segment is just a little more prolonged in the middle.

Tapajos, Ega; five individuals, 3 ♂, 2 ♀; also a female individual from Tapajos, which is much broader than the Ega specimens, so that I think it probable a knowledge of the male might show it to be a distinct species.

2. *Palaminus longicornis*, n. sp. Ferrugineus, antennis, pedibus, palpisque pallide testaceis; capite vertice

emarginato, lineâ verticali bene discretâ; prothorace elongato, basin versus angustato. Long. corp. 2¾ lin.
Mas latet.
Fem.: abdomine segmento 7° ventrali simplice.

Very closely allied to *P. simplex*, but the vertex is more emarginate, so that the fine line which bounds the vertex is not so straight;.the insect is also rather less slender, the thorax is broader in proportion to its length, and has the sides a little rounded and the central space more distinct, and the pubescence of the hind body is more conspicuous.

Ega; a single female individual, which is a little immature.

3. *Palaminus modestus*, n. sp. Ferrugineus, antennis, palpis, pedibusque pallide testaceis, illis elongatis; thorace sat elongato, lateribus rotundatis, basin versus minus angustato. Long. corp. 2½ lin.

Mas: abdomine segmento 7° ventrali producto, apice rotundato, medio excisione parvâ.

Fem. latet.

Antennæ white, slender and very elongate, much longer than head and thorax. Palpi with terminal joint very large. Head rather small, the vertex almost straight; the punctuation similar to that of *P. simplex*. Thorax about as long as broad, a good deal rounded at the sides but not greatly narrowed behind; its surface coarsely and irregularly, not densely punctured, the central space rather conspicuous behind. Elytra broader than the thorax, and nearly one and a half times as long; their punctuation rather coarse, but not dense. Hind body with the pubescence long and conspicuous; the sculpture of the four basal segments coarse, the two apical ones smooth. Legs white; hind tarsi elongate and slender.

In the male the middle portion of the ventral plate of the 7th segment of the hind body is a good deal produced; the hind margin is rounded, and has a small, sharply defined notch in the middle.

Ega; a single male.

Though this species greatly resembles *P. simplex*, it may be very readily distinguished by the male characters; though the female is unknown to me, the different form of the thorax from that of *P. simplex* will no doubt be common to it and the male. The species bears a still greater resemblance to *P. longicornis*, but the head is smaller, the vertex less emarginate, and the thorax less narrowed behind.

4. *Palaminus crassus*, n. sp. Castaneus, antennis, palpis pedibusque pallide testaceis; capite latiore, vertice fortiter emarginato; thorace minus elongato, basin versus fortiter angustato. Long. corp. 2½ lin.
Mas latet.
Fem.: abdomine segmento 7° ventrali simplice.

Antennæ very slender, moderately long, almost white; 3rd joint very slender, a good deal longer than the 2nd; joints 4—8 each very slender, and a little shorter than its predecessor (the three terminal joints are wanting in the individual described). Palpi pale yellow, terminal joint moderately large. Head large, broader than the thorax, and about as broad as the elytra; the eyes distinctly removed from the posterior angles, the vertical line strongly marked, much deflexed in the middle, owing to the emargination of the vertex, and also interrupted in the middle; the surface is coarsely and not densely punctured with distinct, well defined punctures, most numerous between the front part of the eyes, and wanting at the extreme vertex. Thorax rather broad, its length rather less than its width, a little rounded at the sides, and much narrowed behind; the surface sparingly and irregularly punctured with well-defined punctures, these forming a patch on each side the middle; the sides but sparingly punctured, the smooth spaces well marked but not elevated. Elytra one and a half times as long as the thorax, rather paler in colour than the rest of the upper surface, their punctuation moderately coarse and close. The legs are almost white.

Amazons.

The single female individual, the only one I have seen of this species, is very mutilated, and shows no peculiarity of abdominal structure; but the species will be easily recognized by the broad head and the well-defined punctures of the head and thorax, in which respects it resembles some of the black species here described rather than any other pale species I am acquainted with.

5. *Palaminus robustus*, n. sp. Castaneo-testaceus, antennis, palpis, pedibusque pallide testaceis; capite brevi; prothorace subtransverso, crebre minus inæqualiter punctato; elytris crebre punctatis, minus nitidis. Long. corp. fere 3 lin.
Mas latet.

Fem.: abdomine segmento 7° ventrali producto, apice excisione triangulari magnâ, lobis lateralibus acuminatis, leviter recurvis.

Antennæ rather stout (for this genus), almost white, rather short; 3rd joint slightly longer than 2nd; 10th joint about as long as but distinctly stouter than the 9th, 11th a good deal stouter and longer than the 10th, acuminate at the extremity. Palpi pale yellow, last joint not very large. Head very short, broad, almost as broad as the thorax; the vertical line fine, very slightly deflexed in the middle, contiguous at the sides with the hind margin of the eyes, slightly interrupted in the middle; the surface is coarsely and rather closely punctured. Thorax a good deal broader than long, slightly narrower than the elytra, the base and hind angles rounded; the surface rather coarsely and closely punctured, so that the smooth spaces are almost absent. Elytra not quite one and a half times the length of the thorax, rather closely and only moderately coarsely punctured; the punctures near the scutellum dense, at the extremity distinctly more sparing. Hind body broad, darker in colour than the front parts; the two apical segments without imbricate punctures. Legs pale yellow, rather stout and short.

In the female the ventral plate of the 7th segment of the hind body is produced and has a large deep notch in its apical portion; the sides of this plate come to a point at their termination, and are a little curved upwards.

Tapajos; a single specimen.

When I first examined it, I supposed this individual to be a male; but as on dissection I find no trace of the ædeagus, I conclude, with something like certainty, that it is a female.

6. *Palaminus breviceps*, n. sp. Castaneo-testaceus, antennis, palpis, pedibusque pallide testaceis; capite brevi; prothorace transverso, basi sub-truncato. Long. corp. 2½ lin.

Mas latet.

Fem.: abdomine segmento 7° ventrali apice medio excisione sat magnâ.

A rather slender species. The two basal joints of the antennæ are pale yellow, as are, no doubt, also the others, though they are broken off in the specimen described. The palpi are pale yellow, with the terminal joint small.

The head is short, with the vertex scarcely emarginate; the vertical line is fine, contiguous at the sides with the eyes, only slightly deflexed, and interrupted in the middle. The thorax is about as broad as the head, but a little narrower than the elytra, its length considerably less than its width; the sides only slightly curved, and a little narrowed behind, the hind angles obtuse but not rounded; its surface covered with rather coarse punctures, which become wanting at the front angles, but leave no distinct space along the middle smooth; the base in front of the middle slightly bi-impressed. Elytra quite one and a half times the length of the thorax, rather closely and coarsely punctured, with the punctures much more sparing at the extremity. Hind body rather slender, darker in colour than the front parts; the four basal segments with imbricate sculpture. The legs very pale yellow, rather short.

The female has a rather broad, moderately deep notch in the middle of the extremity of the ventral plate of the 7th segment of the hind body.

Amazons, probably Tapajos; a single female.

This species appears rather closely allied to *P. robustus*, but is very much more slender; it has the thorax shorter and the elytra less densely punctured, and in the female the lobes at the sides of the notch of the 7th segment are not acuminate. Though the antennæ are broken, they probably much resemble those of *P. robustus*; the two basal joints which remain are quite as short, but hardly so stout as in *P. robustus*.

7. *Paluminus discretus*, n. sp. Castaneo-testaceus, antennis, palpis, pedibusque pallide testaceis; capite brevi; prothorace transverso, crebre punctato; elytris maculâ suturali fuscâ. Long. corp. 2¼ lin.

Mas latet.

Fem.: abdomine segmento 7° ventrali apice truncato.

Antennæ pale yellow, short, rather stout for this genus: 3rd joint much thinner and a little longer than 2nd; 4—9 similar to one another in thickness, each a little shorter than its predecessor; 10th distinctly stouter than 9th, but scarcely longer, 11th distinctly longer and stouter than 10th. The head is short, as broad as the thorax; the vertical line fine, a little deflexed, but scarcely interrupted in the middle, contiguous with the eyes; it is coarsely and closely punctured. Thorax a little narrower than the elytra, considerably broader than long, the sides a little curved and distinctly narrowed behind; it is coarsely and closely punctured, except at the margins, and has only a very small and narrow smooth space on the middle. Elytra rather paler than the other parts, but with a broad, ill-defined, dark mark on the suture, near the extremity; they are coarsely and rather closely punctured, with the punctures more sparing towards the apex. Hind body rather slender, of a chestnut colour, darker than the front parts; the four basal segments with imbricate sculpture, the next obscurely strigulose, so as to be dull. Legs pale yellow, short and stout.

Rio Purus; a single female, found by Dr. Trail on the 24th September, 1874.

Obs.—Though the resemblance between this species and *P. breviceps* is extreme, they will be easily distinguished, as to the female sex at any rate, by the sexual characters; the thorax in *P. discretus* is rather smaller and more narrowed behind.

8. *Palaminus sinuatus*, n. sp. Testaceo-ferrugineus, antennis, pedibus, palpisque pallide testaceis; prothorace elongato, basin versus angustato; antennis minus elongatis. Long. corp. 2 lin.

Mas: abdomine segmento 7° ventrali sub-producto, apice lato, obsolete trisinuato.

Fem. incog.

Antennæ pale yellow, about as long as head and thorax; the apical joint scarcely longer, but distinctly stouter than the 10th. Maxillary palpi pale yellow, the last joint large. Head rather small, quite as broad as the thorax, but narrower than the elytra, moderately punctured along the middle; the vertex with very few punctures, and distinctly emarginate; the vertical line rather fine at the sides, very close to, but not contiguous with, the eyes, distinctly deflexed and interrupted in the middle. Thorax a good deal narrower than the elytra, quite as long as broad, much narrowed behind, but not curved at the sides; punctured on each side the middle with rather coarse punctures, but with very few punctures at the sides and base, and with a central space along the middle, indistinct towards the front, smooth. Elytra not one and a half times the length of the thorax, rather coarsely punctured; the punctures rather close, except at the extremity. Hind body with the four basal segments with imbricate sculpture, and also with the basal portion of the following segment, with indistinct imbrications. Legs pale yellow, moderately long.

In the male the ventral plate of the 7th segment is very slightly produced, the extremity is broad and faintly tri-sinuate; the dorsal plate of the same segment is much rounded at the extremity.

Tapajos; a single male.

Though this species greatly resembles *P. simplex* and the allied species here described, it may be distinguished from them by its shorter antennæ.

9. *Palaminus apicalis*, n. sp. Rufescens, antennis, palpis, pedibusque testaceis, abdomine apice piceo; thorace infuscato, sub-transverso, basin versus angustato. Long. corp. 1¾ lin.

Mas latet.

Fem.: abdomine segmento 7° ventrali apice emarginato.

Antennæ rather short, about as long as head and thorax, pale yellow; 3rd joint not longer than 2nd, last joint thickened, a good deal broader than the 10th joint and quite as stout as the basal joint. Maxillary palpi pale yellow, their last joint rather small. Head rather small, about as broad as the thorax; the colour reddish, but rather infuscate across the middle; the vertex scarcely emarginate, the vertical line fine, at the sides very near to the eyes, scarcely deflexed and slightly interrupted in the middle. Thorax a little shorter than broad, distinctly narrower than the elytra, curved at the sides and a good deal narrowed behind; the colour reddish, but much infuscate; the punctuation rather fine; the surface with two impressions on the middle towards the base, separated by a slight, raised, smooth space. Elytra about one and a half times as long as the thorax, their punctuation sparing except about the scutellum; they are reddish in colour at the base, pale yellow at the extremity. Hind body rather dark red, with the 7th segment, and the 8th infuscate; four basal segments with imbricate sculpture. Legs pale yellow.

The female has the ventral plate of the 7th segment of the hind body distinctly emarginate at the apex.

Tapajos; a single individual.

10. *Palaminus fragilis*, n. sp. Pallide testaceus, abdomine rufo-testaceo; antennis articulo ultimo incrassato; prothorace transverso, basin versus angustato, lateribus rotundatis, crebre punctato; elytris thorace fere duplo longioribus, basi crebre punctatis. Long. corp. 1⅓ lin.

Antennæ rather short, and for this genus rather stout, almost white; 3rd joint small, finer and rather shorter than 2nd; last joint broader than the 10th, and as long as the 9th and 10th together. Palpi pale yellow, terminal joint small. Head small, with the vertex scarcely visibly emarginate; the vertical line fine and indistinct, at the sides contiguous with the eyes. Thorax a good deal narrower than the elytra, quite as broad as the head, a good deal broader than long, curved at the sides, and a good deal narrowed behind; its surface coarsely punctured, without smooth middle space and only extremely obsoletely bi-impressed. Elytra about twice as long as the thorax, rather closely punctured at the base; the punctures becoming gradually more sparing towards the extremity, where they are nearly entirely wanting. Hind body reddish, darker than the front parts; the four basal segments with imbricate sculpture. Legs almost white, rather short.

Ega; a single individual, whose sex is uncertain; the ventral plate of the 7th segment is very slightly produced, almost truncate at the apex, with the angles rounded.

This is the smallest species I have seen of the genus.

11. *Palaminus niger*, n. sp. Niger, antennis, palpis, pedibusque pallide flavis; thorace elongato, basin versus angustato, sat crebre irregulariter punctato, lateribus parce punctatis; elytris fortiter minus crebre punctatis. Long. corp. 3 lin.

Mas incog.

Fem.: abdomine segmento 7° ventrali medio obsolete emarginato; dorsali medio rotundato-truncato, utrinque distincte emarginato.

A rather slender species. Antennæ long and slender, pale yellow, distinctly longer than head and thorax; 3rd joint very slender, elongate, considerably longer than 2nd; 11th joint slender, only slightly broader than 10th. Palpi pale yellow, only moderately large. Head quite as broad as the thorax, but slightly narrower than the elytra; the vertex a little emarginate, the vertical line fine, at the sides very close to the eyes, only slightly deflexed in the middle; the surface rather coarsely but not closely punctured. Thorax about as long as broad, a good deal narrowed behind but hardly at all curved at the sides, irregularly punctured; the middle with two series of punctures, divergent towards the front, joined together in front of the base, and slightly impressed behind, the front part of the included space bearing large punctures, so that only a small part behind is free from them; outside these central series are a few punctures scattered at the sides. Elytra about one and a half times the length of the thorax, coarsely but not closely punctured, the punctures more sparing towards the apex. Hind body with the four basal segments with imbricate sculpture, the two apical ones smooth. Legs pale yellow, rather long.

The female has the dorsal plate of the 7th segment of the hind body with a curved emargination on each side, between which the hind margin is a little rounded, but slightly truncate in the middle; the ventral plate has the hind margin almost truncate, but with a slight emargination in the middle.

Ega; two female individuals.

12. *Palaminus anceps*, n. sp. Niger, antennis, palpis, pedibusque pallide flavis; thorace basin versus minus angustato, lateribus rotundatis, sat crebre irregulariter punctato, lateribus parce punctatis; elytris sat crebre punctatis. Long. corp. 3 lin.

Mas: abdomine segmento 7° dorsali medio rotundato; ventrali producto, apice subovali, medio excisione parvâ. Fem. incog.

This insect resembles extremely the *P. niger*, but has the thorax rather different, the sides being a little more curved, but less narrowed behind, and the elytra are slightly more finely and closely punctured; these differences are but slight, and it is possible that the single individual I have seen may be the male of *P. niger*.

The dorsal plate of the 7th segment of the hind body is much rounded in the middle, and deeply sinuate on each side; the ventral plate is a good deal produced, and the produced part would have the form of half of an oval plate, were it not for a small notch in the middle at the extremity.

Ega.

Obs.—*P. modestus*, above described, resembles this species, both in the external characters of the male and the structure of the ædeagus, but differs strikingly in colour; it is just possible, however, as all other characters seem to agree, that *P. modestus* is only an immature form of the *P. anceps*.

13. *Palaminus sobrinus*, n. sp. Niger, antennis, palpis, pedibusque pallide flavis; thorace elongato, basin versus angustato, crebre punctato; elytris fortiter punctatis. Long. corp. 2¼ lin.

Antennæ pale yellow, rather slender, moderately long; 3rd joint distinctly longer than 2nd, 11th a little stouter than 10th. Palpi pale yellow. Head about as broad as the thorax; the vertex a good deal emarginate, the vertical line contiguous with the eyes at the sides, strongly deflexed in the middle, and distinctly interrupted by a longitudinal depression; the surface coarsely, distinctly and closely punctured. Thorax quite as long as broad, greatly narrowed behind, but not in the least rounded at the sides, the middle part closely punctured; in front of the base a slightly raised longitudinal smooth space in the middle, the lateral margins sparingly punctured. Elytra scarcely one and a half times the length of the thorax, rather coarsely punctured, the punctures not very close, the apical margin quite smooth. Hind body with imbricate sculpture on the four basal segments, and also on the anterior part of the following segment. Legs pale yellow, rather long.

The structure of the 7th abdominal segment is but little dissimilar in the two sexes, the hind margins being almost simple; the terminal joint of the maxillary palpi is, however, considerably larger in the male than in the female.

Tapajos; two individuals, male and female.

14. *Palaminus puncticeps*, n. sp. Niger, antennis, palpis, pedibusque pallide flavis; thorace minus elongato, basin versus angustato, crebre irregulariter punctato; elytris crebre punctatis. Long. corp. 2¼ lin.

Antennæ pale yellow; 3rd joint slender, a good deal longer than 2nd; joints 4—6 rather short (a good deal shorter than in *P. sobrinus*), the others broken off. Palpi pale yellow. Head slightly broader than the thorax, the vertical line distinct, very close to the eyes at the sides, strongly deflexed in the middle, and interrupted by a longitudinal depression; the surface convex, coarsely, distinctly and rather closely punctured. Thorax not so long as broad, a little rounded at the sides, and narrowed behind; in the middle is a smooth shining space, not reaching the front, and on each side of this the surface is slightly depressed, the depressions coarsely and closely punctured; towards the sides, in the middle, are some other punctures, as also in front of the smooth space, and a few at the lateral margins and base. Elytra more than one and a half times the length of the thorax, distinctly and rather closely punctured, the punctures becoming only a little more sparing towards the extremity, but the apical margin quite smooth and impunctate. Hind body slender; the four basal segments with imbricate sculpture, and the anterior portion of the following segment with similar but less distinct sculpture. Legs pale yellow, rather long.

The male shows little peculiarity in the structure of the 7th segment of the hind body.

Tapajos; a single male specimen.

Obs.—This species greatly resembles *P. sobrinus*, but may be readily distinguished by the shorter thorax; the elytra also are more closely punctured. The structure of the ædeagus is not very dissimilar in the two species, but its size in *P. sobrinus* is double that of *P. puncticeps*.

15. *Palaminus parcus*, n. sp. Niger, antennis, palpis, pedibusque pallide testaceis; thorace sat elongato, basin versus angustato, lateribus leviter rotundatis; elytris sat crebre punctatis, apice late lævigatis. Long. corp. 2½ lin.

Antennæ almost white, rather long, slender; 3rd joint elongate and slender, much longer than 2nd; 11th joint only slightly broader than 10th. Palpi pale yellow, last joint small (in the female). Head large, a little broader than the thorax; the vertex greatly emarginate, the vertical line coarse, at the sides remote from the eyes, greatly deflexed in the middle, and with a narrow interruption; the front part between the eyes coarsely and closely punctured, the vertical portion sparingly punctured, and at the back in front of the line the surface broadly coriaceous and opaque. Thorax about as long as broad, a good deal narrowed behind, and distinctly rounded at the sides; the surface coarsely but irregularly punctured, the punctures consisting of two broadly-separated, rather divergent series, between which, on the front portion, are two other series, and between the middle and the sides are a few other punctures; the two main series are only very slightly impressed. Elytra about one and a half times as long as the thorax, the basal portion rather closely punctured; the punctures become more sparing towards the extremity, and at the apex are quite wanting. Hind body with the four basal segments with imbricate sculpture. Legs almost white, rather long and slender.

In the female the 7th segment of the hind body is simple, the hind margin of the dorsal plate being slightly rounded, while that of the ventral plate (which is scarcely longer) is almost truncate, being scarcely visibly emarginate in the middle.

Tapajos; a single female. Also from the same locality I have another female, which may possibly be a distinct species, the head being a little smaller and the elytra with rather more punctures towards the extremity.

16. *Palaminus pellax*, n. sp. Niger, antennis, palpis, pedibusque pallide testaceis; thorace sat elongato, basin versus angustato, lateribus leviter rotundatis; elytris parcius punctatis, apice late lævigatis. Long. corp. 2½ lin.

Head rather large, with the surface in front of the vertical line scarcely coriaceous. Thorax with two dorsal series of punctures convergent in front of the base; between these in front with a few punctures, and with a few other punctures between the middle and the sides; the main series scarcely impressed, but the portion between them distinctly elevated except in front. The elytra are rather sparingly punctured, the punctures becoming more sparing towards the extremity and altogether wanting at the apex.

In the female the dorsal plate of the 7th segment of the hind body is nearly truncate, being only very slightly rounded in the middle; at the extremity the ventral plate is scarcely longer than the dorsal one, and, like it, is rounded, though rather more distinctly, at the apex.

Amazons; a single female, without special locality.

Obs.— This species resembles *P. parcus* extremely, and differs from the description above given of that species only in the characters here mentioned.

17. *Palaminus fuscipes*, n. sp. Niger, antennis, palpis, pedibusque pallide testaceis, tibiis infuscatis; capite magno;

thorace sat elongato, basin versus sat angustato, parce punctato; elytris parce punctatis, apicem versus lævigatis. Long. corp. fere 3 lin.

Antennæ pale yellow, slender, rather long; 3rd joint very elongate and slender, quite one and a half times the length of the 2nd, 11th only slightly broader than 10th. Palpi pale yellow, last joint moderately large (in the male). Head large, broader than the thorax, about the width of the elytra; the vertex emarginate and depressed in the middle, the vertical line coarse but not much raised, deflexed in the middle and interrupted, at the sides remote from the eyes; the surface coarsely punctured, the punctures moderately close on the front part, sparing towards the vertex, in front of the vertical line a little coriaceous. Thorax about as long as broad, the sides slightly rounded, distinctly narrowed behind; the surface sparingly punctured, the middle space broad and a little elevated behind. Elytra at the sides one and a half times the length of the thorax, even at the base sparingly and not closely punctured, the apical portion impunctate. Hind body with imbricate sculpture on the four basal segments. Legs almost white at the base; the tibiæ and also the extremity of the hind femora infuscate.

External abdominal characters to distinguish the male are almost absent; the dorsal plate of the 7th segment has the hind margin very slightly rounded; the ventral plate is scarcely longer than the dorsal, and is truncate, with the outer angles rounded.

Tapajos; a single male specimen.

18. *Palaminus stipes*, n. sp. Niger, antennis, palpis, pedibusque pallide testaceis, tibiis infuscatis; capite magno; thorace sat elongato, basin versus angustato, parce punctato; elytris fortiter minus crebre punctatis. Long. corp. 3¼ lin.

This species greatly resembles *P. fuscipes*, but is rather larger and much broader, and has the elytra more closely punctured, especially at the extreme base; in all other respects the description above given of *P. fuscipes* will apply to the *P. stipes*.

The female has the extremity of the dorsal plate of the 7th segment of the hind body truncate; the ventral plate has the hind margin very slightly emarginate in the middle.

Tapajos; a single female.

19. *Palaminus sellatus*, n. sp. Piceus, antennis, palpis, pedibusque pallide testaceis, elytris lateribus plagiatim testaceis; prothorace basin versus angustato, lateribus rotundato, parce punctato; elytris minus crebre punctatis. Long. corp. fere 3 lin.

Antennæ rather long and slender, pale yellow; 3rd joint elongate and slender, a good deal longer than 2nd, 11th a little stouter than 10th. Palpi yellow, with the last joint small (in the female). Head large, a little broader than the thorax, and almost as broad as the elytra; the vertex a good deal emarginate, the marginal line moderately remote from the eyes at the sides, in the middle much deflexed and a little interrupted; the surface coarsely punctured, the punctures rather close towards the front, widely separated at the vertex. Thorax broad, very nearly as long as broad, much narrowed behind, a little rounded at the sides, the front angles greatly rounded; the surface sparingly and coarsely punctured, the punctures consisting of two divergent, rather widely separated series along the middle, with accessory punctures between them on the front part, and outside them with a few other punctures; the depression at the hind part of the main series is only slight. Elytra about one and a half times as long as the thorax, pitchy in colour, with a broad straight stripe at each side testaceous; they are rather sparingly and coarsely punctured, the punctures becoming more sparing towards the extremity. Four basal segments of the hind body with imbricate sculpture. Legs almost white, rather long.

The female has the hind margins of the dorsal and ventral plates of the 7th segment of the hind body very slightly rounded, being very nearly truncate.

Tapajos; two female individuals.

20. *Palaminus gracilis*, n. sp. Elongatus, angustus, nigricans, antennis, palpis, pedibus, elytrorumque apice pallide testaceis; capite parvo; prothorace elongato, basin versus sat angustato; elytris parce fortiter punctatis. Long. corp. 2¾ lin.

Antennæ long and slender, much longer than head and

thorax, pale yellow; 3rd joint elongate and slender, a good deal longer than 2nd joint; 11th joint slender, slightly broader than the slender 10th joint. Palpi with the terminal joint small (in the female). Head small, but almost as broad as the thorax; the vertex scarcely emarginate, the vertical line fine, contiguous with the eyes at the sides, distinctly deflexed and a little interrupted in the middle; the surface rather coarsely punctured. Thorax small, rather longer than broad, distinctly narrowed behind; the surface rather coarsely punctured, the punctures not sharply defined; the middle between the dorsal series distinctly elevated behind, the dorsal series indistinct; the sides sparingly punctured, the front angles smooth. Elytra more than one and a half times the length of the thorax, black, with the hind margin narrowly straw-colour; coarsely and sparingly punctured, the punctures more sparing towards the extremity. Four basal segments of hind body with imbricate sculpture. Legs long and slender, pale yellow.

In the female the hind margin of the dorsal plate of the 7th segment of the hind body is rounded in the middle, and distinctly sinuate on each side; that of the ventral plate is truncate, with the outer angles rounded.

Ega; a single female.

21. *Palaminus distans*, n. sp. Robustus, nigricans, antennis, palpis, elytrorumque apice pallide testaceis; capite magno; prothorace parce punctato, evidenter biimpresso; elytris basi dense, apice parce punctatis. Long. corp. 2¾ lin.

Antennæ almost white, only moderately long and slender; 3rd joint distinctly longer than 2nd, 11th hardly stouter than 10th. Palpi pale yellow, last joint rather small (in the female). Head large, rather broader than the thorax, but not quite so broad as the elytra; the vertex much emarginate, the vertical line remote from the eyes at the sides, much deflexed in the middle, and interrupted; the surface coarsely but not closely punctured, the punctures more distant towards the vertex. Thorax not quite so long as broad, rounded at the sides and narrowed behind, only sparingly punctured; the middle part elevated, on each side of the elevation is an irregular series of punctures in a depression; between these series, which are widely separated, are only four or five punctures, and outside them only a very few others. Elytra at the sides quite one and a half times as long as the thorax; they are blackish in colour, with a distinct narrow band of straw-colour at the extremity; they are closely punctured, except at the extremity, the punctures at the base being quite dense. Four basal segments of hind body with imbricate sculpture. Legs rather long, almost white.

In the female the hind margin of the dorsal plate of the 7th segment of the hind body is almost truncate, being straight in the middle, and only slightly longer at the sides than in the middle; the hind margin of the ventral plate is also almost truncate, being only slightly emarginate in the middle.

Tapajos; a single individual.

Stenæsthetus.

This genus was characterized by me a year or two ago, for the purpose of describing an interesting species from Japan, which had no very near known ally. I was therefore much astonished when, on examining the species here described, I found it to be so closely allied to the Japanese insect, that I have not been able to find any characters to distinguish it as a genus therefrom. The Amazonian species possesses the heteromerous tarsi, with the very elongate basal joint to the hind feet of the Japanese insect; and though I have not dissected the mouth of the New World species, its parts, so far as I can see, are quite similar to those of the *S. sunioides*. I have not, however, been able to see the base of the maxillary palpi in *S. illatus*. In my description of the genus (Trans. Ent. Soc. Lond. 1874, p. 79) I neglected to mention the form of the labrum: it is large and simply rounded, without notch or denticulations, and seems quite similar in the two species; in the *S. illatus* the paraglossæ (or possibly the sides of the ligula) project beyond the labrum, and have much the appearance of two slender denticles, so that it might readily be supposed, on a superficial examination, that the labrum was armed with two slender teeth in the middle. As regards the number of joints in the antennæ I am still uncertain whether it be ten or eleven; if the latter number be correct, then there are two stout basal joints, of which the first is short and concealed by the elevation over the point of insertion. The position of the genus is undoubtedly between *Euæsthetus* and *Stenus*, and the occurrence of this form in South America as well as in Japan renders it highly probable that other links will be found between these two dissimilar genera. I may indeed here express my opinion that the interesting genus *Ctenomastax*, recently described by Kraatz, from Spain, should be placed next *Euæsthetus*, and not among the *Pæderidæ*; indeed, the descriptions and figures of Kraatz and Fauvel appear to point out as generic distinction from *Euæsthetus* only a slight difference in the insertion of the antennæ.

1. *Stenæsthetus illatus*, n. sp. Castaneo-testaceus, fere glaber, subnitidus, capite, thorace, elytrisque crebre sat fortiter punctatis, abdomine subtilissime punctulato; thorace subcordato, elytris hujus longitudine. Long. corp. 1⅛ lin.

Antennæ very slender, but with the basal joint stout, it being quite four times as broad, though scarcely so long as, the 2nd; joints 2—8 excessively slender, differing little from one another in length, 9 and 10 very slender, but distinctly broader than the preceding ones, 10th rather longer than 9th. Head short and broad, with the eyes rather broader than the thorax; near the front with two distant foveæ; moderately closely punctured; the punctures, when seen under a high power, are umbilicated, as in *Sunius*, but the interstices are broad, and covered with a very fine, intricate reticulation, which renders the surface nearly opaque; the eyes very convex, moderately large, coarsely facetted, reaching the broad vertex. Thorax a good deal narrower than the elytra, rather longer than broad; the sides a little rounded in front of the middle, narrowed behind the rounded part; the hind angles nearly right angles, not rounded; the surface rather coarsely sculptured, with sculpture similar to that of the head, but the punctures deeper and the interstices narrower; along the middle of the basal part are traces of two longitudinal impressions. Elytra about as long as and rather more coarsely punctured than the thorax. Hind body robust, but with the apical segments very narrow, excessively finely and indistinctly punctured, not shining. Legs slender, pale yellow.

In the male, on the underside, the 4th and 5th segments of the hind body are plicate in the middle, and the elevated part is slightly produced; the 6th has the hind margin slightly emarginate.

Tapajos; one ♂, 2 ♀ individuals; a specimen was also found at Lages, near Manaos, by Dr. Trail, on the 5th January, 1875.

Obs.—This species differs from *S. sunioides* by the larger and more convex eyes, which reach quite to the back of the head.

STENUS.

Although the species of this world-wide distributed genus seem to be everywhere amongst the most numerous of the *Staphylinidæ*, yet only fourteen or fifteen species have been as yet described from the warmer parts of the New World. The twenty-five species here described will help, therefore, somewhat to rectify this disproportion, and indicate that the genus is richly represented in South America, as elsewhere.

Of these species the first eight might, I should have thought, have been properly placed in Erichson's division I. B. Erichson, however, has described three or four species from Columbia (*S. augur*, &c.), which I judge from his descriptions are very closely allied to these eight species, and has placed them in his division I. A. I have therefore left these species without indication as to their position in Erichson's classification of the genus. Species 12 to 18 belong to a group of which the species are numerous in South America, but no species of it occur in Europe. Species 20—25 might be placed in a natural manner between the European *S. cicindeloides* and *S. contractus*.

1. *Stenus inspector*, n. sp. Niger, subopacus, antennis fuscis, basi obscure testaceis, palpis flavis, pedibus testaceis, geniculis infuscatis; fronte excavatâ, vertice angustissime carinato; thorace profunde transversim rugoso-punctato; elytris thoracis longitudine, dense fortiterque punctatis; abdomine gracili, crebre, basi distincte, apice obsolete, punctato, submarginato; tarsis gracilibus, articulo 4° vix lobato. Long. corp. 2¼ lin.

Antennæ moderately long, pitchy, yellowish at the base; 3rd joint a good deal longer than 4th; the three apical joints distinctly stouter, the 10th distinctly longer than broad. Palpi pale yellow. Head with the eyes large, quite as broad as the elytra; the front distinctly excavated, the clypeus abruptly deflexed; the surface densely and coarsely punctured, obsoletely bisulcate, with an extremely narrow carina at the vertex. Thorax a good deal narrower than the elytra, a good deal longer than broad; the sides rounded in front of the middle, and narrowed behind the middle; the surface covered with a very dense, transversely rugose punctuation. Elytra quite as long as the thorax, coarsely, deeply and very closely punctured, the shoulders standing abruptly out from the base of the thorax. Hind body slender and elongate, the basal segment distinctly margined, the following segments each constricted in front of the middle; the constricted part obscurely margined, the apical part not margined; the surface rather closely punctured, the punctuation on the basal segment distinct, but not coarse, on the penultimate segments indistinct; the front part of the two basal segments finely carinate on the middle. Legs yellow, coxæ pitchy, the knees infuscate; the 4th joint of the tarsi small, and scarcely lobed.

In the male the under surface of the 6th segment of the hind body is clothed with fine, pale-yellow pubescence, and is distinctly emarginate at the hind margin; the 3rd, 4th and 5th segments also have the hind margin very obscurely truncate in the middle, the 7th bears a narrow, very elongate notch.

Ega; a single male.

Obs.—I have not referred this species to any of the generally received sections, because doing so would be very likely to create confusion about it. The tarsi might be described as having the 4th joint simple, but this would not be strictly correct, and the same remark would be applicable if the hind body were described as unmargined. The species suggests at first sight an alliance with *S. speculator*.

2. *Stenus obductus*, n. sp. Niger, subnitidus, palpis flavis, antennarum basi pedibusque testaceis, femoribus tarsisque apicem versus infuscatis; fronte bisulcatâ, vertice carinato; thorace transversim rugoso-punctato, elytris fortiter denseque punctatis, interstitiis versus suturam latioribus; abdomine crebre distinctius, apicem versus obsolete punctato, segmento basali marginato. Long. corp. 2¼ lin.

Antennæ rather short, pitchy, with the basal joints yellow; palpi pale yellow. Head with the eyes scarcely so broad as the elytra, densely and coarsely punctured, a little excavate, distinctly bisulcate, and with a narrow central carina. Hind body slender; the basal segment marginate, and the anterior portion of the next following segment finely margined, the others immarginate; the segments rather closely punctured, the punctures on the basal segment quite distinct, but becoming less deep on the following segments, so as to be obsolete on the penultimate segments; the segments bear also a fine, depressed, scanty, ashy pubescence. The legs are long and slender, the hind tarsi elongate and slender, with the 4th joint simple; the coxæ are pitchy black, the femora are yellowish, but somewhat infuscate towards the extremity, and the tarsi become more obscure in colour towards the apex.

In the male the 6th segment beneath is broadly but faintly impressed along the middle; the impressed part is pubescent and densely punctured; the pubescence towards the apex of the segment is dense at each side of the impression, the hind margin scarcely emarginate; the following segment bears a narrow elongate notch.

Ega; two males.

Obs.—This species is closely allied to *S. inspector*, but is undoubtedly distinct; the character of the sculpture is very similar, but is less dense on the thorax and elytra of *S. obductus*; and other less striking differences are conveyed by the two descriptions; the tarsi of *S. obductus* have the 4th joint slender and simple.

3. *Stenus tinctus*, n. sp. Æneo-niger, vix nitidus, palpis flavis, antennis fuscis, basi cum pedibus testaceis, femoribus versus apicem obscurioribus; fronte bisulcatâ et carinatâ; thorace fortiter transversim rugoso-punctato, elytris dense fortiter punctatis, interstitiis versus suturam latioribus. Long. corp. 2 lin.

Antennæ pitchy, yellowish at the base, rather short; palpi pale yellow. Head slightly narrower than the elytra, densely and rather coarsely punctured, a little excavate, and with a distinct shining carina along the middle. Thorax distinctly narrower than the elytra, rather longer than broad, the sides much rounded in front of the middle; the surface covered with coarse and deep transverse rugæ, with one or two of the interstices about the middle rather broader. Elytra rather short and broad, about as long as the thorax; the humeral angles well marked, coarsely and closely punctured, with the interstices near the suture distinct, broader than the external ones. Hind body rather slender, subcylindric; the basal segment margined, the four following ones each with a slight margin on the basal portion; the segments rather closely but obsoletely punctured, the basal one rather more distinctly than the others, and with a carina at the base in the middle, the two following ones with more indistinct carinæ; all the segments with a well-marked, fine, depressed, pale-yellow pubescence. Legs moderately long, reddish-yellow; the femora, as also the tibiæ and tarsi, more obscure in colour in their apical portion; 4th joint of tarsi simple.

In the male the underside of the 6th segment is flattened along the middle and finely pubescent; the 7th segment has a very elongate and narrow notch.

Tapajos; a single male.

Obs.—This species is very closely allied to *S. obductus*, but is more metallic in colour, rather less elongate in form, has the thorax with the sculpture rather coarser, the sides more rounded, the hind body more obsoletely punctured, and the limbs rather shorter and stouter.

4. *Stenus cognatus*, n. sp. Æneo-niger, subnitidus, palpis flavis, antennis fuscis, basi cum pedibus testaceis; fronte bisulcatâ et carinatâ; thorace fortiter transversim rugoso-punctato; elytris fortiter punctatis, interstitiis nitidulis. Long. corp. fere 2 lin.

Antennæ short, yellowish, infuscate at the extremity; palpi pale yellow. Head nearly as broad as the elytra, a little excavate, and distinctly carinate along the middle. Thorax longer than broad, a good deal rounded at the sides; the surface coarsely sculptured, with transverse rugæ, the interstices of which are broad enough to be distinctly shining. Elytra about as long as the thorax, coarsely punctured, with the interstices distinctly shining. Hind body slender, subcylindric; the basal segment margined, the four following ones each with a slight margin on the basal portion; the segments rather closely but obsoletely punctured, the basal one rather more distinctly than the others, and with a carina at the base in the middle, the two following ones with more indistinct carinæ; all the segments with a fine, depressed, pale-yellow pubescence. Legs moderately long, reddish-yellow.

In the male the underside of the 6th segment is flattened

along the middle and finely pubescent, the hind margin slightly emarginate; the 7th segment has a very elongate and narrow notch.

Tapajos; two males.

Obs.—This species bears an extreme resemblance to *S. tinctus*, but is more slender; it is more shining, has the punctures on the elytra rather less crowded, so that the interstices are more shining, and the limbs are still shorter.

5. **Stenus vacillator**, n. sp. Æneo-niger, subnitidus, palpis flavis, antennis fuscis, basi cum pedibus testaceis; fronte bisulcatâ et anguste carinatâ; thorace fortiter transversim rugoso-punctato; elytris fortiter punctatis, interstitiis nitidulis. Long. corp. fere 2 lin.

Antennæ dusky yellow, infuscate towards the extremity, short; 10th joint about as long as broad. Head nearly as broad as the elytra, densely punctured, distinctly excavate, and with a very fine but distinct shining carina along the middle. Thorax rather longer than broad, distinctly rounded at the sides, deeply transversely rugose, the interstices narrow. Elytra about as long as the thorax, rather coarsely and closely punctured, the interstices wider on a space near the suture than elsewhere. Hind body rather closely punctured, the basal segment distinctly, the apical ones obsoletely; the basal segment very finely carinated in the middle, the following two only very obsoletely carinate.

Tapajos; a single female.

Obs.—I have some doubts whether the individual above described be really distinct from *S. cognatus*. It is just the same size as that species, but has the antennæ a little shorter, the vertex more finely carinate, the interstices of the sculpture on the thorax and elytra rather narrower, and the carinæ of the basal segments of the hind body less distinct. These differences, however, are but slight, and it is possible may be sexual or individual rather than specific characters. I have, however, another female from St. Paulo which agrees in these respects with the *S. vacillator*, and is probably conspecific with it; but as it departs very slightly in one or two other respects from the Tapajos individual, I have drawn my description entirely from the latter.

6. **Stenus cursitor**, n. sp. Gracilis, æneo-niger, niti-

dulus, antennis fuscis, basi cum pedibus testaceis, palpis flavis; fronte excavatâ et medio carinatâ; thorace subcylindrico, transversim rugoso-punctato; elytris fortiter punctatis; abdomine crebre, obsolete punctato. Long. corp. 1⅞ lin.

Antennæ rather short, the basal joints yellow, the others infuscate; palpi very pale yellow. Head rather narrower than the elytra, between the eyes coarsely punctured but distinctly shining; a little excavate, with a shining carina along the middle. Thorax much narrower than the elytra, much longer than broad, very little rounded in front of the middle, and but little narrower at the base than in the middle; densely covered with a deep, transversely rugose sculpture. Elytra along the suture quite as long as the thorax, coarsely and closely punctured; the interstices near the suture rather broader than elsewhere. Hind body slender, but a good deal narrower at the apex than the base; the basal segment margined and finely carinate in the middle, the two following segments scarcely carinate, the segments rather closely punctured: the punctures on the ante-apical segments obsolete, the pubescence depressed, fine pale yellow. Legs slender, rather long, yellowish; the femora and tarsi darker towards the apex.

Tapajos; a single female.

Obs.—This species is very closely allied to the *S. cognatus*, but its more slender form, the smaller and more shining area separating the eyes, and the more cylindric thorax, leave me no doubt that it is specifically distinct therefrom. At first sight it suggests, to any one acquainted with the European species, *S. proditor* or *S. impressipennis*. It may be well to give a detailed comparison with the *S. proditor.* Besides the brassy colour of its upper surface and the paler legs and palpi, *S. cursitor* has the head more excavate and the central elevation narrower, polished and impunctate. The thorax is much more cylindric, with deep transverse rugæ for sculpture; the punctures of the elytra are deeper and rather coarser; the hind body is more slender and cylindric, the basal segment only distinctly though finely margined, and the same segment is finely carinate in the middle as in *S. proditor*, but the following segments are scarcely carinate; the punctures on the basal segment are not very dissimilar to those of *S. proditor*, but on the following segments they are more indistinct: the legs are about as long, but are rather more slender, the tarsi being conspicuously more slender.

7. *Stenus fallax*, n. sp. Æneo-niger, dense punctatus, subopacus, palpis flavis, antennis pedibusque testaceis, illis apice infuscatis; fronte excavatâ, medio carinatâ; abdomine sat robusto, crebre, apicem versus obsolete, punctato, segmento basali marginato; tarsis gracillimis, articulo 4º simplice. Long. corp. 2 lin.

Antennæ moderately long, yellowish, infuscate towards the extremity; palpi pale yellow. Head about as broad as the elytra, excavate between the eyes, and with a distinct shining carina in the middle of the excavation. Thorax rather longer than broad, the sides in front of the middle distinctly rounded; the surface coarsely and closely punctured, the punctures somewhat confluent, so as to form transverse rugæ. Elytra about as long as the thorax, deeply, densely and coarsely punctured. Hind body moderately broad, with the basal segment finely margined, and carinate at the base in the middle, the two following segments with less elongate carinæ; the segments rather closely punctured, the punctures much finer towards the apex than on the basal segment. Legs yellow; the femora and tarsi obscured towards the extremity.

In the male the 6th segment of the hind body beneath is broadly impressed along the middle; each side of the impression towards the extremity is densely pubescent, and there is a notch on each side concealed by the pubescence; the following segment bears an elongate narrow notch; the 4th and 5th segments are also a little flattened along the middle.

Tapajos; one male, one female specimen.

Obs.—This species, at first sight, a good deal suggests our common European *S. impressus:* though closely allied in structure to *S. cognatus* and the neighbouring species, it is undoubtedly distinct from them all.

8. *Stenus simulator*, n. sp. Angustulus, niger, vix ænescens, nitidulus, palpis flavis, antennis fuscis, pedibus testaceis, femorum apice obscuriore; fronte excavatâ, medio carinatâ; thorace elytrisque dense fortiter punctatis; abdomine segmento basali marginato, dense sat fortiter punctato, segmentis antepenultimis dense obsolete punctatis; tarsis articulo 4º simplice. Long. corp. 1¾ lin.

Antennæ pitchy, not paler at the base, moderately long; 3rd joint a little longer than 4th; club slender. Palpi pale yellow. Head with the eyes as broad as the elytra, distinctly excavate, and with a shining carina at the bottom of the excavation. Thorax rather narrow, but longer than broad, much narrower than the elytra; the sides distinctly rounded in front and narrowed at the base; the surface densely punctured, the punctures arranged so as to make the interstices assume somewhat the form of transverse ridges. Elytra rather narrow, quite as long as the thorax, densely and coarsely punctured. Hind body slender, the basal segment rather coarsely punctured, carinate in the middle at the base, and finely margined at the sides; the two following segments more indistinctly carinate in the middle, and more finely punctured; the segments towards the extremity very finely punctured, and with a very fine, depressed pubescence. Legs slender, rather long, yellowish; the femora a little darker towards the extremity; tarsi very slender, 4th joint quite simple.

In the male the 5th segment of the hind body on the underside is broadly impressed along the middle before the extremity; the 6th is still more deeply impressed, and has the edges of the impression a little raised, and furnished towards the extremity with a ridge of raised black pubescence; the 7th segment bears a very long and extremely narrow notch.

Tapajos; a single male.

Obs.—This species is conspicuous amongst its close allies here described by the dark basal joints of the antennæ.

9. *Stenus certatus*, n. sp. (Sect. I. A. Er.) Niger, vix ænescens, fere opacus, antennis, palpis, pedibusque testaceis, prioribus apice infuscatis; capite thoracæque dense fortiterque punctatis; abdomine tenuiter marginato, crebre, basi sat fortiter, apice obsolete, punctato. Long. corp. 1⅔ lin.

Antennæ slender, rather long, yellow, the slender extremity infuscate; 3rd joint hardly any longer than 4th. Palpi slender, pale yellow. Head with the eyes large, very nearly as broad as the elytra, distinctly excavate; at the bottom of the excavation with a slightly elevated shining longitudinal space. Thorax longer than broad, a good deal narrower than the elytra, the basal portion distinctly constricted; the surface coarsely and extremely densely punctured; the interstices very small, so that the sculpture is deeply rugose. Elytra slightly longer than the thorax, with the humeral angles prominent and well marked, coarsely, deeply, very closely punctured, scarcely at all shining. Hind body narrowed towards the extremity; all the segments finely margined, the three or four basal ones finely carinate at the base in the middle; the basal segment rather closely and distinctly, the apical ones finely and obsoletely, punctured. Legs unicolorous yellow, very slender, rather long; hind tarsi elongate and slender, clear yellow; 4th joint slender and simple.

Ega; a single female.

Obs.—This species is about the size of *S. incanus*, but is a little more elongate in proportion to the width, the antennæ and legs are longer and more slender; the elongate basal joint of the front and other tarsi separate it abruptly from that species and its allies.

10. *Stenus Traili*, n. sp. (Sect. I. A, Er.) Niger, opacus, dense punctatus, antennarum basi, palpis, pedibusque testaceis, fronte leviter depressâ, fere planâ; antennis distincte clavatis. Long. corp. fere 1½ lin.

Antennæ moderately long, rather slender; joints 3—6 yellowish, the others nearly black; 1st and 2nd joints stout, 3—6 slender and elongate, each a little shorter than its predecessor; 7th joint much broader than 6th, 8th smaller than the contiguous ones, 9th and 10th stout, but each longer than broad; 11th joint small, shorter than 10th. Palpi only moderately long, quite yellow. Head almost as broad as the elytra; the space between the eyes rather depressed, but almost even, very obsoletely bisulcate, evenly and densely punctured. Thorax rather longer than broad, a good deal narrower than the elytra; the sides much rounded in front of the middle, and a good deal constricted behind; it is densely and rugosely punctured and not shining. Elytra a little longer than the thorax, densely and coarsely punctured, quite dull. Hind body finely but distinctly margined, much acuminate, closely punctured; the punctures moderately coarse on the basal segments, quite fine on the penultimate one, 2—4 each with a well-marked carina on the middle of the basal part, and a much shorter one on each side. Legs yellow; tarsi long and slender, the basal joint of the hind one as long as the three following joints together.

In the male there is a moderately large excision at the extremity of the ventral plate of the 7th segment; the 6th is flattened and slightly depressed along the middle, finely punctured and very delicately pubescent, and its hind margin is a little emarginate; the hind margin of the 5th segment is obscurely emarginate.

Ananá; a single ♂ found by Dr. Trail on 6th September, 1874.

11. *Stenus pedator*, n. sp. (Sect. II. A, Er.) Niger, dense punctatus, subopacus, palpis, pedibus, antennisque testaceis, his apice nigricantibus; capite elytris fere latiore, fronte sat excavatâ; abdomine fortiter, minus crebre punctato, subnitido. Long. corp. 2 lin.

Antennæ long, slender, yellow, with the three or four apical joints blackish; 3rd joint elongate, twice as long as 2nd, and a good deal longer than the 4th; 7th and 8th joints slender and elongate, scarcely at all thicker than the preceding ones, 9—11 also slender and elongate, but distinctly stouter than the others; 11th almost as long but scarcely so broad as 10th. Palpi elongate, pale yellow. Head with the eyes very large, a little broader than the elytra; the space between the eyes is distinctly depressed, but obsoletely bisulcate, rather coarsely but not evenly punctured, with a small, shining, smooth space at the vertex in the middle. Thorax a good deal narrower than the elytra, a little longer than broad; the sides a good deal rounded in front, and distinctly narrowed behind the middle; it is coarsely, very densely and quite rugosely punctured, so as not to be shining. Elytra broad, scarcely longer than the thorax, very coarsely, densely and rugosely punctured, not shining. Hind body rather slender, with the sides finely but distinctly margined; it is rather coarsely and distinctly and not closely punctured; it is distinctly shining, and there are no carinæ in the impressions on the base of the segments. The legs are yellow, the tarsi elongate; the hind tarsus has the basal joint very long, quite as long as the three following together; the 2nd joint is also elongate, and about half the length of the basal one, the lobes of the 4th joint are elongate and slender. The under surface is shining and coarsely punctured.

In the male the femora are rather stout, and the hind tibiæ at their apex are a little incrassate, and with a minute tooth or tubercle on the inside; the ventral plate of the 7th segment has a rather large notch, the 6th is flattened along the middle, and finely but sparingly pubescent on the flat part, and has the hind margin a little emarginate.

Rio Purus; a single male, captured by Dr. Trail on the 25th October, 1874.

12. *Stenus ventralis*, n. sp. (Sect. II. B, Er.) Elongatus, plumbeo-niger, sat nitidus, albido-pubescens, antennis, palpis, pedibusque pallide flavis; thorace elytrisque crebre fortiter punctatis, his thoracis longitudine; abdomine parcius subtiliter punctato. Long. corp. $2\frac{1}{3}$ lin.

Antennæ pale yellow, elongate and slender, longer than head and thorax; 3rd joint more than twice as long as 2nd, and a good deal longer than 4th; joints of the club elongate and slender. Palpi pale yellow, elongate. Head fully as broad as the elytra, slightly depressed between the eyes, the central part very indistinctly elevated; the punctuation moderately fine, not dense, rather more sparing about the middle. Thorax a good deal longer than broad, subcylindric, but distinctly broader in the middle than at the extremities; the surface closely and rather coarsely punctured. Elytra just about as long as the thorax, but a good deal broader; the shoulders distinct, the sides slightly curved; the punctuation rather coarse, a little coarser than on the thorax, rather close. Hind body elongate and cylindrical, the basal segment finely margined, each segment with a long white pubescence, which is most distinct on its basal portion; the basal segments finely but distinctly, rather sparingly punctured, the apical ones quite finely and sparingly. Legs pale yellow, elongate, rather slender, the lobes of the tarsi broad.

In the male, on the underside, the 7th segment of the hind body bears a deep, narrow notch; the 6th is more closely punctured and pubescent along the middle than at the sides; the 5th is depressed along the middle before the apex, the depression impunctate but scarcely shining; the 4th and 3rd have similar but not such deep depressions, while the basal one is smooth and shining in the middle at the extremity.

Tapajos; one male, one female.

13. *Stenus extensus*, n. sp. (Sect. II. B, Er.) Elongatus, plumbeo-niger, sat nitidus, albido-pubescens, anten-

nis, palpis, pedibusque pallide flavis; thorace crebre sat
fortiter punctato, medio carinâ angustâ; elytris thoracis
longitudine, crebre fortiter punctatis; abdomine obsolete
punctato. Long. corp. 2⅓ lin.

Antennæ pale yellow, elongate and slender, longer than
head and thorax; 3rd joint more than twice as long as
2nd, and a good deal longer than 4th. Head about as
broad as the elytra; the front not excavate, but a little
depressed on either side between the eyes and the middle
part, which is shining and impunctate. Thorax a good
deal narrower than the elytra, much longer than broad,
cylindric, the sides nearly straight, a little narrowed
towards the base and slightly towards the front; moderately coarsely and closely punctured, with an abbreviated,
shining, narrow space along the middle. Elytra about as
long as the thorax, rather coarsely punctured, the punctuation moderately close; the shoulders but little prominent.
Hind body elongate, the basal segment margined finely,
each segment only obscurely constricted at the base; quite
obscurely punctured, with a distinct, silvery, depressed,
long pubescence; its under surface finely and sparingly
punctured. Legs very pale yellow, elongate and slender;
the lobes of the tarsi broad.

Tapajos; a single female.

Obs.—This species bears an extreme resemblance to
S. ventralis, but is undoubtedly distinct; *S. extensus* is
rather the more slender of the two, and has the thorax
more cylindric and distinctly carinated along the middle;
it is best distinguished, however, by the punctuation of
the hind body, which is finer than in *S. ventralis*, a difference which is very easily perceived when the undersides
are compared.

14. *Stenus genalis*, n. sp. (Sect. II. B, Er.) Elongatus, angustus, niger, metallescens, antennis, palpis, pedibusque pallide flavis; fronte planâ, thorace cylindrico,
dense punctato, fere opaco; elytris thoracis longitudine,
dense fortiterque punctatis; abdomine crebre fortiterque
punctato, fere nudo. Long. corp. 2½ lin.

Blackish, with a leaden-green tinge. Antennæ elongate
and slender, yellow; palpi pale yellow. Head broad, quite
as broad as the elytra; the space between the eyes broad,

flat, and scarcely depressed below the margin of the eyes; densely, evenly and rather coarsely punctured; on the underside the genæ are very broad and very densely and coarsely punctured. Thorax greatly narrower than the elytra, nearly twice as long as broad, cylindric, the sides not at all rounded, deeply, very densely and rather coarsely punctured, the punctuation rather coarser at the basal margin than at the front. Elytra narrow, but with the shoulders well marked and prominent, densely, deeply and coarsely punctured; behind the scutellum depressed, and the punctuation there rather finer and denser; that towards the hind margin rather coarser and more sparing, so that that part is more shining than the base. Hind body elongate and narrow; the segments closely and rather coarsely punctured, the 6th smooth towards the hind margin, the 7th sparingly and obsoletely punctured. Legs pale yellow, the tarsi moderately slender.

The male has a very deep excision on the ventral plate of the 7th segment; the 6th is broadly impressed along the middle, and there extremely finely and densely punctured, and bearing a fine, pale pubescence, and its hind margin is a little cut away in the middle; the 5th segment is more finely punctured along the middle than elsewhere.

Pará and Tapajos; several specimens.

15. *Stenus Paræ*, n. sp. (Sect. II. B, Er.) Elongatus, angustus, niger, leviter metallescens, antennis, palpis, pedibusque pallide flavis; thorace dense fortiterque punctato; elytris thoracis longitudine, fortiter punctatis, nitidulis; abdomine fortiter, minus crebre punctato. Long. corp. 2$\frac{1}{3}$ lin.

Head quite as broad as the elytra, the front closely and coarsely punctured, but still shining, a little depressed in the middle. Thorax much longer than broad, a good deal narrower than the elytra; the sides only slightly narrowed towards the front, but distinctly contracted behind the middle; densely covered with a coarse, almost rugose punctuation, but with the interstices distinctly shining. Elytra about as long as the thorax, with the shoulders not very prominent, covered with a coarse punctuation, which becomes more sparing towards the hind margin, where, however, it is still quite distinct, the interstices quite shining. Hind body slender and cylindric, the three or four basal segments rather coarsely and closely, but not densely punctured; almost without pubescence. The punctuation of the metasternum coarse.

Pará; a single female, collected by Mr. Smith.

Obs.—Though this species is closely allied to *S. genalis*, it is undoubtedly quite distinct, the sculpture being considerably coarser and more sparing.

16. *Stenus nigricans*, n. sp. (Sect. II. B, Er.) Elongatus, angustus, niger, nitidus, antennis, palpis, pedibusque flavis; prothorace subcylindrico, dense punctato; elytris thoracis longitudine, crebre fortiter punctatis; abdomine fortiter crebre punctato. Long. corp. 2½ lin.

Antennæ yellow, long and slender, the three apical joints distinctly incrassate. Palpi elongate, yellow. Head even broader than the elytra, the space between the eyes a little depressed, almost flat, the middle being very obsoletely elevated; it is black and shining, moderately coarsely and not densely punctured. Thorax a good deal narrower than the elytra, much longer than broad, distinctly rounded at the sides, so that it is slightly narrowed both in front and behind; it is covered with a dense, rather coarse, almost rugose punctuation. Elytra scarcely so long as the thorax, narrow, the shoulders not very prominent; they are black and shining, coarsely punctured, the punctuation at the base dense, more sparing at the apex; at the extreme base is some delicate white pubescence. Hind body cylindric, slender, only the basal segment margined, the four basal segments rather coarsely punctured, and with a white pubescence at the extreme base of each. Legs clear yellow.

The male has a broad and deep excision on the ventral plate of the 7th segment of the hind body; the 6th segment is a little depressed near the hind margin, and there finely punctured, the hind margin being distinctly emarginate.

Pará; two individuals (♂ and ♀), collected by Mr. Smith.

Obs.—This species bears a great resemblance to *S. excisus* and *S. Puræ*; but it is more slender than *S. excisus*, and has the elytra more closely punctured, and the punctuation of the hind body considerably coarser. It is rather smaller than *S. Puræ*, is blacker, and more finely punctured, and has the pubescence at the base of the abdominal segments distinct.

17. *Stenus excisus*, n. sp. (Sect. II. B, Er.) Elongatus, angustus, niger, nitidus, pedibus, palpis, antennisque pallide flavis, his apice fuscis; prothorace subcylindrico, sed medio distincte dilatato; elytris hoc vix longioribus, fortiter punctatis, apice fere lævigatis; abdomine minus fortiter punctato, segmentis singulis basi albido-pubescentibus. Long. corp. 2¼ lin.

Antennæ slender and elongate, quite as long as head and thorax, pale yellow, with the three or four apical joints infuscate, the three apical ones distinctly thickened; 3rd joint more than twice as long as 2nd. Head with the eyes almost broader than the elytra; the space between the eyes slightly depressed, almost even, rather coarsely and closely punctured, yet distinctly shining. Thorax elongate and narrow, yet distinctly contracted behind the middle, so that the sides in the middle appear a little prominent; it is also slightly narrowed towards the front; it is densely and coarsely punctured and yet shining. The elytra are rather narrow; they are scarcely longer than the thorax, their sides are a little rounded, the humeral angles quite prominent; they are coarsely and not closely punctured and shining, the punctures being fine and sparing at the hind margin; at the extreme base is a distinct white pubescence. The hind body is slender and cylindric, with only the basal segment margined; the four basal segments are distinctly, but not coarsely nor densely punctured, the apical ones obsoletely punctured; each segment has at the extreme base a distinct, whitish pubescence. The legs are long and slender, very pale yellow; the tarsal lobes elongate and slender.

In the male the ventral plate of the 7th segment has a very broad and deep incision; the 6th segment is a little flattened towards the extremity and finely punctured, and with a fine, pale pubescence; its hind margin is a little emarginate.

A single male was found by Dr. Trail on the 5th November, 1874, but no special locality has been sent me.

18. *Stenus laticeps*, n. sp. (Sect. II. B, Er.) Niger, subnudus, antennis, palpis, pedibusque testaceis; fronte leviter depressâ, fortiter punctatâ, medio glabrâ; thorace cylindrico, dense fortiter, profundeque punctato; elytris latis, fortiter crebre punctatis, nitidulis; abdomine crebre fortiter punctato. Long. corp. 2¼ lin.

Antennæ yellow, elongate and slender; 3rd and 4th joints of about the same length. Palpi pale yellow. Head broad and short, about as broad as the elytra; the eyes separated by a broad space, which is distinctly depressed, and closely and coarsely punctured, but in the middle the punctures become sparing, so as to leave an irregular longitudinal shining space; on the underside the genæ are very broad, and densely and coarsely punctured. Thorax much longer than broad, greatly narrower than the elytra, subcylindric; very slightly rounded at the sides, very densely, coarsely and deeply punctured, so that the interstices are rugose and very narrow. Elytra about as long as the thorax, broad, outstanding, the humeral angles strongly marked; the surface coarsely and deeply punctured, the punctures rather close, but the interstices broad and shining. Hind body cylindric and elongate; the segments coarsely and closely punctured, the 6th towards the extremity sparingly and finely, the 7th obsoletely punctured. Legs yellow, elongate, moderately stout; the knees reddish.

Pará; a single female.

Obs.—This species, as well as *S. genalis*, is remarkable from the broad, very densely and coarsely punctured genæ.

19. *Stenus tricolor*, n. sp. (Sect. II. B, Er.) Elongatus, angustulus, nitidus, viridi-æneus, abdomine rufo-testaceo, apice abrupte nigro; antennis rufis, basi cum palpis pedibusque flavis. Long. corp. 2½ lin.

Antennæ elongate and slender, pale yellow at the base, darker towards the apex; palpi pale yellow. Head broad, not excavate, but the middle slightly elevated, and between this and the eye with a small depression on either side; the antennal tubercles elongate, the punctuation sparing and irregular. Thorax longer than broad, narrower than the elytra, subcylindric, but distinctly broadest in the middle, and the basal portion slightly contracted; shining, rather coarsely and not sparingly punctured, the basal portion with a smooth space along the middle. Elytra slightly longer than the thorax, rather narrow and elongate; the shoulders rectangular and sharply marked; the colour shining-green like the thorax; the punctuation coarse, moderately close. Hind body elongate, slender and cylindric; reddish, shining; the basal segment distinctly, the others obsoletely punctured, the 7th and 8th quite black; segments 3—5 a little constricted at the base. Legs yellow, rather long; lobes of tarsi strongly developed.

In the male, on the underside, the segments of the hind body are each flattened on the middle, and the 7th bears a very deep and rather broad notch.

Of this elegant species only a single male was found at Tapajos.

20. *Stenus heres*, n. sp. (Sect. II. B, Er.) Niger, sat nitidus, antennis, palpis, pedibusque testaceis, his geniculis, illis clavâ, infuscatis; prothorace subcylindrico, dense fortiterque punctato; elytris dense fortiterque punctatis, thorace longioribus; abdomine crebre fortiter, apicem versus subtiliter, punctato. Long. corp. 1½ lin.

Antennæ elongate, yellow, with the club nearly black; 3rd joint distinctly longer than 4th. Palpi yellow. Head broad, about as broad as the elytra, rather coarsely punctured; the middle longitudinally a little elevated, and between this and the eyes a small depression on each side. Thorax longer than broad, much narrower than the elytra, subcylindric, but the sides a little curved and slightly contracted towards the base; very densely and coarsely punctured, so as to be rugose. Elytra elongate and outstanding, much broader than the thorax; the shoulders well marked and rectangular, very coarsely and closely punctured, but the interstices quite broad enough to be shining. Hind body cylindric, rather coarsely and closely punctured, very finely and very scantily pubescent; segments 3—5 much constricted near the base. Legs yellow; the base of the tibiæ and the apical portion of the femora broadly infuscate; lobes of the tarsi long; punctuation of metasternum and under face of hind body very coarse; genæ rather coarsely but not densely punctured.

Ega; a single individual, which I believe to be a female.

21. *Stenus cerritus*, n. sp. (Sect. II. B, Er.) Æneoniger, nitidulus, parcius albido-pubescens, fortiter profundeque punctatus, antennis pedibusque testaceis, palpis pallide flavis; abdomine parcius sat fortiter punctato. Long. corp. 2½ lin.

Antennæ yellow, elongate; 3rd joint much longer than 4th, club elongate, 10th joint twice as long as broad

Palpi elongate, very pale yellow. Head with the eyes large, rather smaller than the elytra, distinctly excavate; at the bottom of the excavation with a smooth, carina-like space. Thorax much longer than broad, subcylindric, but distinctly narrowed towards the front, and a little towards the base; shining brassy, with fine, pale hairs; coarsely, deeply and rather closely punctured. Elytra quite as long as the thorax, with the shoulders well marked and prominent; coarsely, deeply and closely punctured, shining brassy, with a fine, pale, scanty pubescence. Hind body cylindrical, not margined, shining, sparingly, moderately coarsely punctured, with a fine, pale, elongate pubescence. Legs yellow, rather long; 4th joint of tarsi bilobed, the lobes rather long and narrow.

The male characters are extraordinary: the hind legs are deformed; the femora are incrassate, the lower margin thickened near the middle, abruptly contracted near the base; the tibiæ are also thickened, and furnished near the middle with an angular prominence on their inner face, and below this prominence the inner face is partly sliced off; the basal joint of the tarsus is also distinctly dilated. On the underside of the hind body the basal segment has the hind margin thickened in the middle, but much emarginate, so as to form a broad notch, with rather prominent edges; the next segment has a smaller but shining notch, the 4th and 5th segments are also shining in the middle, in front of the hind margin; the 6th is broadly emarginate, and the 7th is emarginate at the extremity, the emargination being continued forwards as a narrow, deep fissure.

Tapajos; one male, two female individuals.

22. *Stenus Batesi*, n. sp. Plumbeo-niger, breviter albido-pubescens, dense punctatus; antennis palpisque testaceis, illis apice infuscatis; pedibus fuscis, tibiarum basi testacea; capite coleopteris multo angustiore, bisulcato, crebre punctato; prothorace maculâ medio lævi; elytris thorace longioribus, dense fortiterque punctatis; abdomine elongato-conico, dense punctato. Long. corp. fere 2 lin.

Antennæ moderately long, and rather stout, dark yellow; the club, which is long in proportion to the other part, more obscure. Head rather small, much narrower than the elytra, not excavate, but with a small space on the middle elevated, and a depression on either side of this; rather closely punctured, except on the elevation. Thorax rather longer than broad, much narrower than the elytra, the basal portion distinctly contracted, and the sides narrowed towards the front; the surface closely and moderately coarsely punctured, with a small spot behind the middle free from punctures. Elytra largely developed, much longer and broader than the thorax, slightly depressed within the prominent shoulders, rounded and contracted towards the extremity, densely and coarsely punctured. Hind body broad at the base, and gradually narrowed towards the extremity; rather closely and moderately coarsely punctured, with a fine, distinct, depressed pubescence. Legs pitchy, the base of the tibiæ yellow; tarsi pitchy-yellow, slightly paler at the base. Genæ rather sparingly punctured.

The male has a very broad notch at the extremity of the 7th segment; the 6th segment is flat along the middle, and finely pubescent, scarcely emarginate at the hind margin.

Tapajos; a single male.

23. *Stenus collaris*, n. sp. (Sect. II. B, Er.) Niger, nitidus, glaber, antennarum basi, palpis pedibusque rufescentibus; fronte areâ mediâ latâ lævi; prothorace medio ampliato, fortiter sat crebre punctato, disco lævi; elytris fortiter sat crebre punctatis. Long. corp. fere 2 lin.

Antennæ only moderately long, pitchy, with the two basal joints yellow, and the following ones intermediate in colour; 3rd joint a good deal longer than 4th. Palpi elongate, yellow, the last joint dusky yellow. Head broad, but narrower than the elytra, only sparingly punctured, a broad space in the middle being quite smooth and even, outside this the punctures are but sparing; the front is not excavate, but there is an impression on each side of the smooth central place; the antennal tubercles are very small. Thorax about as long as broad, much narrower than the elytra, a good deal broader in the middle than at the front and base; coarsely but rather sparingly punctured, the punctures most numerous near the front margin; a broad space on the middle impunctate. Elytra broad, the shoulders prominent, rather longer than the thorax, coarsely but sparingly punctured. Hind body broad at the base, but much narrowed towards the extremity; segments 2—5 transversely depressed near the base in the middle, and a little contracted at the sides; rather sparingly but coarsely punctured, the apical segments with only sparing and fine punctures. Legs dark yellow; tarsi rather long, with the lobes broad. Genæ coarsely and rather closely punctured; underside of hind body very coarsely punctured.

Tapajos; a single individual, which is I believe a female.

24. *Stenus parviceps*, n. sp. (Sect. II. B, Er.) Niger, nitidus, supra parcius albido-pubescens, antennis palpisque testaceis, illarum clavâ obscuriore; pedibus infuscato-rufis, femoribus tibiisque basi quam apice dilutioribus; capite thorace vix latiore, bisulcato; elytris thorace longioribus, crebre fortiter punctatis; abdomine apicem versus angustato, crebre fortiter punctato. Long. corp. 2 lin.

Antennæ moderately long, yellow, with the club darker; 3rd joint a good deal longer than 4th. Palpi yellow, the apical joint darker than the preceding one. Head much narrower than the elytra, and scarcely broader than the thorax, not excavate, but distinctly bisulcate; the surface irregularly punctured, as when viewed from the front, the middle and some spaces near the eyes appear like shining spots. Thorax scarcely longer than broad, greatly narrower than the elytra, the sides not rounded but a little narrower at the base than in front; the surface coarsely, moderately closely punctured, the punctures absent from a very small space behind the middle. Elytra much longer and broader than the thorax, the shoulders rectangular and a little elevated, coarsely and closely but not densely punctured. Hind body a good deal narrowed towards the extremity, rather coarsely punctured; the punctures on the apical segments much finer than at the base, the punctures rather close, the pubescence very fine and scanty. Base of femora reddish, extremity infuscate; front coxæ reddish, hind ones pitchy; tibiæ yellowish, infuscate towards the extremity; tarsi infuscate-yellow: under surface deeply, coarsely and densely punctured, opaque and with the white pubescence elongate and conspicuous.

The male on the underside exhibits a notch at the extremity of the 7th segment; it is, however, much concealed by the dense, fine, elongate, pale pubescence, which covers the middle of the segment; the 6th segment is also densely pubescent along the middle, and its hind margin is slightly notched at the extremity.

Amazons; a single individual, without more special locality.

Obs.—I have also a specimen from Ega, which I believe to be merely a variety of this species; the chief difference it exhibits from the individual above described consists in the absence of any smooth space on the thorax.

25. *Stenus proximus*, n. sp. (Sect. II. B, Er.) Niger, supra nitidus, parcius albido-pubescens, antennis palpisque testaceis, apice infuscatis; pedibus infuscato-rufis, femoribus tibiisque basi quam apice dilutioribus; capite thorace vix latiore, bi-impresso; elytris thorace longioribus, crebre fortiter punctatis; abdomine crebre fortiter punctato. Long. corp. $1\frac{1}{5}$ lin.

Antennæ reddish with the club, dusky; palpi yellow, but with the front half of the last joint distinctly darker. Head small, not excavate, with two impressions between the eyes, which can scarcely be called sulci, as they do not reach the vertex, and are also abbreviated in front by the well-marked antennal tubercles; only sparingly punctured, the more elevated portions appearing as smooth spaces. Thorax much narrower than the elytra, rather longer than broad, slightly curved at the sides towards the front, coarsely and rather closely somewhat irregularly punctured. Elytra longer than the thorax, distinctly impressed within the prominent shoulders, coarsely but not closely punctured. Hind body rather coarsely and moderately closely punctured, much more finely at the extremity than on the basal segments.

Amazons (probably Tapajos ; a single male.

Obs.—This species is excessively closely allied to *S. parviceps*, and differs therefrom only in slight characters; it is a little smaller than *S. parviceps*, and has the antennæ distinctly shorter; the sulci on the head are less distinct, being more fovea-like; the thorax is slightly narrower, the elytra are rather shorter and rather more coarsely and less closely punctured; the punctuation of the hind body is not quite so coarse and deep. The male characters seem scarcely to differ.

MEGALOPS.

Many points of the structure of these remarkable insects remain to be ascertained, before the position and affinities of the species can be satisfactorily decided on. Erichson describes the antennæ as ten-jointed. I find them, however, to be certainly eleven-jointed, the basal joint being short and very stout, and much concealed by the prominence above its point of insertion. Erichson has also described the tarsi as five-jointed, and the 4th joint to be minute. On inspection, however, the tarsi appear at first to be only four-jointed, but a more careful examination reveals the fact that the tarsi are really five-jointed, and that the 4th joint is not minute, but consists of a very small basal and articular portion, to which are attached two long slender lobes, which are so closely applied to the 5th joint as only to be detected by bending or lifting up the terminal joint.

Nothing is known as to the structure of the labrum, which is quite invisible in the species. It is probable, however, that it is concealed under the largely-developed horny clypeus, and that it is moveable; and that the two long spines which appear to proceed from the front of the clypeus are in reality appendages from the labrum. The sexual characters have hitherto escaped observation. I have pointed them out in the following description of *M. spinosus*, but should add that, though in *M. spinosus* the antennæ are similarly formed in the two sexes, in some of the other species there is a remarkable sexual disparity in the structure of the apical joints.

Seven species of the genus have already been described: two from Northern America, three from South America, one from Australia and one from South Africa. The species, however, are undoubtedly more numerous in South America than elsewhere, for I have thirteen species from thence in my collection, while the only other species I have seen is the Australian one.

1. *Megalops spinosus*, n. sp. Niger, nitidus, antennis pedibusque testaceis, illis clavâ fuscâ; thorace transversim quadrisulcato, sulco secundo medio vix, tertio sat late, interrupto; elytris ante medium striolâ obliquâ impressâ. Long. corp. 2 lin.

Antennæ yellow, darker towards the extremity, with the club fuscous; 3rd joint twice as long as 2nd, 4th about as long as 2nd; 5—7 each shorter than its predecessor; 8th small, 9th a good deal broader than 8th, bead-like; 10th rather strongly transverse, 11th moderately large, slightly broader than 10th, as long as 9th and 10th together, obtusely pointed. Palpi yellow. Head broad; the eyes very large, even broader than the elytra; the clypeus armed in front with two very elongate, pitchy-yellow spines, which are ciliated internally, the horny clypeus deeply impressed and separated by a straight depressed line from the front; the front with elevations and depressions so placed as to form a central elevated space, surrounded, except at the summit in the middle, by a broad, irregular depression; also with a fine depression along the inner margin of the eyes, in which are a few punctures. Thorax broad, with four transverse furrows in which are large punctures; the first of these grooves is placed near the front margin, whose course it follows; the second extends in a nearly straight line across the thorax, so that it is nearer to the front one at the sides than in the middle,—it can scarcely be said to be interrupted in the middle; the third furrow is the broadest and is distinctly interrupted in the middle, the hinder one is placed close to the base; the sides appear a little waved and have two angular projections near the front. Scutellum emarginate-truncate at the apex, bearing two foveæ. Elytra broad, broader than long, about as long as the thorax, deeply impressed at the base for the thorax; each near the shoulder with an oblique stria, sharply limited on the inner but not on the outer side. Legs yellow; coxæ castaneous.

Ega; two specimens, both of which appear to be females.

Obs.—Besides the two females above described, I have also two males of this species, which I obtained from the collection of Mr. E. W. Janson, where they were labelled, "Pará, Brazil." These two males have the ventral plate of the 7th segment of the hind body slightly emarginate on each side the middle at the extremity; and the preceding segments are slightly flattened along the middle, and furnished there with a very fine and scanty short pubescence. The structure of the antennæ is quite the same as in the females.

2. *Megalops impressus*, n. sp. Niger, nitidus, antennis pedibusque testaceis, illis clavâ fuscâ; thorace grosse punctato, minus distincte sulcato; elytris disco striolâ profundâ impressâ. Long. corp. 1⅜ lin.

Antennæ with the two basal joints yellow, the others

more obscure; joints 6—8 small, 9th transverse, 10th much broader than 9th, strongly transverse, 11th rounded, rather large. Palpi yellow; clypeus with two elongate spines, its front with an emargination on each side. Head broad and short, with impressions placed much as in *M. spinosus*, but the depressions formed as it were by confluent punctures. Thorax with very coarse punctures covering the greater part of its surface; a series behind the front margin, a second series separated from the front one by a rather elevated space; along the middle with a third, broad, confused, double series, interrupted in the middle, also with a basal series, and with additional punctures (not extending across the middle) in front of the basal series. Elytra broad and short, the sutural stria very deeply impressed at the base, across the middle with a deep oblique impression, and near the inner edge of this with two obsolete punctures. Hind body with the impressions at the base of the segments large and distinct. Legs yellow; coxæ darker.

Villa Nova; a single female, found under chips.

OSORIUS.

About eighteen species of this genus have been described, eight from warm America, one from North America, and the others from the warm parts of the Old World. I here describe seven Amazonian species, and though this seems a considerable addition to the South American species, it is, in comparison with the undescribed species, but small; for I find the specimens of the genus from South America extant in my own collection must be referred to about forty species. The genus is one of excessive difficulty to the student, from the extreme resemblance of the species to one another; and it is not until careful examinations and comparisons are made, that the characters distinguishing the species from one another are seen and appreciated. The structure of the species indicate very sedentary habits; the cohesion or attachment of the different parts of the body together is but slight, so that these insects drop to pieces in our collections with only too great ease. It has been observed that some of the species live in burrows in decaying wood, but it is not indicated whether they follow the borings of other insects, or make the burrows for themselves. The almost complete absence of external characters to distinguish the sexes is worthy of

remark, as is also the simple but peculiar form of the
œdeagus, this organ scarcely varying, moreover, in the
different species.

1. *Osorius stipes*, n. sp. Niger, sat nitidus, capite
coriaceo, subopaco, clypeo antice emarginato, et in medio
prominulo; thorace parce punctato, medio breviter canaliculato; elytris strigulosis. Long. corp. 7—8½ lin.

Head with the surface coriaceous, and sprinkled with
distinct punctures; the clypeus emarginate in front, so
that the anterior angles form blunt projections, but not at
all spinose, and also obtusely prominent in the middle;
the temples, over and behind the eyes, with coarse distinct
rugæ, and some small rough elevations. Thorax much
broader than long; the sides much narrowed behind and
distinctly sinuate in front of the hind angles, which are
distinct and nearly right angles, but a little obtuse; the
sides distinctly impressed in front of the hind angles; so as
to make the lateral margin appear strongly raised there;
the surface entirely coriaceous, but more finely so than the
head, and with distinct though sparing punctures, and
with a short, fine channel along the middle. Elytra distinctly longer than thorax, entirely covered with shallow
irregular rugæ; on the underside the prominence of the
prosternum is very marked, the mesosternum distinctly
carinate in front of the coxæ. The hind body with very
few setæ, the apical segment coarsely strigose, especially
at the sides, and with strongly marked tubercles.

Pará, Ega; eight individuals.

Obs.—This is the largest species of the genus yet
known; the structure of the front of the head readily distinguishes it from *O. ater*.

2. *Osorius nitens*, n. sp. Cylindricus, niger, nitidus,
antennis pedibusque piceis; capite pernitido, parce sat
fortiter punctato, clypeo utrinque emarginato; thorace
nitido, coriaceo, parce fortiter punctato, angulis posterioribus obtusis, minus prominulis; elytris nitidis, rugu-
losis, et parce obsolete punctatis; abdomine supra lævigato. Long. corp. 5 lin.

Front of clypeus distinctly notched on each side, the
lateral angles only slightly more prominent than the
middle. Head very shining, and with the front part not
in the least coriaceous; the surface sparingly sprinkled with

distinct punctures. Thorax about as long as its breadth at the base; the sides gradually narrowed from the front to the base, and not sinuate in front of the hind angles; the base distinctly curved near the hind angles, so that these are obtuse, the lateral margin strongly raised on its posterior half; the surface coriaceous, but shining, sprinkled with distinct punctures, with a very short indistinct channel on the middle. Elytra slightly longer than the thorax, shining, but distinctly rugulose, and with some rather coarse, but obsolete punctures. Hind body above shining, and with one or two setigerous punctures on each segment; its under surface with rather coarse, sparing, setigerous punctures; hind portion of the apical segment sparingly punctured, but longitudinally smooth along the middle.

St. Paulo; two individuals.

3. *Osorius simplex*, n. sp. Cylindricus, nitidus, nigro-piceus, pedibus rufescentibus; capite parce sat fortiter punctato, clypeo antice subtruncato, angulis prominulis; prothorace parce punctato, angulis posterioribus obtusis; elytris subrugulosis et parce obsolete punctatis; abdomine supra lævigato. Long. corp. 4¼ lin.

Clypeus almost straight in the middle in front, but with the angles thick and prominent, but not at all spinose. Head shining, rather sparingly but distinctly punctured, the antennal tubercles well marked, and the eyes distinctly prominent; the punctures are wanting in front of the vertex, and are wanting about an irregular longitudinal space along the middle. Thorax slightly longer than it is broad at the base, a little narrowed from the front to the base; the surface shining, sprinkled with punctures like the head, indistinctly channelled along the middle; the hind angles distinctly obtuse, but not far from right angles. Elytra a little longer than the thorax, shining, with indistinct rugulose impressions and elongate punctures. Hind body shining, rather slender, impunctate above; on the underside the 7th segment bears a large smooth impression in the middle at the extremity, on each side of which it is sparingly sculptured, and furnished with scanty hairs, the punctures quite at the side forming obscure distant rugæ; the 6th segment coarsely punctured, the punctures rather close on the middle; the preceding segments also with coarse punctures on the middle.

Ega; a single male.

Obs.—This species is closely allied to *O. nitens*, but is rather smaller and distinctly more slender; the clypeus is not prominent in the middle, and the antennal tubercles are more distinct; it is possible that the paler colour may be only the result of the immaturity of the individual described.

4. *Osorius integer*, n. sp. Niger, nitidus, capite thoraceque crebre, fortiter punctatis, illo clypeo antice emarginato, hoc angulis posterioribus subrectis; elytris haud dense punctato-rugulosis; abdomine supra parcius et obsolete punctato. Long. corp. fere 5 lin.

Clypeus emarginate in front, the angles not produced. Head rather coarsely and moderately closely punctured, with a smooth space in front of the finely-punctured vertex; black and shining, not coriaceous. Thorax about as long as it is broad at the base, only a little narrowed from the front to the base; the hind angles distinctly obtuse, but not far from right angles; the surface shining black, rather coarsely and not very sparingly punctured, and with traces of a short channel on the middle. Elytra a little longer than the thorax, shining, with a distinct but neither dense nor deep sculpture, consisting of indefinite rugæ, and sparing, ill-defined, elongate punctures. Hind body black; above sparingly sprinkled with obsolete punctures, beneath with the 7th segment in the middle coarsely punctured, and bearing a fine elongate pubescence, the punctures at the sides more sparing, but coarse, and passing into shallow rugæ; 6th segment coarsely punctured about the middle, and the preceding segments also with some coarse punctures about the middle.

Ega; a single male.

Obs.—This species at first sight exactly resembles *O. nitens*, but the clypeus is not prominent in the middle; the head and thorax are more distinctly punctured, and the punctuation of the hind body is different.

5. *Osorius solidus*, n. sp. Cylindricus, nitidus, piceus, antennis pedibusque rufis; capite thoraceque fortiter punctatis, hoc angulis posterioribus rotundato-obtusis; elytris fortiter punctatis, nitidulis; abdomine parce obsoleteque punctato. Long. corp. 4 lin.

A rather narrow and parallel species. Mandibles bidentate, the upper tooth on the right one very large. Clypeus with the front angles distinctly prominent; surface of head shining, rather coarsely punctured, with a smooth space in front of the vertex; over the eyes the punctures become strigose. Thorax about as long as it is broad in front; the sides gently narrowed to the base, the angles obtuse, and, owing to the base being distinctly curved near the angles, appearing somewhat rounded; the surface shining, rather coarsely and not very sparingly punctured. Elytra scarcely longer than the thorax, shining, with rather coarse, elongate, distinct punctures. Hind body above with a few obsolete punctures, which are most numerous and most visible on the 6th and 7th segments; these segments are besides very finely strigose, so as not to be shining. The sculpture of the ventral plate of the 7th segment similar to that on the dorsal plate, viz., some coarse but obsolete, sparing, elongate punctures, the surface besides being finely and densely strigose, so as to be opaque; the 6th segment beneath is also scarcely shining, and with some coarse, scattered punctures. Mesosternum strongly carinate.

St. Paulo; a single individual, which I consider to be a female.

Obs.—This species is remarkable by reason of the comparative large development of the upper tooth on the right mandible.

6. *Osorius affinis*, n. sp. Angustulus, nigro-piceus, antennis pedibusque rufis; clypeo antice subtruncato; capite thoraceque coriaceis, subnitidis, sat fortiter punctatis, hoc angulis posterioribus obtusis; elytris minus discrete punctatis; abdomine segmentis 6° et 7° minus profunde punctatis. Long. corp. 3⅔ lin.

Head rather small, with the eyes distinctly convex; the clypeus nearly truncate in front, the antennal tubercles distinct; the surface distinctly coriaceous, so as to be but little shining, with coarse but shallow, elongate punctures, which pass into rugæ over the eyes, and are wanting in front of the vertex. Thorax about as long as it is broad in front, the lateral margin fine throughout; the sides gently narrowed towards the base, slightly sinuate in front of the hind angles, which are therefore distinct, and are obtuse; the surface coriaceous, and sprinkled with rather coarse but obsolete punctures. Elytra shining, rather sparingly sprinkled with ill-defined, somewhat coarse punctures. Hind body shining above; the 6th and 7th segments with shallow, rather sparing punctures; sculpture of the 7th segment on the under side almost similar to that of the upper side; 6th with sparing coarse punctures. Legs reddish, with the femora pitchy red.

St. Paulo; a single individual, which I consider to be a female.

Obs.—This species is closely allied to *O. solidus*, and has the mandibles similarly formed, but is readily distinguished by the different sculpture.

7. *Osorius oculatus*, n. sp. Piceus, cylindricus, antennis pedibusque rufis, oculis majoribus, prominulis; capite, thorace, elytrisque fortiter sat crebre punctatis, nitidulis; abdomine dense asperato-punctato, opaco. Long. corp. 2⅛ lin.

Antennæ red; 7th joint abruptly larger than the preceding ones. Clypeus nearly straight in front, the angles very slightly prominent; surface of head shining, coarsely punctured, the punctures wanting along the middle and at the vertex. Thorax about as long as broad, coarsely, deeply and rather closely punctured, with a sharply-defined longitudinal space along the middle impunctate; the sides slightly narrowed from the front to behind the middle, and thence more abruptly to the base; the lateral margin very fine, the hind angles obtuse and indistinct. Elytra rather longer than the thorax, coarsely and moderately closely and deeply punctured. Hind body above densely punctured, and with a distinct, rough, pale pubescence; beneath coarsely punctured, but more sparingly than on the upper side, and therefore more shining. Legs red; hind tibiæ slender, bearing three spines.

Ega; three individuals.

Obs.—This species at first sight greatly resembles the North American *O. latipes*, but it is considerably smaller, and may be at once distinguished by the larger and more convex eyes. Laporte has described a species about this size from Columbia; but the few words of his description (*Osorius pygmæus*, Etudes Ent. p. 130) are so meagre as to render the identification of his species extremely difficult.

HOLOTROCHUS.

Six species are at present placed under this generic name, five of which inhabit tropical America, the other being found in Madagascar. I here describe six other species; of these the *H. durus* appears to be somewhat allied to the *H. volvulus* figured by Erichson. The species I have described as *H. syntheticus* differs from the *H. durus* by some structural peculiarities, among the more interesting of which is the formation of the apical segments of the hind body. Attention being paid to this character and to the facies of the species an affinity with *Lispinus* is strongly suggested. The other two species, viz., *H. pubescens* and *H. subtilis*, when I first examined them, at once suggested to me a relationship with an insect which has been one of the unsatisfactory ones in the classification of the *Staphylinidæ*, viz., *Phlœocharis subtilissima*; and, on comparing the *H. pubescens* with *P. subtilissima*, I find such a primâ facie resemblance in the structure of the thorax and middle body as to lead me to think that the natural connection of *Phlœocharis subtilissima* will be found to be with this group of *Holotrochus*. Indeed, I may say, the variety of facies and of certain structural characters which exist in *Osorius* and its allies and in these species of *Holotrochus*, suggest to me that the accurate study of these insects will be found to suggest an improved arrangement of some of the *Piestini*, *Oxytelini* and *Phlœocharini*; and I shall not be at all surprised if it be ultimately considered that we have here preserved for us some of the more primitive forms of the *Staphylinidæ*. It will not improbably be suggested that I ought to have established a new genus for *H. syntheticus* and *H. pubescens*, but after the examination of several undescribed intermediate American forms which exist in my collection, I have considered it better not to do so. As regards the two last species here described, viz., *H. clavipes* and *H. Fauveli*, I think it highly probable that they will prove to be closely allied to *Ancæus megacephalus*, Fauvel. In establishing the genus *Ancæus*, M. Fauvel, to distinguish it from the *Oxytelini*, pointed out the hidden and retractile 7th segment of the hind body. The structure of that segment is, however, subject to so much variation both in the *Oxytelini* and *Piestini* (as they are at present limited) that this point throws but little light on the affinities of M. Fauvel's insect. M. Fauvel considered the *Anœus megacephalus* to be most allied to *Lispinus*; but the two insects I have described as *H. clavipes* and *H. Faureli* differ in a highly important respect from the species of *Lispinus*, inasmuch as they have the front coxæ exserted and not covered or separated by any process similar to that which is so conspicuous in *Lispinus*. I must not be understood as implying that the two insects I here allude to will be ultimately considered congeneric with the other *Holotrochi* here described, for I consider that in the present state of our knowledge of these *Staphylinidæ* it is quite impossible for us to decide where there really occur those separations and gaps between species which warrant the formation of genera.

1. *Holotrochus durus*, n. sp. Niger, glaber, nitidus, parce sed distincte punctatus, antennis piceis, pedibus piceo-rufis. Long. corp. $3\frac{1}{3}$ lin.

Mas: abdomine segmentis ventralibus 6 et 7 profunde impressis.

Antennæ stout, shorter than head and thorax, insertion near the front of the eye in a large cavity, greatly overlapped by the side of the clypeus; 1st joint much concealed, and thick, 2nd joint small, 3rd a good deal longer than 2nd, dilated towards the extremity; joints 4—10 transverse, 11th rather narrower than 10th, obtusely pointed; the six basal joints are pitchy, the others red, but this colour is much obscured by the pubescence. Head much narrower than the thorax; clypeus greatly rounded in front; the surface shining and rather finely and sparingly but quite distinctly punctured. Thorax scarcely so long as broad, almost broader than the elytra; the sides a little rounded towards the front, nearly straight behind the front, and thence only very slightly narrowed to the base; the hind angles almost right angles; the surface is shining black, rather sparingly but distinctly punctured; it is transversely convex, the sides margined, the base closely applied to the elytra and not margined; near the hind angles are traces of a longitudinal impression. Elytra slightly longer than the thorax, with a well-marked sutural stria moderately finely and sparingly punctured, with an indistinct longitudinal series of four or five larger punctures along the middle. Hind body with segments 2—5 of equal width, each segment smooth in the middle, obscurely punctured at the sides, less shining than the front parts of the body. Legs dark reddish. Prosternum prominent, but not carinate in the middle in front; mesosternum with a prominent piece between the middle coxæ, which are therefore distinctly separated; metasternum smooth, shining black, deeply channelled. In the male the ventral plate of the 6th segment bears a large and deep, almost horseshoe-shaped impression, which is surrounded by a kind of margin, and has a peculiar granular pubescence along its sides; the 7th segment bears a similar but more elongate impression, the bottom of which is covered by the peculiar granular pubescence; the apical portion of this segment is produced in the middle as a broad lobe; the punctures on the under face of the hind body in this sex are coarser than in the female.

The female lacks the impressions described above, and the produced lobe of the 7th segment is narrower and more pointed.

Amazons; three females, one male. One of these specimens is indicated as being from Ega.

Obs.—The structure of the 8th segment of the hind body and of the ædeagus in this species are peculiar; the dorsal and lateral plates of the former are formed much as in *Osorius*, but each lateral plate bears a pencil of elongate delicate hairs; the ventral plate appears to be altogether absent; the ædeagus is complicated in its structure and laterally asymmetrical, and the missing ventral plate of the 8th segment appears to me to be attached to one side of the ædeagus as a lateral appendage thereof.

2. *Holotrochus syntheticus*, n. sp. Piceus, antennis testaceis, pedibus rufis; capite, thorace, elytrisque pernitidis, glabris, parcius sat fortiter punctatis, thorace versus angulos posteriores acute rectos foveâ magnâ; abdomine apicem versus attenuato, subtiliter pubescente, fere opaco. Long corp. 2—2¼ lin.

Antennæ yellow; 1st joint rather long and stout, not much concealed by the clypeus, 2nd and 3rd joints subequal, rather slender, 4, 5 and 6 small, not differing much from one another, the 6th hardly so long as broad; joints 7—11 much larger than the preceding ones, 7—10 scarcely differing from one another, each distinctly transverse, 11th scarcely so broad as 10th, obtusely pointed. Head small, much smaller than the thorax; the eyes prominent: the clypeus rounded; the surface pitchy, very shining, rather sparingly but distinctly punctured, the vertex smooth. Thorax strongly transverse, the sides slightly curved, about as broad at the hind as at the front angles; the hind angles sharply-marked right angles, much more elevated than the front ones; the lateral margin very fine; close to each hind angle is a large broad and deep impression; the surface is shining, and it is distinctly, not closely punctured, the punctures are most numerous about the middle, nearly wanting at the sides, and there is a very indistinct longitudinal space along the middle smooth. Elytra much longer than the thorax, very shining, rather coarsely and sparingly punctured, each with a well-marked sutural stria. Hind body elongate, conical cylindric, being much narrowed towards the extremity; the surface nearly opaque, being finely pubescent, and covered with a peculiar obsolete punctuation. Legs reddish, the tibiæ bearing hair-like spines, which are most distinct on the intermediate legs. Prosternum short, with a very fine tubercle in front of the coxæ; mesosternum with a sharply-elevated laminar carina; metasternum with a short, coarse channel, shining, sparingly punctured.

In both sexes the ventral plate is produced at the extremity, so as to form an angular projection; this projection is more elongate and the angle at its extremity more acute in the female than in the male.

Tapajos; five individuals.

Obs.—This species departs widely in its facies from *H. durus*, and in the structure of its hind body approaches to *Lispinus*; the dorsal plate of the 8th segment has not the peculiar box-like structure found in *H. durus* and in the species of *Osorius*; the ventral plate appears to be absent as in *H. durus*, the under face of the segment being formed by the folded lateral plates, which are ample, and have quite lost the very hard, spinous character of the same parts in *H. durus*.

3. *Holotrochus pubescens*, n. sp. Ferrugineus, densius pubescens, obsolete punctatus, haud nitidus; prothorace transverso, angulis posterioribus fere acutis. Long. corp. 1⅓ lin.

Antennæ yellowish, rather short; 1st joint stout and rather long, 2nd joint oval, 3rd joint about as long as 2nd, slender at the base; 4th joint smaller than the others, 5th and 6th about equal, bead-like; 7—11 broader than the preceding ones, 7—10 rather strongly transverse; 11th joint as broad as the 10th, obtusely pointed. Head much narrower than the thorax, dull reddish; punctuation very indistinct, but with a well-marked, fine, rather long, yellowish pubescence; eyes rather prominent. Thorax as broad as the elytra, rather strongly transverse, very slightly curved at the sides; the front angles more depressed than the hind ones, the base a little emarginate, so that the hind angles project backwards; the surface clothed with a fine, dense pubescence, but with scarcely visible sculpture. Elytra a good deal longer than the thorax, with similar pubescence and obsolete sculpture. Hind body cylindric, scarcely narrowed till the 7th segment, which is very retractile. Legs reddish, the tibiæ with fine spines, which are most distinct on the intermediate legs.

Tapajos; ten individuals.

Obs.—Notwithstanding the great difference in appearance between this species and the preceding (*H. syntheticus*), they appear to be structurally closely allied. In certain individuals of *H. pubescens* the ventral plate of the 7th segment of the hind body is somewhat prolonged and acuminate in the middle; judging from what is the case in *H. syntheticus*, I consider these to be females.

4. *Holotrochus subtilis*, n. sp. Ferrugineus, subtiliter sat dense pubescens, subnitidus, obsolete punctatus; prothorace transverso, angulis posterioribus fere acutis. Long. corp. 1¼ lin.

This species is extremely closely allied to *H. pubescens*, but is readily enough distinguished, on comparison, by the much shorter and more inconspicuous pubescence and the less opaque surface; it is also a little smaller, and has the antennæ a little shorter; in other respects it appears scarcely to differ from *H. pubescens*.

Ega; three individuals.

5. *Holotrochus clavipes*, n. sp. Piceus, angustulus, sat nitidus; capite, thorace, elytrisque dense subtilissime longitudinaliter strigosulis, et parce subtiliter punctulatis; capite magno, mandibulis porrectis; thorace basin versus angustato, ad angulos posteriores foveolato; pedibus sordide testaceis. Long. corp. (abdomine extenso) 1½ lin.

Antennæ pitchy, stout, very short; 1st joint concealed by the mode of insertion, joints 2—10 each shorter than broad, 11th joint obtusely pointed, rather lighter in colour, and scarcely so broad as 10th. Mandibles porrect, conspicuous, crossed in repose. Head elongate; the front angles of the elongate clypeus rounded, the middle slightly emarginate; on each side, near the front, is a large depression; the surface very finely strigose, so as to be but little shining, and with a few fine, distant punctures. Thorax about as long as broad, distinctly narrowed towards the base, but with the sides not curved; the lateral margin excessively fine, and only distinct on the posterior part; close to each hind angle is a rather large, but not sharply defined depression; the hind angles are obtuse, the sculpture of the surface is similar to that of the head, and there is a fine, abbreviated channel along the middle. Elytra rather longer than the thorax, and about as broad as it is at the base, with a well-marked sutural stria; the sculpture similar to that of the head and thorax, but the scattered punctures excessively fine and indistinct. Hind body almost without sculpture. Legs short, pitchy yellow, the hind femora extremely stout, the four basal joints of the tarsi very short. Under surface smooth and without sculpture.

Amazons; a single individual, without indication of any special locality.

6. *Holotrochus Faureli*, n. sp. Testaceo-castaneus, angustulus, nitidus, glaber; capite magno, mandibulis porrectis; thorace basin versus angustato, ad angulos posteriores foveolato. Long. corp. 1 lin.

Antennæ yellowish, rather short, a good deal of the basal joint exposed; 2nd joint stout, but a good deal more slender than the basal one, 3rd slightly shorter and more slender than 2nd; 4—8 small, differing but little from one another, the 7th and 8th, however, distinctly transverse; 9th and 10th distinctly broader than the preceding ones, rather strongly transverse; 11th obtuse at the extremity. Head elongate, with the mandibles not large but prominent; the clypeus with the front angles rounded and with an impression near these; the eyes small but distinct and rather prominent; the surface almost without sculpture. Thorax about as long as broad, distinctly narrowed behind; the sides slightly sinuate in front of the hind angles, which are nearly right angles, and not rounded; within each is a small impression; the surface almost without sculpture. Elytra longer than the thorax, each with a fine sutural stria, almost without sculpture. Hind body cylindric, with a few upright setæ. Under surface impunctate. Legs short; hind femora incrassate.

Amazons; a single individual without special locality. I have named this species in honour of M. Albert Fauvel, of Caen, whose labours on the *Staphylinidæ* are well known to all interested in this family of *Coleoptera*.

BLEDIUS.

The species of this well-known genus here described are seven in number, and suggest no special remark; only one species was found by Mr. Bates, the other six being discovered by Dr. Trail. Only three or four species have been previously described from South America, yet it is very probable that the genus is numerously represented there, for these insects are very retiring in their habits and little likely to come under the notice of collectors, except special search be made for them.

1. *Bledius albidus*, n. sp. Pallide testaceus, obsolete punctatus, subnitidus; capite castaneo, bituberculato. Long. corp. 1½ lin.

Antennæ very pale yellow, slender; basal joint as long as the three or four following ones together; 2nd more than twice as long as 3rd; the four or five apical joints stouter than the others, each of them about as long as broad. Mandibles elongate, their upper edge with a tooth near the base, and beyond the middle with a long spine-like tooth directed forwards and upwards. Head darker than the rest of the insect, castaneous, with the eyes black; the clypeus much deflexed; close to the eye on each side is a prominent tubercle; the punctuation is quite obsolete. Thorax a little narrower than the elytra, not so long as broad, nearly straight at the sides, with the hind corners oblique; it is almost white, except that the fine depressed basal margin is black in the middle; it is extremely finely channelled, and finely and obsoletely punctured. Elytra short, but a little longer than the thorax, pale yellow, finely and indistinctly punctured. Hind body pale yellow, impunctate. Legs very pale yellow; the front tibiæ very broad below the middle, abruptly contracted at the apex.

Jurua; a series of individuals, captured by Dr. Trail on the 3rd November, 1874.

Obs.—This very distinct little species can be readily identified by the perpendicular front part of the head. I do not observe any indications of sexual differences.

2. *Bledius rarus*, n. sp. Niger, nitidus, antennis, pedibus, elytrisque testaceis, his disco late infuscato; prothorace sat crebre fortiterque punctato, medio canaliculato. Long. corp. 1½ lin.

Antennæ yellow, elongate; basal joint long, equal to the three following together; 2nd joint a good deal longer than 3rd, 6—10 each longer and slightly broader than its predecessor, 10th a good deal longer than broad, 11th pointed, longer than 10th. Head black, with a few indistinct punctures between the eyes, which are large, very prominent, and coarsely facetted. Thorax shining black, nearly straight at the sides till behind the middle, and thence a good deal narrowed to the base, so that the hind angles are very obtuse; it is not quite so long as broad, and the surface is rather coarsely but not closely punctured, and has a distinct channel along the middle. Elytra rather short, a little longer than the thorax, yellowish, with a large, common, dark patch on the middle, rather coarsely and moderately closely punctured, a little shining. Hind body a good deal narrower at the base than near the extremity, shining black, with the apex paler, almost impunctate, the basal segment not so shining as the others. Legs very pale yellow; anterior and middle coxæ a little infuscate.

Ega; a single individual.

3. *Bledius addendus*, n. sp. Testaceus, capite nigricante, antennis articulis 7—10 leviter transversis; thorace canaliculato, sat crebre minus profunde punctato; elytris thorace paulo longioribus, crebre punctatis; abdomine fere impunctato. Long. corp. 1¾ lin.

Antennæ yellow, slender, moderately long, distinctly thickened towards the extremity; 1st joint about as long as the four following; 3rd and 4th joints slender, of 5—10 each is distinctly broader than its predecessor, the 6th about as long as broad, the following ones a little transverse. Mandibles reddish, elongate and curved, but unarmed. Head with the clypeus black, the vertex blackish-red; it is opaque, without any distinct sculpture; the eyes very convex. Thorax yellowish, about as broad as the elytra, a little shorter than broad; the sides quite straight till behind the middle, then gradually narrowed to the very obtuse hind angles; along the middle is a distinct, rather deep channel, and the surface bears shallow, moderately coarse and rather distant punctures, and is a little shining. Elytra a little longer than the thorax, yellow, densely and moderately finely punctured, the hind margin of each separately much rounded. Hind body yellow, coriaceous, but with no distinct punctures. Legs yellow and rather stout; the spines on the four anterior tibiæ elongate.

Rio Solimoes; a single individual, found by Dr. Trail on the 11th October, 1874.

4. *Bledius simplex*, n. sp. Obscure testaceus, capite nigro, elytris fusco-testaceis, lateribus testaceis; prothorace transverso, subtiliter canaliculato, minus distincte punctato; elytris thorace longioribus, crebre subtiliter punctatis; abdomine impunctato. Long. corp. vix 1½ lin.

Antennæ yellow, rather short, gradually thickened from the 3rd joint to the extremity; joints 6—10 rather strongly transverse. Mandibles reddish, slender and curved, only moderately long, unarmed. Head black, with the eyes very convex, without sculpture and quite unarmed. Thorax reddish, only slightly shining, much broader than long, nearly as broad as the elytra, the hind angles very obtuse; it has a very fine channel along the middle, and is indistinctly and rather sparingly punctured. Elytra distinctly longer than the thorax, yellow, with the sutural portion broadly infuscate; they are finely and rather closely punctured, the hind margin of each separately much rounded. Hind body quite impunctate. Legs pale yellow, short; the tibiæ rather slender.

Rio Solimoes; a single individual, captured by Dr. Trail, 11th October, 1874.

5. *Bledius muticus*, n. sp. Testaceus, sat nitidus, capite nigricante; thorace vix transverso, obsolete canaliculato, parce obsoletequo punctato; elytris crebre subtiliter punctatis. Long. corp. ⅞ lin.

Antennæ yellow, rather stout; all the joints from 3—10 short, and each stouter than its predecessor, so that the penultimate ones are strongly transverse. Mandibles not elongate. Head black, without distinct punctuation, quite unarmed, with the eyes very convex. Thorax slightly narrower than the elytra, only a little broader than long, the hind angles quite rounded; it is yellowish and distinctly shining, only sparingly and very obsoletely punctured, and with a very obsolete channel along the middle. Elytra distinctly longer than the thorax, yellowish, finely and moderately closely punctured. Hind body yellow, impunctate. Legs yellow, only moderately stout.

Rio Madeira; two individuals, captured by Dr. Trail 25th May, 1874. They were attracted by light.

6. *Bledius similis*, n. sp. Fusco-testaceus, capite abdominisque apice summo nigricantibus, pedibus pallidis; prothorace subtransverso, obsolete punctato, subtiliter canaliculato, angulis posterioribus sinuatis; elytris subtiliter sat crebre punctatis. Long. corp. 1 lin.

Antennæ yellowish; all the joints from 3—10 short, and each distinctly stouter than its predecessor, so that though the 3rd joint is both short and slender, the penultimate ones are rather strongly transverse. Head black. Thorax reddish, blackish towards the front margin, distinctly shorter than broad, only slightly narrower than the elytra, sparingly and indistinctly punctured, channelled along the middle; it is a little shining: the sides are nearly straight, till near the hind angles, when they are a good deal narrowed, in such a way as to make the hind angles appear a little sinuate. The elytra are rather narrow, a good deal longer than the thorax, of a dirty-yellowish colour, finely and indistinctly punctured. The hind body is yellowish and impunctate, the extreme apex a little blackened. The legs are pale yellow, rather slender.

Rio Purus; a single individual, found by Dr. Trail on the 25th October, 1874.

Obs.—This minute species resembles extremely the *B. muticus*, but has the thorax differently formed.

7. *Bledius modestus*, n. sp. Piceus, elytris dilutioribus, antennis pedibusque testaceis; prothorace latitudine haud breviore, lateribus rotundatis, canaliculato, sat evidenter punctato; elytris crebre subtiliter punctatis. Long. corp. $\frac{7}{6}$ lin.

Antennæ yellow, short; all the joints from 3—10 short and each distinctly broader than its predecessor, so that the penultimate ones are rather strongly transverse. Mandibles and palpi yellow, head nearly black, with the antennal tubercles rather strongly marked, and the eyes very convex. Thorax distinctly narrower than the elytra, as long as broad, the sides a little rounded, and the base and hind angles rounded; it is of a pitchy colour and distinctly shining, with a channel along the middle, moderately closely but not very distinctly punctured. Elytra of an obscure-yellowish colour, distinctly longer than the thorax, finely and rather closely punctured. Hind body pitchy, curved at the sides and evidently contracted at the base; it is shining and almost impunctate. Legs pale yellow, rather slender.

Rio Madeira, a single individual found by Dr. Trail on the 25th May, 1874; it was attracted by light.

Obs.—This is the smallest species of *Bledius* I have seen; it is closely allied to *B. muticus* and *B. similis*, but is rather more slender in form, and has the sides of the thorax more curved.

TROGOPHLŒUS.

Five species of this genus are here described, and only one other was previously known from the Continent of South America; the genus is one of almost universal distribution, and likely to prove numerous in species even in the tropics. The paucity of species as yet known from South America is pretty certainly, therefore, only the result of neglect on the part of collectors; several species are known from Cuba, and species of the genus are numerous in Chili.

1. *Trogophlæus mundus*, n. sp. Niger, dense punctatus, pedibus testaceis, antennarum basi fusco-testaceo; elytris subtestaceis; prothorace basi minus discrete bi-impresso. Long. corp. 1 lin.

Antennæ moderately long, a good deal thickened towards the extremity, blackish, the basal joint yellowish; 3rd joint shorter and more slender than 2nd; 4th joint small, 6—8 small and transverse, 9—11 distinctly broader than the others. Palpi infuscate, the 3rd joint broad. Head nearly as broad as the thorax, the eyes large, and occupying nearly all the side of the head; antennal tubercles elongate and sharply elevated, the surface depressed on the inside of each, so that the front of the head bears two longitudinal impressions; the surface of the head very finely punctured, dull. Thorax a good deal narrower than the elytra, a good deal broader than long, much narrowed towards the base; the surface densely punctured, and near the base with two indistinct impressions extending towards the hind angles, and with scarcely any traces of longitudinal impressions. Elytra brownish, the base rather infuscate, densely and finely punctured, much longer than the thorax. Hind body black, very densely punctured, with a very fine, short, delicate, pale pubescence. Legs, including the coxæ, pale yellow; middle coxæ elongate, and separated by a very narrow space.

Ega; a single individual.

Obs.—This species, at first sight from its small size and comparatively narrow form, suggests a comparison with our *T. corticinus* and *T. pusillus*, but its structural characters show it to be more nearly allied to *T. obesus*. A second specimen from the same locality has the palpi, the base of the antennæ and the thorax paler in colour; but I consider it probable that these differences are only the result of the immaturity of the specimen.

2. *Trogophlœus breviceps*, n. sp. Brevior, latiusculus, piceus, antennis, palpis, pedibus, abdominisque apice testaceis; capite brevi, lato; thorace fortiter transverso, longitudinaliter bi-impresso; elytris subnitidis, minus dense punctatis. Long. corp. $1\frac{1}{4}$ lin.

Antennæ yellowish, longer than head and thorax; 1st joint very elongate, 3rd much more slender than 2nd, 8th about as long as broad, 9th and 10th a little transverse. Head very short and broad, but a little narrower than the thorax, the eyes, however, leaving a distinct prominent space behind them at the sides; the antennal tubercles short, the surface dull and extremely finely punctured. Thorax a good deal narrower than the elytra, about twice as broad as long, distinctly narrowed towards the base; the surface dull, extremely indistinctly punctured, with two distinct longitudinal impressions along the middle, which do not reach to the front, and near each side with a more indistinct broad impression. Elytra much longer than the thorax, neither altogether finely nor densely, but still indistinctly punctured, a little shining. Hind body broad, the apex yellowish; the segments indistinctly and sparingly punctured, and a little shining. Legs bright yellow; under side of head and thorax obscure reddish; middle coxæ scarcely contiguous; under face of hind body densely and finely punctured, dull.

Ega; a single individual.

3. *Trogophlæus latifrons*, n. sp. Rufus, capite fusco-rufo; prothorace transverso, lateribus subdenticulatis, dorso minus distincte bi-impresso; elytris sat fortiter punctatis; abdomine obsolete punctato, apicem versus nitidulo. Long. corp. 1½ lin.

Antennæ reddish, rather long, a good deal thickened towards the extremity; 3rd joint about as long as, but more slender than 2nd; 4—6 subequal, rather small, bead-like; 7 and 8 broader than the preceding, rather transverse; 9 and 10 broader than the preceding, also transverse. Head broad and short, but distinctly narrower than the thorax, very dull, more obscure in colour than the other parts; the eyes large, but with a distinct prominent space at the sides behind them, the antennal tubercles short and strongly elevated, but without any distinct impression on the inner side of each. Thorax very broad and short, twice as broad as long, a good deal narrower than the elytra; the sides distinctly narrowed towards the base, and each with three or four minute setigerous prominences, giving them the appearance of being denticulated; the surface reddish, very opaque, very densely and indistinctly punctured, with two indistinct longitudinal impressions on the middle, and outside these scarcely visibly impressed. Elytra broad, much longer than the thorax, rather distinctly and somewhat closely punctured, a little shining. Hind body large, very indistinctly punctured, the basal segment nearly dull, the others more shining, so that the 6th is quite shining. Legs clear yellow; under surface unicolorous red, with the hind body densely and finely punctured and pubescent.

Ega and Tapajos; two individuals.

4. *Trogophlæus hilaris*, n. sp. Rufus, capite fusco-rufo; antennis elongatis, articulis nullis transversis; prothorace transverso, lateribus subdenticulatis, dorso minus distincte bi-impresso, angulis anterioribus minus rotundatis; elytris dense minus fortiter punctatis; abdomine apicem versus nitidulo. Long. corp. 1¾ lin.

Antennæ elongate, formed almost as in *T. latifrons*, but with joints 4—10 each slightly longer, so that 4—6 are less bead-like, and 7—10 not transverse. Head very short and broad, with a very minute fovea on the vertex in the middle. Thorax twice as broad as long, the sides slightly rounded towards the front, and distinctly narrowed

towards the base, and with two or three fine denticles; the
disc with two indistinct longitudinal impressions. Elytra
much longer than the thorax, densely and indistinctly, but
not altogether finely punctured.

Tapajos and Ega; several specimens.

Obs.—This species, though very closely allied to *T.
latifrons*, is larger, and may be readily distinguished by its
more elongate antennæ. My description is drawn entirely
from one of the Tapajos specimens, for I am not at all
sure that I have not before me two or three very closely
allied species; two of the individuals are considerably
darker in colour, so that the term "piceo-rufo" would more
correctly describe them, and one of these dark individuals
shows, near the front angles of the thorax, a patch of
peculiar elongate pubescence; which may, however, only
be wanting from the other individuals on account of their
being rubbed.

5. *Trogophlœus vicinus*, n. sp. Rufus, capite fusco-
rufo; antennis elongatis, articulis nullis transversis; prothorace transverso, lateribus subdenticulatis, versus angulos
anteriores bene rotundatis, dorso minus distincte bi-impresso; elytris dense minus fortiter punctatis; abdomine
apicem versus nitidulo. Long. corp. 1¾ lin.

This species is extremely closely allied to the *T. hilaris*,
but it has the thorax a good deal more rounded towards
the front angles, and the part of the head behind the eyes
is less distinct and less prominent. In other respects I see
scarcely anything to distinguish the two species.

Amazons; a single specimen, without special locality.

APOCELLUS.

This genus I consider one of the most interesting of the
Staphylinidæ of the New World. It at present contains
six species, found both in North America and South
America. The facies of the species is greatly that of the
Falagria forms of the *Aleocharini*, and the structure of
the thorax is scarcely dissimilar from what may be found
in some of the *Aleocharini*; the genus, therefore, affords
us a connecting link between the *Oxytelini* and *Aleocharini*. The *A. planus* I here describe is of special interest
as indicating in a certain manner what are the changes
that have taken place in the head of the *Aleocharini*, so
as to give rise to the appearance of a different insertion of

the antennæ. These changes may be briefly summed up as follows: in *Osorius* the labrum appears to be attached to the edge of the front of the head, but is, in fact, attached to a membrane underneath the front of the head, and is, therefore, mobile; in *Apocellus* this membrane intervenes in a very visible manner between the front of the head and labrum; for I consider the large semi-corneous portion of the head, intervening between the transverse suture (very visible in *Apocellus*) and the labrum, to be clearly the homologue of the membrane above mentioned in *Osorius*, and which is very visible in many of the *Staphylinini*. In the *Aleocharini* the transverse suture above alluded to has disappeared, and the antennal tubercles can scarcely be recognized.

1. *Apocellus planus*, n. sp. Testaceus, metasterno, abdomine, elytrisque plus minusve infuscatis; capite, thorace, elytrisque opacis, dense subtilissime strigosulis; abdomine nitidulo, fere impunctato. Long. corp. 1½ lin.

Antennæ yellow, rather stout and very elongate, longer than head, thorax and elytra; 2nd and 3rd joints slender and elongate, the latter a good deal the longer; 10th joint much longer than broad, 11th distinctly longer than 10th. Head yellow, the part in front of the transverse suture between the antennal tubercles more shining and less corneous than the other parts; the antennal tubercles strongly elevated, the front rendered opaque by very fine, indistinct, strigose sculpture, the middle with an indistinct fovea. Thorax small, only about half as broad as the elytra, about as long as broad, distinctly narrowed towards the base, very finely margined at the sides and base, subquadrate, but with the angles rounded; the surface rendered opaque by a very fine, indistinct, strigose sculpture. Elytra distinctly longer than the thorax, yellowish, but somewhat infuscate, especially towards the pleuræ; opaque, densely and finely strigose. Hind body broad, a little curved at the sides, shining yellowish, slightly infuscate, especially towards the extremity, almost impunctate; its under face as well as the metasternum more distinctly infuscate. Legs elongate, pale yellow.

Ega; three individuals, of doubtful sex.

2. *Apocellus lævis*, n. sp. Castaneo-testaceus, nitidulus,

impunctatus, antennis pedibusque testaceis; prothorace basi
medio impresso; abdomine basi angustato. Long. corp.
1⅜ lin.

Antennæ yellow, quite as long as head and thorax, a
good deal thickened towards the extremity; 3rd joint distinctly longer than 2nd, and a good deal longer than 4th,
which is slender and longer than broad; of 6—10 each is
slightly longer and distinctly broader than its predecessor,
each a little longer than broad; 11th long, a good deal
longer than 10th. Head chestnut yellow, the hind angles
very rounded, the clypeus large. Thorax narrow, scarcely
so broad as the head, and not much more than half as
broad as the elytra; it is longer than broad, quite convex,
and with an impression at the base in the middle. Elytra
quadrate, about as long as the thorax, of a chestnut colour,
rather darker than the head and thorax. Hind body
broad, but much narrowed at the base, so that its sides are
greatly rounded; it is of a yellowish colour, with the apical
segments a little infuscate. The legs are long and slender,
the four hind femora very slightly infuscate towards the
extremity.

Manaos; three individuals, captured by Dr. Trail in
August, 1874. They were attracted by light.

Obs.—An undescribed species from Rio de Janeiro is
very closely allied to *A. lævis*, but is rather smaller, has
joints 4—6 of the antennæ longer, and the hind body more
contracted at the base.

OMALIUM.

The insignificant little species I here describe with this
generic name is interesting, as being the only representa-
tive of the *Omalini* yet detected in the Amazons. Indeed,
the group seems to be extremely poorly represented in
tropical America, only two species of it having as yet
been described from those parts, and scarcely any others
existing, so far as I know, in collections; it is, however,
quite possible that this paucity may prove not to be so
complete as these facts would suggest, for our knowledge
of the smaller species of tropical *Staphylinidæ* is still so
very fragmentary, that no generalization as to an extensive group can with propriety be more than hinted at.

1. *Omalium nanum*, n. sp. Subopacum, depressum,
nigrum, antennarum basi pedibusque testaceis; protho-

race transverso, dorso obsolete bi-impresso. Long. corp. ⅞ lin.

Antennæ short, the five basal joints yellow, the others black; 2nd joint short and stout, 3rd very slender, 4th and 5th similar to one another, very small; 6—10 strongly transverse, the 6th much broader than the preceding ones, 7th broader than 6th, and 8th than the 7th; 8—10 very similar to one another, 11th short. Palpi yellow. Head greatly narrower than the thorax, only half as broad as the elytra; black, indistinctly but not altogether finely punctured, almost dull; ocelli small but distinct. Thorax about twice as broad as long, distinctly narrower than the elytra; the sides rounded and rather more narrowed in front than behind; on the middle are two very obsolete, large impressions; its punctuation is obsolete, but it is scarcely shining. Elytra twice as long as the thorax, their outer hind angle rounded, the sutural one almost rectangular; they are rather closely but quite indistinctly punctured, the punctuation at the apex becoming even finer and more indistinct than at the base; they are black and almost opaque. Hind body black, dull, its punctuation excessively fine, the lateral margins broad. The legs are yellow, but somewhat infuscate; they are short and slender; the tarsi are very short; the metasternum has a deep channel on its hinder part.

Two individuals of this species were found by Dr. Trail on the 5th November, 1874, but he has sent me no special locality.

Piestus.

The insects of this genus are confined to the warm parts of the New World, though it must not be forgotten that the North American and European genus *Siagonium* approaches them very closely. The genus comprises about eighteen described species, and I here add five to that number. Of these five, two—viz., *P. validus* and *P. frontalis*—belong to the group of large species having the head armed with horns, and the mandibles much developed; *P. rectus* belongs to the *P. minutus*, Er., group; while *P. rugosus* and *P. aper* have as their only described near ally, *P. angularis*, Fauv.

The species of the genus have, many of them, an apparently wide range in South America, and their discrimination from one another in a satisfactory manner is a matter of great difficulty. The characters by which the sexes may be distinguished are extremely slight; in some species the antennæ are very elongate in some male specimens, but in other individuals scarcely differ at all from those of the female, while the structure of the ædeagus shows scarcely any variation in very dissimilar species.

1. *Piestus validus*, n. sp. Niger, nitidus, subdepressus, abdominis apice piceo; antennis setosis, articulis 1, 3, 4 et 5 setis intus densioribus; fronte bispinosâ, spinis approximatis; prothorace sat crebre obsolete punctato. Long. corp. 6 lin.

Antennæ blackish, rather stout, $2\frac{1}{2}$ lin. in length, clothed with tawny, elongate setæ. These setæ are specially long and dense on the inner side of the 1st, 3rd and 4th joints, while the 5th joint is less setose than the 4th, but more than the 6th. Front of head armed with two moderately long acuminate spines; behind these the head in the middle is depressed, the depression shaped somewhat like a narrow V; the spines at nearly half their distance from the base are only separated by a width of about $\frac{3}{10}$ of a line; the surface is sparingly and finely punctured. Thorax 1 lin. in length, $1\frac{3}{8}$ lin. in breadth, channelled along the middle, transversely a little convex; the surface very shining, and with rather numerous, but obsolete punctures. Elytra shining black, $1\frac{1}{4}$ lin. in length, their greatest breadth just that of the thorax, viz., $1\frac{3}{8}$ lin., each with 5 deep striæ, and outside these with indications of a 6th stria sufficiently well marked at the extremity. Hind body black, not very shining; the segments punctured at the base and sides of each, the basal segment nearly entirely coriaceous; the extremity pitchy yellow, the paler colour commencing on the hind part of the 6th segment. Legs black, with the tarsi pitchy.

Pebas; three specimens, 2 ♂, 1 ♀.

Obs. I.—There are several species of *Piestus* mixed in descriptions and collections under the name of *P. bicornis*. I have not, however, seen the above-described species from any other locality than this of the Upper Amazons; its large size, together with its distinctly punctured thorax, distinguish it from all the closely allied forms.

Obs. II.—I have ascertained the sexes of this species by dissection, without which they cannot be distinguished. Of the two males, one has the thorax remarkably developed, it being considerably broader than the elytra and more arched transversely. As this is probably an extreme sexual characteristic, my description has been made from the smaller male, which, in the development of its thorax, quite resembles the female.

2. *Piestus bicornis*, Ol., Er.
Pará; two female specimens. Var. *Oxytelinus*, Lap.; seven specimens from Ega. I am very doubtful whether this so-called variety be not rather a distinct species.

3. *Piestus spinosus*, Fab., Er.
Pará and Ega; six individuals.

4. *Piestus frontalis*, n. sp. Depressus, rufescens, nitidus, antennis elongatis; capite vertice bi-impresso, fronte spinis duabus brevibus distantibus armatâ; prothorace medio canaliculato, et punctato; elytris punctato-striatis. Long. corp. $4\frac{1}{2}$ lin.

Antennæ elongate and rather slender, $3\frac{1}{2}$ lin. in length; the three basal joints red, the others pitchy. Mandibles moderately long, greatly curved, each with a very long transverse tooth on the inner side; in the middle and on the upper side with a tooth some distance from the extremity; on the right mandible this tooth elongate and spine-like. Head shining dark red, without punctures; the front armed on each side with a short straight horn, the distance between the two horns being about $\frac{3}{8}$ lin.; within the antennal elevations the surface is depressed on each side, the two depressions are not connected by a channel. Thorax strongly transverse, $\frac{3}{4}$ lin. in length by $1\frac{1}{4}$ in width; very shining red, with a channel along the middle, and the disc with a few distinct punctures. Elytra $1\frac{3}{8}$ lin. in length by fully $1\frac{1}{8}$ in width, each with five deep punctured striæ, and with traces of a 6th stria externally. Hind body broad, infuscate-red, with the extremity paler, coarsely but not closely punctured. Legs red. Hind margin of prosternum in the middle almost straight; mesothoracic keel very obsolete.

Ega; a single individual, of doubtful sex.

Obs. I.—Though closely allied to *P. spinosus*, this species is rather smaller, and is very readily distinguished by the differences of the mandibles, of the frontal horns, and of the vertex.

Obs. 11.—It is only after a good deal of hesitation that I have decided to consider this insect a new species. *P. capricornis*, Lap., Er., must be closely allied to it if not the same species. Laporte's figure represents an insect with less transverse thorax, and his words, "abdomen finement ponctué," are singularly inapplicable to the species I have described above as *P. frontalis*. Erichson's description of *P. capricornis* agrees much better with the *P. frontalis*. Nevertheless, though Erichson gives a detailed description of the mandibles of his species, he omits any notice of the remarkable transverse spine-like tooth which exists on the inner side of each mandible in *P. frontalis*, this being one of the most striking characteristics of the species.

5. *Piestus rectus*, n. sp. Rufescens, antennis (basi exceptis), capite abdomineque piceo-rufis; capite vertice unifoveolato, antice utrinque curvatim lineato; abdomine segmento 6° toto dense punctato. Long. corp. 3 lin.

Antennæ 1¾ lin. in length, the basal joint red, the two following pitchy red, the rest pitchy; the 1st joint not swollen on the upper side, but with a rough spot bearing a few long hairs, the other joints with sparing setæ. Head on the upper side blackish, shining, the middle part nearly on a level with the antennal elevation, on the inner side of this latter is a curved impression; these impressions are much abbreviated behind, so that they are far from meeting; the vertex is quite flat and bears a distinct fovea in the middle; the surface is sparingly sprinkled with fine punctures. Thorax strongly transverse, reddish, very shining, channelled along the middle; the surface with excessively obsolete, sparing, fine punctures, only to be detected on a careful examination with a high power; the hind angles obtuse, the impression near the hind angle small. Elytra just as broad as the thorax, about ⅘ lin. in length and ¾ lin. in breadth, shining red, with the suture and outer and hind margins indistinctly blackish, each with five moderately fine striæ, which are indistinctly punctured. Hind body broad and parallel, pitchy, rather coarsely and closely punctured; segments 2—4 each with a smooth space in the middle behind, the punctures on the corresponding part of the 5th segment more sparing than at the base, the 6th evenly and distinctly punctured throughout; its extreme hind margin reddish, as are the following segments; on the underside the hind body is pitchy red, paler towards the extremity, and all the segments 2—6 are coarsely and rather closely punctured; on segments 3 and 4 and base of 5 the punctures are almost wanting on a small space on the middle. Legs red.

Ega; four individuals, probably all females.

Obs. I.—Besides the specimens above described, I have another individual from the same locality (Ega), which I believe to be the male of *P. rectus*; it is rather smaller than the individuals described, but the antennæ are $2\frac{1}{4}$ lin. in length, the head is longitudinally depressed along the middle, and the front part is scarcely shining, being minutely strigose; the 1st joint of the antennæ is more densely penicillate, and the 2nd and 3rd joints have also a dense long pubescence on the inner face.

Obs. II.—I believe the most closely allied described species to be *P. pennicornis* and *P. plagiatus*, Fauv. I have a fourth still undescribed species in my collection from Rio, which is considerably larger, but has the head similarly formed and, as well as the antennæ, manifesting similar sexual disparities.

6. *Piestus minutus*, Er., Fauv.

Pará and Ega; four individuals. These individuals indicate a peculiar race, which I have not seen from any other locality; they are small (about two lines in length), and broad in proportion to their width; the elytra are nearly black in mature individuals.

7. *Piestus pygmæus*, Lap., Er.

Pará and St. Paulo; numerous specimens. This is another very puzzling species, owing to the variations it exhibits, and it is quite possible that two or three closely-allied species are mixed under this name.

8. *Piestus sulcatus*, Grav.

Pará, Lages, Ananá; five specimens, which agree exactly *inter se* and only differ slightly from individuals from Bahia.

9. *Piestus rugosus*, n. sp. Piceus, pedibus abdominisque apice rufescentibus, opacus; prothorace dense rugoso, lateribus pone medium obtuse dentatis; elytris lineis 6 elevatis; abdomine dense punctato. Long. corp. $2\frac{1}{2}$ lin.

Antennæ 1½ lin. in length, rather stout, pitchy black, with the basal joints rather paler; of joints 4—10 each is a little longer than its predecessor. Head with two curved lines on the front part, which meet so as to form an acute angle on the middle of the head; antennal callosities stout, but not greatly elevated; the whole surface densely rugose, quite opaque. Thorax strongly transverse; the sides rounded and distinctly narrowed in front, behind the middle with a short stout tooth, and behind this tooth cut away, the hind angles being distinctly marked and just rectangular; the whole surface is densely and coarsely rugose, quite dull; there is no distinct impression along the middle, and in front of the hind angles is a scarcely definite longitudinal impression, reaching nearly to the front margin. Elytra longer than the thorax, their sculpture on each consisting of a sutural, and six other fine, elevated, longitudinal lines; the broad spaces between these lines are peculiarly opaque, and are marked with some indistinct transverse marks, which probably represent obsolete coarse punctures. Hind body quite dull, densely and rather finely punctured; the apical segments and the hind margin of the 6th segment reddish; under surface not so dull as the upper; metasternum covered with coarse punctures and with a fovea at the extremity in the middle. Legs dull red.

Ega; a single individual, whose sex is unknown.

10. *Piestus aper*, n. sp. Opacus, piceus, pedibus rufis; capite mutico, longitudinaliter striguloso; prothorace lateribus trisinuatis, dense fortiterque punctato; elytris lineis 6 elevatis; interstitiis transversim rugosis; abdomine dense fortiterque punctato, apice dilutiore. Long. corp. 1¾ lin.

Antennæ 1⅛ lin. in length, stout, pitchy, the basal joints obscure red; 3rd joint scarcely so long as 2nd, 4th a good deal stouter than and quite as long as 3rd, 5—10 differing little from one another in length, each slightly stouter than its predecessor, 11th a little longer than but scarcely so broad as 10th. Head without impressions or lines, the antennal callosities only being present; the surface densely covered with longitudinal rugæ. Thorax broader than long, the front margin a little sinuate on each side the middle, so that the front angles are somewhat prominent; the sides each with two notches, so that they exhibit three obtuse projections, and behind these much cut away at the hind angles; the surface covered with coarse dense punctures, so that the interstices are very narrow. Elytra rather broader and a good deal longer than the thorax, each with seven fine, raised, longitudinal lines, of which one is close to the suture and indistinct, and the outer one most distinct; the spaces between these lines marked with transverse lines, which are irregular, and represent a coarse confluent punctuation. Hind body coarsely punctured, the punctuation becoming more indistinct on the apical segments, which also are paler in colour. Metasternum coarsely, closely and deeply punctured throughout. Legs reddish; tarsi short, pubescent beneath, the basal joint of each distinctly broader than the others.

St. Paulo; two individuals, of uncertain sex.

HYPOTELUS.

Only four species have yet been described as forming part of this genus; one from Kansas, and the others from tropical America; to these I now add a fifth species. It is probable that the genus, like *Piestus*, is one of the forms characteristic of the South American fauna.

I feel considerable doubts whether the species should at present be separated from *Piestus*. The two genera are extremely close in their general structure, and the only characters given to distinguish them are drawn from the trophi; but as the oral organs are subject to an excessive difference of development within the bounds of the genus *Piestus*, and as the comparison of these parts rests, so far as I know, on the examination of a single species of each genus, it is clear enough that the individuality of the genera has not yet been properly determined.

1. *Hypotelus micans*, n. sp. Piceo-rufus, nitidus, antennis pedibusque rufo-testaceis; elytris stramineis, suturâ epipleurisque infuscatis; capite thoraceque sat crebre, distincte, elytris parcius obsoletiusque punctatis. Long. corp. 1 lin.

Antennæ reddish, not thickened at the extremity; 3rd joint much shorter than 2nd; 4th joint small; 5th to 10th broader than the preceding joints, differing little from one another, each subquadrate. Head rather small, dark reddish, shining, rather sparingly but quite distinctly punctured, and with two large, well-defined impressions. Thorax a good deal broader than long, rounded at the sides, and a good deal narrowed behind, similar in colour to the head; the surface shining, rather sparingly but quite distinctly punctured, with an excessively fine abbreviated impressed line on the middle. Elytra a good deal longer than the thorax, sparingly and indistinctly punctured, pale shining-yellow; the suture infuscate, and the pleuræ piceous. Hind body pitchy, with the extremity reddish; the segments scarcely punctured, but finely strigose at the sides and base. Legs reddish-yellow.

Ega; a single individual.

Isomalus.

This genus at present consists of about twenty-five described species found in all the warm parts of the world; the genus *Chasolium*, Cast., is apparently not sufficiently distinct from *Isomalus* to be at present adopted with advantage. The species are of the most difficult character, in so far as their discrimination from one another is concerned. I am not able to point out any external characters to distinguish the sexes; in certain species some males have the head very large, but it seems that this extra development in other individuals of the same sex and species disappears entirely.

1. *Isomalus agilis*, n. sp. Latiusculus, parallelus, perdepressus, nitidus, fere lævis, rufo-castaneus; capite oblongo; thorace lateribus antice rectis, postice rotundatis, margine anteriore utrinque impresso; elytris thorace paulo longioribus. Long. corp. 2½ lin.

Antennæ slender, shorter than head and thorax, red, scarcely thickened towards the extremity; 10th joint almost as long as broad; 11th elongate, fully twice as long as 10th. Palpi red. Head as broad as thorax, oblong, the sides quite parallel, shining red, almost impunctate. Thorax quite as long as broad; the sides in front nearly straight, behind the middle rounded and narrowed towards the base; the front margin on each side bears a rather large impression, and at the hind angle on each side are two indistinct depressions, one in front of and external to the other. Elytra a little longer than the thorax, red, very shining, a little infuscate at the extremity and sides, each with a fine puncture on the middle. Hind body reddish, shining, impunctate.

Ega; three individuals, one of which bears a label, "under bark, nimble."

2. *Isomalus dubius*, n. sp. Perdepressus, piceus, nitidus, fere lævis, antennis pedibusque testaceis; thorace basin versus fortiter angustato; elytris disco unipunctatis. Long. corp. 1½—1¾ lin.

Antennæ rather longer than head and thorax, slender, reddish, slightly thickened towards the extremity; 3rd joint a little longer than 2nd; 10th joint quite as long as broad, 11th a good deal longer than 10th. Head as broad as the thorax, flat, impunctate, very shining, pitchy, with the mandibles and parts of the mouth red. Thorax as broad as the elytra; the sides rounded at the front angles, greatly narrowed behind, with an indistinct denticle behind the middle, the base very narrow; shining piceous, almost without sculpture, the front margin on either side bearing two approximate punctures. Elytra narrowed towards the shoulders, a good deal longer than the thorax, shining, pitchy, each with a puncture on the middle. Hind body broad, flat, piceous, paler at the extremity, impunctate, bearing a few upright setæ. Legs reddish.

Ega; three individuals, which vary in size, in the development of the head, and in the length of the antennæ.

Obs.—This species is extremely closely allied to *I. pallidipennis*, Fauv., but is a little larger, has the elytra concolorous, and the antennæ a little longer.

3. *Isomalus tenuis*, Fauv.

St. Paulo; a single individual. The species was described by Fauvel from a specimen coming from Venezuela; the St. Paulo exponent does not differ from an example in my collection from Santa Rita, Brazil, so that the species appears to possess a wide range.

LISPINUS.

This genus as at present constituted consists of about forty described species, which occur pretty generally in the warmer parts of the globe; though not yet recorded from Australia, I can state that species of the genus occur even there. The distinction of the species from one another appears to me likely to be a point of great difficulty, owing to the extreme general resemblance that exists between numerous species. The genus *Holosus* of Motschoulsky and Kraatz appears to be excessively closely allied to *Lispinus*, and scarcely justifies the adoption for it of a distinct generic name in the present state of our knowledge of the exotic *Staphylinidæ*. The South American species of *Lispinus* appear to me to belong to two groups, easily distinguished from one another by the form of the head; in *L. striola* and its allies the antennal tubercles are quite obsolete, and the front of the head is very evenly and distinctly margined, and its upper surface convex, while in the other group the antennal tubercles are distinct, and the front of the head is scarcely or not at all margined. *L. simplex* and *L. lætus*, here described, belong to this latter group, while all the other species I here describe belong to the first-mentioned group, the *L. depressus* making, however, a considerable approach to the second group.

1. *Lispinus striola*, Er.

Pará; a series of specimens.

This is, apparently, one of the species of *Staphylinidæ* having a great geographical range in the warm parts of the New World.

2. *Lispinus catena*, n. sp. Elongatus, angustulus, parallelus, nigro-piceus, nitidus; antennis, pedibus, abdominisque apice rufis; thorace basin versus leniter angustato, lateribus haud sinuatis, basi utrinque impressione angustâ; capite thoraceque sat crebre et subtiliter punctatis; elytris parcius punctatis, disco striolâ minus discretâ e punctis elongatis formatâ notatis. Long. corp. fere 2½ lin.

Antennæ reddish; joints 7—10 transverse, 3rd slightly longer than 2nd. Head rather finely and moderately closely punctured, on the disc with two larger distant punctures. Thorax not quite so long as broad; the sides only a little narrowed towards the base, on each side near the hind angles with a narrow, elongate impression, reaching quite half-way to the front; on the middle is an excessively fine channel; the surface is rather finely and moderately closely punctured. Elytra a little longer than the thorax, only a little less shining, sparingly but not altogether finely punctured, the punctures being rather elongate; along the middle these punctures are crowded together and confluent, so as almost to form an impressed stria; the sutural stria deep. Hind body narrow and elongate, cylindric, blackish, with the extremity reddish; above finely and indistinctly punctured, beneath the oblique striæ on the sides of each segment are quite distinct. Legs, including the coxæ, red.

Ega; a single specimen, which I believe to be a female, and an individual of a slightly different variety from Pará.

Obs.—This species may readily be distinguished from *L. striola* by the more setose surface, by the more slender joints of the base of the antennæ (2—5), and by the less ample elytra and hind body; its nearest described ally is probably *L. quadripunctulus*, Fauvel.

3. *Lispinus apicalis*, n. sp. Niger, nitidus, antennis piceis, pedibus abdominisque apice summo rufis; thorace elytrisque crebre sat fortiter punctatis; abdomine punctis magnis leviter impressis, puncto singulo parte posteriore omnino deficiente, segmento apicali fortiter profundeque punctato. Long. corp. 2¾ lin.

Antennæ rather stout; 3rd joint nearly twice as long as 2nd, 7—10 each a little transverse. Head a good deal narrower than the thorax, distinctly, moderately closely punctured. Thorax a good deal broader than long, the sides nearly straight, only slightly curved; it is covered with rather coarse, moderately close punctures, bears an extremely fine channel on the disc, and an elongate impression near the outer angles, which extends considerably more than half-way to the front. Elytra a good deal longer than the thorax, rather coarsely punctured; the punctures at the outside, however, quite fine, and at the lateral margins wanting. Hind body a little flattened, blackish, with the hinder part of the 7th segment red; the segments on the upper side bear a peculiar obsolete punctuation, the hinder part of an obsolete coarse puncture being entirely wanting; the 7th segment, however, offers a striking contrast to the preceding ones, for it is deeply and distinctly punctured: the under surface of the hind body bears numerous fine, raised, oblique striæ.

Pará; a single individual (I believe a female), collected by Mr. Smith.

4. *Lispinus terminalis*, n. sp. Niger, sat nitidus, an-

tenuis piceo-rufis, pedibus abdominisque apice summo rufis; thorace elytrisque crebre sat fortiter punctatis; abdomine punctis magnis leviter impressis, puncto singulo margine posteriore omnino deficiente, segmento apicali fortiter punctato. Long. corp. 2¾ lin.

Antennæ dark red, moderately stout; 3rd joint much longer than 2nd, 7—10 slightly transverse. Head rather finely punctured. Thorax broader than long, straight at the sides and not narrowed behind; rather closely punctured, with a narrow smooth space along the middle, and an elongate impression near each outer angle, which reaches more than half-way to the front. Hind body punctured as in *L. apicalis*, but the punctures on the 7th segment not quite so deep and coarse.

Pará; a single individual (I believe a female), collected by Mr. Smith.

Obs.—This species is extremely closely allied to *L. apicalis*, but is not quite so large, and the sculpture of the dorsal plate of the 7th segment is a good deal less coarse. It is very similar in size and form to *L. striola*, but has the 3rd joint of the antennæ longer. The elytra are without the discoidal stria of *L. striola*, and their sutural portion is more coarsely punctured, and the sculpture of the hind body is different.

5. *Lispinus punctatus*, n. sp. Nigro-piceus, nitidus, fortiter punctatus, antennis piceis, pedibus abdominisque apice rufis; capite lato; prothorace transverso, basin versus angustato, basi utrinque impresso. Long. corp. 2 lin.

Antennæ stout and short; joints 4—6 broader than long, 7—10 distinctly broader than the preceding, strongly transverse; 11th joint short, rather paler than the 10th. Head broad and short, with a small impression on each side, near the front; rather coarsely punctured, the punctures towards the vertex finer and more sparing, over the eyes almost passing into striæ. Thorax much broader than long, a good deal narrowed towards the base, and a little sinuate in front of the hind angles; shining black, coarsely and rather closely punctured, with an excessively fine channel on the middle, close to each hind angle, with a broad but not sharply limited impression, the front part of which forms a narrow, indistinct channel. Elytra a good deal longer than the thorax, coarsely, deeply, and rather closely punctured with elongate punctures. Hind body shining black, red at the extremity; on the upper surface distinctly punctured along the middle and at the sides with oblique striæ, the basal segment broadly opaque at the base; its under surface shining and with shallow, oblique striæ. Legs red.

Ega; a single male.

6. *Lispinus cognatus*, n. sp. Nigro-piceus, nitidus, fortiter punctatus, antennis piceis, pedibus abdominisque apice summo rufis; prothorace transverso, basin versus angustato, basi utrinque impresso. Long. corp. 1¾ lin.

Antennæ short, rather stout; joints 4—6 a good deal broader than long, 7—10 strongly transverse. Head rather large, with a very small impression on each side near the front; the surface shining, moderately closely and finely punctured. Thorax rather strongly transverse, distinctly narrowed towards the base, the surface rather coarsely and closely punctured, and at each hind angle with an ill-defined, not very large impression. Elytra a good deal longer than the thorax, coarsely and closely punctured. Hind body cylindric, but a little depressed; on the upper surface along the middle segments 2—5 are rather coarsely but indistinctly and not closely punctured, at the sides the punctures become obscure oblique striæ; the 6th segment finely and sparingly punctured; the front portion of the basal segment with a moderately broad opaque space at the base. Legs red.

Ega; a single individual, which I believe to be a male.

Obs.—This species is very closely allied to *L. punctatus*, but is a little smaller, and the head and thorax are narrower in proportion to their length, and the impressions on the thorax are narrower in their basal portion, and the antennæ are not so stout.

7. *Lispinus modestus*, n. sp. Angustulus, subparallelus, piceus, pedibus abdominisque apice rufis; minus nitidus, crebre sat fortiter punctatus, abdomine segmento 6° sat crebre punctato. Long. corp. 2 lin.

Antennæ short, not much thickened towards the extremity; joints 4—6 rather stout, the 6th distinctly transverse; 7—10 broader than the preceding, rather strongly transverse. Head distinctly narrower than the thorax, rather closely and finely punctured, and with a distinct small impression on each side towards the front. Thorax a little broader than long, the sides a little narrowed towards the base, not at all sinuate; the surface moderately finely and closely punctured, near each hind angle, with an elongate impression. Elytra distinctly longer than the thorax, their punctuation rather coarser than, and not quite so close as, that of the thorax. Hind body depressed, cylindric, but little shining; all the segments, including the 6th, distinctly punctured along the middle, the punctures at the sides becoming shallow striæ; on the under face the striæ along the sides are more deep and distinct, the punctures along the middle sparing. Legs red.

Ega; a single individual, which I believe to be a female; also two specimens from Pará, collected by Mr. Smith.

8. *Lispinus planus*, n. sp. Rufo-piceus, depressus, parallelus, fere opacus, antennis pedibusque rufis; capite thoraceque subtiliter sat crebre punctatis; elytris parcius punctatis; thorace basi utrinque profunde longitudinaliter impresso; lateribus basin versus subsinuatis. Long. corp. fere 1¾ lin.

Antennæ reddish, rather slender at the base, distinctly thickened towards the extremity; joints 4—6 not at all transverse, 7—10 distinctly broader than the preceding, each a little transverse. Head narrower than the thorax, rather short; the surface finely and not closely punctured, and with a small punctiform impression on each side the middle towards the front. Thorax broader than long; the sides scarcely narrowed till near the base, where they are a little contracted; the surface finely and not closely punctured, close to each hind angle with a large longitudinal impression, which is sharply limited both on the inside and the outside; it extends more than half-way to the front, and is opaque at the bottom, because of a very dense, fine sculpture. Elytra a good deal longer than the thorax, rather sparingly punctured with elongate but not deeply impressed punctures. Hind body flattened, pitchy red, red at the extremity, sparingly and indistinctly punctured. Legs red. Metasternum with an ill-defined longitudinal impression along the middle.

Ega; one male and one female individual.

Obs.—This species is, no doubt, allied to *L. linearis*,

Er.; it is probable that the *L. linearis* of Fauvel (Not. Ent. ii. p. 47) is a distinct species from that of Erichson, and I have three or four other very closely allied species from different parts of Brazil.

9. *Lispinus depressus*, n. sp. Rufescens, depressus, parallelus, fere opacus; capite thoraceque subtiliter sat crebre, elytris parcius punctatis; thorace basi utrinque profunde longitudinaliter impresso, lateribus basin versus subsinuatis. Long. corp. 1½ lin.

This species resembles extremely *L. planus*, but is rather smaller, narrower and paler in colour; the antennæ are a trifle more slender, and the punctuation of the upper surface is just a little finer. I should have considered it a small pale variety of *L. planus*, had it not been that the metasternum possesses only very indistinct traces of any impression along the middle.

Ega; a single female individual.

10. *Lispinus simplex*, n. sp. Piceus, depressus, subopacus, parce obsoletissime punctatus, antennis pedibusque testaceis; elytris disco impunctatis. Long. corp. ⅞ lin.

Antennæ yellowish, a good deal stouter at the extremity than at the base; 3rd joint shorter and more slender than 2nd, 4—6 small; 7—10 broader than the preceding joints, short, and though not very broad decidedly transverse. Head rather small, margin of the clypeus very indistinct, antennal tubercles distinct, eyes rather prominent; the surface nearly dull from being finely coriaceous, and also showing a few excessively fine punctures. Thorax transverse, a little rounded at the sides and with the basal portion distinctly contracted; the surface with a sculpture similar to that of the head, with an excessively fine channel along the middle, and with a broad, short, very ill-defined impression at each hind angle. Elytra much longer than the thorax, finely coriaceous and dull, each with a distinct puncture on the disc, and with a few very obsolete punctures. Hind body almost impunctate, dull, the extremity and the hind margin of each segment yellowish. Legs yellow.

Ega; two individuals, of doubtful sex.

11. *Lispinus lætus*, n. sp. Angustulus, depressus, rufo-

testaceus, nitidus, parce obsoletissime punctatus; prothorace lateribus rotundatis, basin versus angustato. Long. corp. ⅞ lin.

Antennæ reddish, rather slender; 3rd joint shorter and smaller than 2nd, 4—8 differing little from one another, not transverse, 9 and 10 broader than the preceding joints, transverse. Head rather small, clypeus almost immarginate, eyes prominent, antennal tubercles distinct; the colour is infuscate-red, the surface is shining, but very finely coriaceous, and with a few excessively fine punctures. Thorax transverse, a good deal rounded at the sides and narrowed behind, shining reddish, with a very few extremely obsolete punctures, and with an indistinct impression near each hind angle. Elytra much longer than the thorax, bright yellowish-red, almost impunctate. Hind body infuscate-reddish, the extremity and the hind margin of each segment brighter, impunctate, the sides of each segment coriaceous. Legs bright yellow.

St. Paulo; two individuals, sex unknown.

THORAXOPHORUS, Motsch. (*Glyptoma*, Er.)

Under this generic name there are mixed in collections and entomological works two distinct genera; of these, the one to which *Glyptoma crassicorne*, Er., and its allies belong, is allied to *Lispinus*; while the other, containing *Thoraxophorus corticinus*, Motsch., and its allies, is one whose natural affinities are much more obscure to me. Of the two new species here described, one, *T. opacus*, is allied to *G. crassicorne*, while the other, *T. crassus*, is an ally of *T. corticinus*. Of these two genera or groups, the one containing *G. crassicorne* is peculiar to South America, while the other is found in South America, North America, Europe and East India.

1. *Thoraxophorus opacus*, n. sp. Elongatus, niger, opacus, pedibus piceis; antennis crassiusculis, elongatis, articulo 2° solum transverso; elytris unicostatis; abdomen cylindricum. Long. corp. 1⅔ lin.

Antennæ stout, about 1 lin. in length, blackish; 2nd joint smaller than the others, not so long as broad; 3rd a little longer than 4th, quadrate, 4th nearly as long as broad; 5—10 differing little from one another, each a little longer than broad; 11th joint pointed at apex. Head quite as broad as thorax; the surface dull, velvet-like, coarsely but quite indistinctly punctured, the vertex with two indistinct elevations. Thorax narrower than the elytra, rather longer than broad; the sides straight from the front angles till just behind the middle, thence suddenly narrowed to the base; the surface dull and velvet-like, with a broad impression along the middle, most distinct on the front part; the lateral portions with coarse, indistinct punctures. Elytra longer than the thorax, each with a sharply-defined, longitudinal line down the middle; the surface dull and quite obsoletely punctured. Hind body cylindrical, elongate, dull and velvet-like, the basal segments with some coarse rugosities, the apical ones almost without sculpture. Legs pitchy.

Ega; a single individual.

Obs.—A species very closely allied to the above, and apparently as yet undescribed, is not uncommon in Brazil; it has the antennæ shorter, the penultimate joints being transverse, and the thorax has the sides a little rounded near the front angles.

2. *Thoraxophorus crassus*, n. sp. Ferrugineus, opacus, breviusculus; antennis brevibus, articulis transversis; capitis angulis posterioribus acutis; prothorace transverso lateribus dentatis; elytris tricostatis; abdomine conico, segmentis longitudinaliter lineato-striatis. Long. corp. 1 lin.

Antennæ short and stout, reddish, basal joint dilated on the inner side, towards the extremity; 2nd joint short, broad, bead-like; 3rd joint rather shorter and narrower than 2nd, 4—6 short, similar to one another; 7—10 broader than the preceding ones, strongly transverse; 11th short, pointed at apex. Head with the clypeus margined in front, the antennal tubercles joined by a fine curved line parallel with the clypeus; along the middle two strongly raised, longitudinal marks, which are connected in front; the hind angles projecting and quite acute, the surface depressed between the elevations. Thorax strongly transverse, but narrower than the elytra, the front angles prominent, the sides with two or three angular projections; along the middle two strongly raised elevations, which are connected in front, and appear to project over the front margin. Elytra short and broad, longer than the thorax, each with three elevated longitudinal lines, and the surface between these reticulated. Hind body short, convex, pointed; the segments extremely finely margined at the sides, covered with fine, straight, longitudinal, elevated lines. Legs slender.

St. Paulo; a single individual.

Obs.—I have also in my collection an individual of this species, which was captured by Mr. Squires near Rio de Janeiro.

LEPTOCHIRUS.

I find that considerable confusion exists both as to the generic characters and the specific forms represented by the name *Leptochirus*. There are, it seems to me, two distinct forms confounded under the generic name, viz., one in which the anterior coxae are separated by a well-developed process of the prosternum, and another in which this process is absent; *L. scoriaceus*, and all the Amazonian species here described, belong to the first of these groups, as do also *L. laticeps* and other species from the tropics of the Eastern hemisphere. On the other hand, a large number of the Eastern species, such as *L. ebeninus* and *L. mandibularis*, have the front coxae comparatively elongate and exserted, and the division between their cavities quite concealed, so that it is probable they will ultimately be considered generically distinct; it is to this latter group that the *L. bicornis*, Fauv., from Mexico, should be referred.

The species of the genus as yet described from the New World are only seven in number, but they are in reality more numerous than has been supposed, for I have at least a dozen undescribed species from this part of the globe in my own collection. The species, are, however, very difficult to distinguish, from their great similarity, and demand a careful study before their distinctions can be satisfactorily elucidated. I can only satisfy myself as to four distinct species from the Amazons, two of which I here describe as new.

1. *Leptochirus fontensis*, n. sp. Nigerrimus, nitidus, fronte bi-impressâ, clypeo oblique declivo, utrinque acute tuberculato, pone tuberculam bipunctato; abdomine supra fere impunctato. Long. corp. (extenso) 8 lin.

Antennae 2 lin. in length; 10th joint scarcely so long as broad. Head with the two frontal impressions without punctures, or rather each with two very indistinct punctures; clypeus descending obliquely from the front, with an acute tubercle on each side, and behind this with two coarse punctures on each side. Thorax scarcely 1¼ lin. in length by 1½ in breadth. Elytra just as long as thorax, with their inflexed margin rather coarsely punctured. Hind body slender; on the upper side, at the base of each segment, there are no punctures in the middle; on the under side the basal segment is without punctures in the middle; the following segments rather coarsely punctured at the base; the 6th segment more sparingly than the the others; metasternum with only ten or twelve not very distinct punctures on each side, external to the middle coxæ.

Fonte Boa; two individuals, communicated by Dr. J. W. H. Trail, who captured them there on the 17th October, 1874; and also found at Ega by Mr. Bates.

Obs. I.—This species, in the structure of its head and mandibles, is very closely allied to the Brazilian *L. scoriaceus*, Germ, but is much smaller and especially narrower; the antennæ are considerably shorter, and the fine dense punctuation found at the base of each of the two or three front segments, on the upper side of *L. scoriaceus*, is wanting.

Obs. II.—I have drawn up the above description entirely from one of the Fonte Boa individuals, which is, I believe, a female; the individuals from Ega, of which several are before me, are slightly smaller, and show a slightly greater development of the punctures on the under side of hind body; but these differences are only very slight, and I believe all the individuals from both localities to be of one species.

2. *Leptochirus brunneoniger*, Perty; Fauvel.

Ega; two individuals.

This appears to be a widely distributed, and yet but little variable, species. I see no difference between these Ega specimens and individuals from near Rio de Janeiro, and the species is recorded by Fauvel as Mexican. I have not myself seen any specimens from the north of the Amazons; on the other hand, an individual in my collection is labelled "Peru."

3. *Leptochirus latro*, n. sp. Niger, nitidus, thorace sanguineo; mandibulis brevibus; antennis crassiusculis; vertice canaliculato, canaliculâ antice profundâ, clypeo

parte posteriore elevatâ, anteriore abrupte declivâ; abdomine subtus magis punctato. Long. corp. 6 lin.

Antennæ short and stout, 1½ lin. in length; the 10th joint hardly so long as broad. Mandibles short and stout, their upper edge strongly sinuate near the base. Head with the longitudinal channel as deep in front as behind; the hind part of the clypeus elevated to the level of the vertex, and on the same plane with it, but quite distinctly marked out therefrom, the front part of the clypeus at right angles to the hind portion; the upper part of the head is black, with the neck obscurely reddish, beneath blackish-red. Thorax rather broad, red, 1 line in length, and quite 1¼ in breadth. Elytra narrower than the thorax, 1⅜ lin. in length, and just about the same in breadth, quite black. Hind body black, moderately stout; on the under face, the segments 3—6 are coarsely punctured over a large space on each side the middle. Legs black, with the tarsi reddish.

Ega; a single individual, of the male sex.

Obs.—This species is closely allied to both *L. brunneo-niger* and *L. maxillosus*. The structure of the head and mandibles is almost the same as in the former species, but *L. latro* is smaller and has the thorax shorter, and the elytra black. The structure of the head distinguishes the species from *L. maxillosus*, and the shorter antennæ distinguish it from both the species alluded to.

4. *L. maxillosus* (Fab.).

Ega; St. Paulo, about a dozen examples.

I find great difficulty about this species; indeed, it seems to me probable that two or three variable and yet closely-allied species may be confounded in collections under this name. M. Fauvel has separated one form and given it the name of *L. proteus*, but I have not been able to satisfy myself as to which of the forms his description refers. This author gives a figure of the front of the head of an insect he calls *L. maxillosus* (Notices Ent. pt. ii. pl. i. fig. 2), but I have seen no specimen at all like his figure.

These Amazonian individuals vary much in size, colour, and also somewhat in the front of the head; but as I cannot see that these characters indicate distinct species, I think it best merely to record them as *L. maxillosus*.

TURELLUS, n. gen.

Antennæ clavatæ, 9-articulatæ.
Tarsi omnes 4-articulati.

Labrum ample, corneous; front margin rounded and serrated, the serrations being about twenty in number. Mandibles elongate, slender, acutely pointed, strongly curved, with an elongate pointed tooth on the middle of the inner side of each. Maxillary palpi elongate; 1st joint not observed, 2nd joint rather long and slender, two or three times as long as broad; 3rd joint similar to 2nd, but nearly twice as long; 4th joint large, longer than 3rd, elongate oval, but a good deal dilated on the inner side about the middle.

Antennæ inserted at the sides of the front, as in *Oxytelus*, short, 9-jointed; the 1st joint stout, joints 2—7 small, each a little shorter than its predecessor; 8th joint strongly transverse, 9th joint forming a pointed oval club, terminating in setæ. Head shaped as in *Oxytelus*; the eyes rather small, but prominent. Prothorax strongly transverse, the base cut away on each side at the hind angles; beneath, the horny portions of the prosternum and sides occupy nearly the whole space, so that the front coxæ are nearly entirely covered; they are small, quite contiguous, placed quite at the hind part of the thorax, the openings of their cavities being apparently reduced to two minute, circular, contiguous (? confluent) spaces, without lateral prolongations. Middle coxæ minute, nearly contiguous, subglobose. Elytra longitudinally costate. Hind body short and broad, much narrowed to the extremity, rather strongly margined, composed of six visible segments. Legs rather small, tibiæ slender and simple; tarsi short, 4-jointed, the three basal joints short, and about equal to one another, the 4th joint rather longer than the three others together. Form of the whole insect—short and broad, flat on the upper side.

This minute insect, for which I have been obliged to find a new generic name, is perhaps the most interesting of the *Staphylinidæ* discovered by Mr. Bates. Only a single individual was brought back by Mr. Bates, and I have therefore been able to expose its characters only in a very imperfect manner; the lower lip and its appendages and also the maxillæ and base of the maxillary palpi have quite eluded my observation, and also I have been able to see only in an imperfect manner the structure of the hind parts of the prosternum. Nevertheless, it is evident that the insect is one of the most anomalous of the *Staphylinidæ*, and that the determination of its nearest allies will be a matter of difficulty. The structure of the anterior coxal cavities, I anticipate, will be found to be very close to that which obtains in *Thoraxophorus corticinus*; from that insect its trophi, however, remarkably separate it and appear to indicate a relationship with the anomalous genus *Evæsthetus*. I am unacquainted with the North American genera *Edaphus* and *Stictocranius*, and am unable to guess what relationship it may bear thereto. Its nearest allies I cannot, therefore, at present point out.

1. *Turellus Batesi*, n. sp. Obscure rufescens, opacus, antennis pedibusque testaceis, elytris tricostatis. Long. corp. ⅚ lin.

Antennæ yellow, about as long as head. Head a good deal narrower than the thorax, the front part in the middle rather largely depressed; the depressed part triangular, and limited in front by the slightly raised and curved edge of the clypeus, and on each side by an obscure raised line proceeding from the point of insertion of the antennæ; these two lines converge about the middle of the head, but do not quite meet with one another; the surface is opaque and dull, but has no distinct sculpture. Thorax quite as broad as the elytra, quite twice as broad as long; the front margin a little sinuate on each side, the front angles not in the least deflexed, a little rounded, the short sides nearly straight; the hind margin a good deal cut away on each side, so as to leave a gap between the thorax and elytra at the sides; the surface not quite even, but with two indistinct, distant, longitudinal elevations; between these, at the base, are three very minute and indistinct foveæ; the surface quite dull, but with no distinct sculpture. Elytra broad, a good deal longer than the thorax, each with three raised, longitudinal lines, the inner one of which is the most distinct, and is placed about midway between the suture and side; the second is placed quite at the side, and the third, or outer one, which is near the second, is placed on the deflexed lateral portion of the wing case; the pleural portion of the wing case under this third line is abruptly inflexed and rather broad. The surface is dull, but without distinct sculpture. Hind body dull, without distinct sculpture.

Ega; a single individual, whose sex I do not know.

www.ingramcontent.com/pod-product-compliance
Lightning Source LLC
Chambersburg PA
CBHW030427300426
44112CB00009B/885